SOUND IN HISTORY

Emma Dillon, *Series Editor*

THE ATTENTIVE EAR

Sound, Cognition, and Subjectivity in the Long Nineteenth Century

Edited by Francesca Brittan
and Carmel Raz

PENN

UNIVERSITY OF PENNSYLVANIA PRESS

PHILADELPHIA

This book is freely available in an open access edition thanks to the University of Pennsylvania Press's participation in the University Press Library Open program, an initiative of Paradigm Publishing and the De Gruyter eBound foundation. Read this and many other quality open access books at UPLOpen.com.

Published by
University of Pennsylvania Press
Philadelphia, Pennsylvania 19104–4112 USA
www.pennpress.org

EU Authorized Representative: Easy Access System Europe—Mustamäe tee 50, 10621 Tallinn, Estonia, gpsr.requests@easproject.com

Publication of this book in open-access form has been made possible by the Max Planck Institute for Empirical Aesthetics.

Printed in the United States of America on acid-free paper
10 9 8 7 6 5 4 3 2 1

Cataloging-in-Publication record for this book is available from the Library of Congress.

Hardcover ISBN 978-1-5128-2994-5
eBook ISBN 978-1-5128-2995-2
OA Electronic ISBN 978-1-5128-3001-9

CONTENTS

PART III. ECSTASY

PART IV. MEASUREMENT

Introduction: Sound Attentions

Francesca Brittan and Carmel Raz

> . . . it often comes about, being occupied with another thought, we do not in conversation seem to have heard even ourselves.
>
> —Augustine, ca. 389[1]

> . . . when the mind is strongly occupied in contemplating certain objects, it is in no way aware of the impression that certain bodies make on the organ of hearing, although the impression is quite strong. But no perception comes from it, if the soul takes no knowledge of it.
>
> —Gottfried Wilhelm Leibniz, 1704/1765[2]

> . . . While two persons are engaged in interesting discourse, the clock strikes within their hearing, to which they give no attention. What is the consequence? The next minute they know not whether the clock struck or not. Yet their ears were not shut.
>
> —Thomas Reid, 1788[3]

> It is a well known fact that when the mind is strongly preoccupied and attention engaged, many vague feelings and ideas flit about consciousness without leaving in memory any durable traces of their visit. And more than this, nervous processes which would under ordinary circumstances yield distinct and strong impressions,

> remain during such mental preoccupation without any appreciable effect. Thus moderate sounds fail to have any disturbing effect on a mind deeply absorbed in reading.
>
> —James Sully, 1874[4]

Perception depends on attention. From Augustine, writing ca. 389, to Leibniz and Reid in the 1700s, to Sully in the late nineteenth century, philosophers and psychologists have made this claim, often with recourse to sonic examples. The unattended sound, they point out, remains "unheard." Sensation may be present—we may be awash in acoustic vibration—and yet, as Leibniz puts it, "no perception comes from it," thus an epistemological gap is created: "The soul takes no knowledge of it." Acknowledgment and understanding occur only when we stretch toward a sound, honoring the "tendere" at the Latin root of attention. But to do so, we must withdraw our focus from other stimuli—from our visual surroundings, our book, the tangle of thoughts in our head, the chatter in the room. To attend fully to the choir on stage, we must allow other voices to retreat into faint or non-perception. To listen attentively to one instrument in the orchestra means "making it dominant," allowing all others to fade in our awareness such that they "produce only vague sensations."[5]

Attention, then, is contingent on inattention. And both, historically, are caught up with psychologies and philosophies of listening: Etymologically, "to listen" overlaps directly with "to attend," "to accompany," "to mind," and even "to obey."[6] The cognitive questions arising from this are manifold: In what ways does listening cultivate, mold, or discipline attention? The ethical queries are equally obvious: To which sounds should we listen? To which voices? instruments? landscapes? And what happens to those that fall into inattention and therefore nonperception? Add to these a spate of economic and empirical concerns: How much psychic bandwidth do we have? How much giving over—or stretching toward, or minding—can we do? There are also social-environmental quandaries: In noisy or chaotic environments, can we truly attend to anything? What happens when, as English physician John Mason Good puts it, "the attention is suffered to wander to something of stronger attraction; or the imagination is left to the play of its own capricious ideas"?[7] This leads to even trickier conundrums: Is there an unconscious attention? And what are we to make of ambient, involuntary, or passive forms of noticing?

Interest in these issues has galvanized an outpouring of literature on attention over the past few decades, including work across the fields of psychology, cognitive science, philosophy, history, economics, art history, and literary criticism. Studies unpacking the attentional politics of our own moment have intertwined with a growing body of scholarship interrogating modes of focus back through classical antiquity. In some accounts, attention emerges as a neural or cognitive process, in others as a physical act, an ethical stance, a commodity, a psychic web, or a form of purposeful selfhood.[8] Attention's relationship to both consciousness and volition remains undecided, as do the particularities of its mechanisms (bottlenecking, filtering, spotlighting, selecting, enhancing, inhibiting, etc.).[9] Rather than a stable concept, "a faculty of some already formed subject," as Jonathan Crary puts it, attention has tended to operate as a "sign," a discursive trope by which "new conditions of subjectivity [are] articulated."[10] Culturally constructed and historically bounded, attention is also profoundly mediated, channeled through, and defined by the technology of the moment—material, digital, scientific, social. As such, it is difficult to apprehend in any pure or abstracted form; indeed, apperception (the act of paying attention to attentiveness itself) involves an overriding of enculturated habit and a vertiginous calling into question of the nature of perception itself. We might think of such an effort as thinking about thinking—contemplating the psychic glue that holds us in relation to ourselves, to one another, and to the systems of interaction, exchange, and care that sustain or constrain us.

Key work in the arenas of art history and literary criticism has demonstrated how modes of alertness determine, and arise from, ways of seeing and reading—but even more obviously (as our opening quotes suggest), conceptions of attention are marked by experiences of hearing and listening. We can choose to look away or put down our book, but our defenses against sound are more porous (we have no earlids, as has often been pointed out), thus questions of attentional agency are especially urgent in sonic environments. This observation was made explicit in 1866 by German philosopher Hermann Ulrici in his text *Leib und Seele: Grundzüge einer Psychologie des Menschen*:

> This possibility of preventing certain sensory perceptions from entering our consciousness, of annulling them, as it were, is of particular importance for [the sense of] hearing, since we cannot voluntarily shut our ears (as we can close our eyes), to avoid the effects of the outside world. This is perhaps why attention is even more important in the domain of auditory perception.[11]

Over the course of the long nineteenth century the role of audition in studies of alertness and distraction became increasingly central, particularly in experimental domains. Psychologist Carl Stumpf, writing in 1883, positioned this outcome as inevitable: "It is only natural that attention, on which all mental activity ultimately depends, develops its nature and effect particularly richly in auditory judgments, and that it is here that a theory of attention can best be developed."[12] In and outside of Europe, Stumpf's prognostication held true: Listening became a key index of attentional depth and subtlety as well as a testing ground for functional, structural, psychophysical, and social theories of focus/unfocus. As imperial expansion accelerated through the late century, pedagogies of auditory attentiveness emerged as powerful tools for enforcing cognitive receptivity and therefore mechanisms of valuation and control. Widening webs of communication and exchange generated interactive networks within which competing modes of auditory alertness—heedfulness, passivity, absorption, vigilance—reinforced and challenged one another with increasing insistence. In what follows, we unpick several strands of this mesh, offering meditations on some of the most prominent attentional orientations of the long nineteenth century. These are not exhaustive—many others exist—nor are they sealed off from one another. Rather than a static taxonomy, attentional regimes existed within a fluid system of attraction, investment, and mutual modulation whose complexities remain to be fully explored. Here, we begin to imagine how this ecology operated—essentially, what it might look like to rethink auditory culture ca. 1780–1930 in cognitive terms—offering a series of starting points, which are also invitations to extension, reexamination, and new inquiry.

Concentration

Among the most characteristic attentional attitudes of the period 1780–1930 was that of concentration. Urgent requests for narrowed and sustained auditory focus resonate across the period, often inextricable from spiritual, intellectual, and aesthetic imperatives. Heinrich Wilhelm Wackenroder's famous insistence on the "true" way of listening as entailing "the complete surrender of the soul to this sweeping stream of sensations; in the distance and detachment from every disturbing thought and from all foreign sensory impressions" (1792) is echoed in Hermann von Helmholtz's call for "undisturbed and concentrated [auditory] attention" (1877) and, a decade later, in music theorist Salomon Jadassohn's definition of "earnest listening" (*Anstrengung des Hörens*) as "the shutting off and guarding one's self as perfectly as possible against all other impressions" (1888).[13] This fetishization of concentration can be traced to Enlightenment philosophy and aesthetic theory, which placed the focused mind at the epicenter of rational, productive, and fully volitional selfhood. As Georg Friedrich Meier put it in 1733, attention was "as it were, the hand with which the soul grasps a dark idea in its depths, raises it up, and thereby brings it into the light of day."[14]

Mid-century debates across Britain and continental Europe over the nature and extent of attention embraced a variety of modalities and gradations, but general trends can be identified. Among these was a gradual shift from John Locke's notion of attention as an involuntary or automatic function (the mind passively receiving ideas that "offer themselves") to Dugald Stewart's conception of attention as a voluntary and active power ("the effort of the mind, to detain the idea or the perception").[15] Conceptions of multifocal attention (the notion of paying attention to several things simultaneously) were gradually overtaken by a valorization of unifocal attention (the steady concentration on a single object or train of thought). In the expanding media and information markets of the late eighteenth century, distractions (musical entertainments, the popular press, advertisements, public fairs), posed an increasing threat to this narrowed focus. Partly in response, strident warnings against mental wandering became common, including James Beattie's 1783 claim that distraction produced the "absent" man and Stewart's insistence that "attention consists partly (perhaps entirely) in the effort of the mind, to detain the idea or the perception, and to exclude the other object that solicits its notice."[16]

The process of selection valorized by Beattie, Stewart, and others was often explained in terms of listening. As early as 1782, Thomas Reid compared focalized attention to the action of isolating *either* the bass or the treble of a two-voice tune; and Stewart himself argued that "in a concert of music, a good ear can attend to the different parts of the music separately."[17] Musical analogy extended seamlessly into compositional theory. As Matthew Riley has shown, both Johann Georg Sulzer and Johann Nikolaus Forkel highlighted attentional management as a key aspect of compositional craft in their aesthetic writings. These theorists described musical structures (phrases, sentences, periods) as tools for the careful calibrating and sustaining of listener focus, and offered advice for the use of surprising or unusual musical effects to refresh flagging alertness.[18] Similar advice was dispensed in the Scottish context by music theorist Walter Young, who observed in 1790 that "a considerable variety, both of movement and of measure . . . when skilfully managed, enliven the music, surprise the hearer, and excite his attention."[19] Such strategies might be seen as reflections of an "enlightened" monofocal regime—technologies of attentional capture in line with the goals of sustained and reasoned thought. But equally so, they were symptoms of the erosion of such a regime in the face of the widening diversions of the public sphere, and in this sense, operated as savvy guides to an emerging attention economy whose strategies of focalization were also bids for cognitive investment.[20]

The cult of auditory concentration was shored up—indeed, reached an apex—with the rise of romantic idealist philosophy. No longer just an act of discipline, musical absorption was now recast as a form of theological obeisance with a sublime payoff. Alongside Wackenroder, E. T. A. Hoffmann drew on the language of both analysis and philosophy to describe musical attentiveness: "The captivating siren song" of Beethoven's varied musical phrases, as he put it, "entices me [to go] deeper and deeper."[21] Once completely absorbed, the organic mysteries of Beethovenian construction were revealed and, along with them, a glimpse of the divine macrocosm. This line of discourse was extended, famously, through A. B. Marx and Eduard Hanslick, for whom listening was an act of reasoned contemplation predicated on sustained alertness. Truly aesthetic audition, as Hanslick argued in his landmark *Vom Musikalisch-Schönen* (1854), was to be separated from mere passive receptivity. The latter was the mark of the *Liebhaber*, the musical amateur whose mind moved only on the surface of the work, distracted by superficial features and indistinct emotional reactions. The *Kenner* (the devotee), on the other hand, gave his mind fully to the deep intricacies of musical form and thus to higher meaning and metaphysical truth.[22] Revelation (access to beauty, spiritual insight, and substance) came only at the cost of full attentive investment.

This brand of formalism, and the regime of auditory vigilance on which it relied, exercised a powerful sway over the following several generations, its attentional economics inextricable from German nationalism and cultural-imperial ambition. Dogged insistence on totalized musical focus echoes down through psychological treatises, music theoretical tracts, "appreciation" guides, and school tutors well into the twentieth century, exemplified by Theodor Adorno's exhortation that "everything has to be drawn into the field of attention lest the meaning of the musical work should be missed altogether.[23] To listen properly was to give full attention and, in turn, full attention could be cultivated by listening properly—a relentlessly circular logic. Not confined to Europe and America, the concentrative ideal also infiltrated burgeoning European colonies as far afield as Bengal (see **Richard David Williams**), bringing with it an insistence on musical alertness as a purely mental phenomenon divorced from embodied or kinesthetic forms of attending.[24] Over time, the ubiquitousness and stridency of demands for musical concentration began to suggest a certain resistance (even implausibility) and, as commentators as early as the 1770s had pointed out, ethical dubiety.

Almost as soon as the regime of concentration was established, it became the subject of critique across philosophical, literary, and scientific arenas. As Natalie M. Phillips has shown, burgeoning Gothic literatures emanating first from England, then from continental Europe and America, dramatized the perils of hyperfocus, linking it with atrophied cognition, peripheral attentional blindness, and erosion of moral and ethical awareness. From the 1760s onward—and with particular urgency through the 1780s and 1790s—authors of Gothic novels tied monofocus to monomania, tracing the obsessive tendencies of their antiheroes to forms of pathological attentional narrowing. "In a 180-degree turn," as Phillips puts it, "the traditional mark of reason and mental discipline became an emblem of cognitive inflexibility."[25] Composers, too, made this connection, most famously Hector Berlioz, whose Gothic antihero in the *Symphonie fantastique* concentrates so fully on an "ideal" melody that he is driven first to solipsism, then madness.[26] Musical focus here leads to what Garrett Stewart calls "attentive surfeit disorder," a damning excess of concentration.[27] Much earlier, another Frenchman—Denis Diderot—had hinted at the same problem in his article on "Distraction" for the *Encyclopédie* (1754), which derided unwavering attention as a form of stupidity in which the mind "rests on . . . the same idea."[28]

The cognitive perils and moral-political pitfalls of selective attention were noted not just by English and French critics but also by Germans themselves. Friedrich Schiller, in his treatise *Über die ästhetische Erziehung des Menschen* (*On the Aesthetic Education of Man*, 1795), argued for the importance of liberated contemplation, positioning the "free play" of the mind as an antidote to the relentless concentration (or "specialization") associated with industrial labor. For him, the unconcentrating of the mind was a form of enlarging, a sensuous opening to beauty that "consummate[ed]" one's "humanity."[29] His ideas were extended by Friedrich Schlegel, whose novel *Lucinde* (1799) rejected both "concentrated reason" and the culture of utility with which it was associated, hailing cognitive idleness or "intentional . . . passivity" as the true source of knowledge and aesthetic insight.[30] These attitudes constituted forms of cognitive disobedience or reversal, moments (to borrow Crary's formulation) in which "institutionally competent attentiveness veers into something vagrant, unfocused, something folded back against itself."[31] They also had musical ramifications, including in the harmonically wandering, repetitious writing of Schubert, whose work embraced an alternative (Romantic) attention directly at odds with "enlightened" cultures of effortful focus.[32]

German doubts around the healthfulness and ethics of single-mindedness continued to surface through mid-century, including in the writings of philosopher Hermann Lotze, a figure straddling the transition from German idealism to scientific positivism. In his 1852 *Medicinische Psychologie*, he noted that "someone tuning the strings of a piano with the most strained attention to his task has a minimum of self-consciousness," arguing that such absorption should surely be tempered by cognitive "flux" for thoughtful consideration to be possible. Indeed, for Lotze, "prolonged narrowing of the thought process," no less than "prolonged distraction," constituted "the first stages of a disturbance of the soul."[33] By this point in the 1850s, even the most earnest concentrators had begun to falter, faced with the drastically elongated works of Schumann, Wagner, and Verdi. To many, the prospect of sustaining rapt absorption during the whole of *Tannhäuser* or *Lohengrin* was not just exhausting but suspect, suggesting abdication of wider social and political awareness and, in extreme cases, even a kind of catalepsy. Musical works of this length (along with the thick Victorian novels that served as their literary counterparts) elicited, instead, an intermittent or cyclic attention, which began to be positioned as a more realistic alternative to single-mindedness.[34] The notion of "dynamic" focus, which, as **Ewan Jones** observes, had its roots in eighteenth-century aesthetic theory, emerged as a standard construct in psychological literature of the late nineteenth century. Théodule Ribot, in his *Le psychologie de l'attention* (1889), asserted that voluntary attention was by nature "precarious and vacillating" (*précaire et vacillante*), and William James, in his landmark *Principles of Psychology* (1890), argued that it could be sustained "for no more than a few seconds at a time," operating as "a repetition of successive efforts."[35]

By the early twentieth century, as Benjamin Steege has demonstrated, philosophy of a phenomenological bent had launched an active campaign against absorbed, hermetically sealed concert listening. Heinrich Besseler, in his "Grundfragen des musikalischen Hörens" ("Fundamental Issues of Musical Listening," 1925), argued that giving over total mental bandwidth to a concert performance meant estranging the ear from wider soundscapes, producing a kind of willful mishearing—a silencing of community and a capitulation to capitalist forces of attentive extraction. Besseler's contemporary, philosopher Günther Stern, in his "Zur Phänomenologie des Zuhörens" ("On the Phenomenology of Listening," 1927), went further, arguing that true musical listening was not a matter of either attention or distraction (since one was enfolded within the other) but a radical "*gehen-lassen*" ("letting go"), an eradication of selection in favor of total inclusion.[36] The writings of these young critics hardly marked the end of Hanslickian absorption, but they opened the door to a burgeoning discourse around attention, intention, and (musical) value that would—as **Steege**'s chapter in the present volume argues—extend across the Atlantic.

Collectivity

In January 1838, in one of the earliest demonstrations of the electric telegraph, Samuel F. B. Morse tapped out the message "Attention, the Universe!," transmitting it along ten miles of wire emanating from an exhibition space at New York University (see Figure 0.1).[37] His experiment anticipated a new kind of networked alertness—a collectivity hardwired into the global psyche by mid-century, when cables linked urban centers across several continents. No longer an individual investment, attention was rerouted by the wired web along collective arrays, producing what philosopher Yves Citton calls "transindividual flows."[38] In telegraphic (and other mass media) systems, he argues, attentiveness and nonattentiveness were produced and defined via reciprocal modulation: "Through 'me,' it is always 'our' collective thinking and feeling [νους/nous] who is paying attention." Or, to put it differently, "I" pay attention to what "we" attend to collectively—attention operates as an interaction within a distributed psychic ecology.[39]

Of course, the "I," "me," and "we" in Citton's claims are loaded, begging questions around the politics of collectivity and the control mechanisms of (unequal) psychic "distribution." These issues were addressed by pioneering social psychologists of the late nineteenth century, including Gustave Le Bon and Gabriel Tarde, who initiated theories of group mind ("esprit collective") and crowd behavior as well as notions of a "media" or "opinion" public.[40] Such collectives were marked, as Tarde argued, by forms of attentional virality and capture. In the case of crowds, "their attention and also their inattention is always stronger and more persevering than would be that of each of the

Figure 0.1. Violet Oakley, Samuel F. B. Morse, study for "Attention the Universe by Kingdoms Right Wheel" with original Morse code. 1939. Ink on paper, 33.655 cm x 27.94 cm. Gift of the Violet Oakley Memorial Foundation. https://www.pafa.org/highlighted-collection/violet-oakley-memorial-foundation.

individuals who make them up, if he were alone." Members of the opinion public—consumers of print media, telegraph dispatches, and (by the early 1900s), wireless news—were subject to even deeper fascination: "The very fact that their attention is concentrated on a single object, in a kind of collective monoideism, brings them close to the state of dreaming or hypnosis."[41] Power, often disguised, remote, or invisible, was concentrated in the individual(s) riling the crowd or the authors/operators of newly powerful and far-reaching media flows, who manipulated attentional dynamics in the interest of producing compliant, productive, and lucrative networks.[42]

As in the case of selective attention, the mechanisms of collective alertness were strongly associated with forms of auditory conditioning, including (most obviously) modes of telegraphic and radio listening, which bound military forces, governments, and special interest groups into rapt auditory networks.[43] But music and sound cultures—particularly performing bodies—also operated as crucial metaphors and models in the theorization of attentional collectivity. From the intracranial to the geopolitical, networked alertness was framed in terms of listener-performer thrall; indeed, the group consciousness or public "harmonization" theorized by Tarde and Le Bon had its origins in early nineteenth-century phrenological constructions of the "orchestral" mind. As Francesca Brittan has shown, these originated with Franz Joseph Gall and Johann Spurzheim, who first conceived of the brain as a collective organ, an interactive constellation of physical faculties comparable to the instruments (or instrumental sections) of an orchestra. According to phrenological theory, each of the twenty-seven faculties of mind (including those responsible for calculation, vision, language, memory, etc.) was endowed with an individual attentional capacity stimulated via external cues and modulated by the natural propensities/powers of the faculty itself. But these separate attentions were bound into a collective under the watchful eye of a central intelligence, a volitional "I" or mind, conceptualized as a harmonizing power (the "leader" of Tarde and Le Bon's crowd).[44] As phrenological theory spread through the mid-century, this instrumental-harmonic model proliferated, appearing in its most literal form in the work of Spanish phrenologist Mariano Cubí y Soler. In Cubí y Soler, the volition was figured as an orchestral conductor (a "supreme and sovereign attention") wired via a system of "electric telegraphy" (charged nerve fibers) to all individual brain instruments. Acting as a central surveillance node—a receiver, signaler, and enforcer—the volition ensured that all local attentional faculties played in time and tune. Cubí y Soler's "I" controlled, while also being constituted by, a collective alertness, its overarching focus indistinguishable from the web of overlapping, continuously modulating individual attentions of the multiplayer mind.[45] The result was an enforced cognitive unity troped as collective harmony.

The metaphor of the conducted brain orchestra was constructed by and through the attentive dynamics of actual orchestras, which modeled the centrally monitored collectivity described by phrenologists. Nowhere is this clearer than in the writings of nineteenth-century conductors, who described the act of orchestral leadership as an explicitly attentional power. Hector Berlioz, in his *Memoirs*, offers a clear example. When the leader of a large orchestra is on the podium, Berlioz writes,

> his attention embraces everything; his eye is everywhere; with a glance he signals . . . entries to voice and instrument, up, down, to the right, to the left; with his right arm he flings forth terrible chords that burst in the distance like harmonious projectiles; then, on a pedal point, he arrest all this movement; rivets the attention of all; suspends every arm, every breath, listening for a moment to the silence . . . and then gives freer scope than ever to the whirlwind he has subdued.[46]

In this account, attentional subjugation is inextricable from violence, rendering individual players mere extensions of conductorly impulse. Reviewers were quick to understand the political implications of such a model, describing Berlioz as a general and his instrumentalists as an obedient army.[47] And it was not just mesmeric capture but quasi-telegraphic infrastructure that governed the attentional collective of the orchestra. To hold his thousands-strong group together, Berlioz employed a newfangled "electric baton," a device linking his own conductorly stick via wires to subsidiary batons arrayed around vast performance spaces. Rhythmic transmissions sent from the central podium along these wires kept far-flung sections of the ensemble (strings, winds, brass, chorus, percussion) in unison, producing a tightly controlled unit, much as Cubí y Soler's "volition" networked spatially separated brain organs into attentive harmony. These interrelated musico-neural webs were obvious refractions of nineteenth-century telegraphic networks—imperial forms of collectivity rooted in attentional hegemony and extraction.[48] We might think of them (paradoxically) as atomizing collectives, holding out the promise of psychic wholeness while privileging consolidated control, thus draining cognitive along with cultural and material resources.

Systems of collectivizing extraction took other guises across the long nineteenth century, including the quasi-inclusive "global listening" regimes excoriated by **Nicholas Mathew** and the musico-temporal synchronizations described by Jacob Olley. In the former case, fantasies of universal attentiveness derived from the construction of an "omniaudience," a fantasized auditory network generated by European consumption of "world" music excerpts offered in proto-ethnomusicological collections. The presentation of these snippets in Western staff notation might be compared to conversion into Morse code, which was then sent along the European wire in the form of musical "messages" reinforcing dominant auditory-cognitive modalities. Synchronizations, as described by Olley, operated similarly, generating the illusion of a temporally unified collective in which attention was timed and defined by the ticking of Greenwich Mean Time. Even more obviously, as Erin Johnson-Williams and Roe-Min Kok have pointed out, pedagogical institutions including the Associated Board of the Royal Schools of Music (ABRSM) created instructional networks across South and East Asia, holding generations of students hostage to carefully curated repertories. The very definition of musical "attention"—as a solo, memorized performance of set pieces and technical exercises—was enforced by examiners fanning out from Britain along routes already well-trodden by military and commercial enterprise.[49] What Johnson-Williams describes as the "auditory surveillance" of British examiners was also a form of cognitive molding, an inculcation of global physical and mental "concentration" in line with Euro-German ideals of auditory focus.[50]

Dehumanizing attentional collectives—often operating as cognitive technocracies—eroded, while often seeming to generate, community. Rather than forms of acknowledgement or inclusion, their wired alertness bound individuals into vampiric systems siphoning off cognitive capital to controlling powers. Characterized by "overhearing," "listening to" and "listening in," they relied on "hungry" attention (to adapt Dylan Robinson's formulation).[51] Their extractive regimes have had lasting legacies, including what American sociologist David Riesman described as early as 1950 in *The Lonely Crowd* as the "other-directed person"—a networked individual who "must be able to receive signals from far and near" even when "the sources are many, the changes rapid." What this character-type internalizes "is not a code of behavior but *the elaborative equipment needed to attend to such messages* and occasionally to participate in their circulation"; in other words, a form of cognitive remaking such that individuality is subsumed into voracious attentive collectivity.

Among the results, as Riesman argues (and as research on contemporary social media has increasingly confirmed), is "a diffuse anxiety," a sense of colonized or commodified attentional consciousness always listening outward, never inward, always listening "to," never "with."[52]

In more life-giving guises, collective attention could reverse this psychic evisceration. Telegraphy itself, with its new language of taps and clicks, generated surprising forms of intimacy among operators who learned one another's auditory signatures. Likewise, radio and telephonic devices could sustain social ties over distances via newly ethereal attentive techniques (see **Brian Kane**).[53] But more obviously, attentional community was related to liveness—to the forms of rhythmic entrainment and emotional induction associated with shared physical experience. Tarde, elaborating on these distinctions, noted that while the media public was the most modern and powerful collective of the age, the older body of the crowd had capacities for connectedness that went beyond mere simultaneity or linkage. The distinction lay in "the mutual contagion of feelings among the assembled individuals" and their broader gestural entrainment (synchronization).[54] Crowds witnessing theatrical events, music festivals, or religious celebrations generated a crucial form of "social union" in which "all dissent is momentarily erased in the communion of a single desire, the desire to see one another, to embrace one another, to sympathize." Rather than a public, physical gatherings generated a community, a multitude of individuals joined sympathetically, whose "love and joy are hidden in the recesses of the heart" such that "long after the party, there remains an extra sympathetic and conciliatory disposition which translates into a thousand unnoticed forms in the gestures, words, and relationships of everyday life."[55]

Citton, contemplating constructive forms of collectivity, emphasizes their capacity to effect "a particular quality of attention rooted in care—which is to say consideration of the vulnerability of the other, of our solidarity and our responsibility towards them." Noncoercive attentional groups involve forms of collective individuation—the sense that the whole animates, enlarges, and resonates rather than subsuming the individual. Examples of this phenomenon across auditory/musical domains of the long nineteenth century are too numerous and complex to unpack in full here but might include the collective forms of kinesthetic, multisensory musical attention of Javanese villagers at an 1815 wedding celebration (see **Day and Weiss**). In European performance contexts, they can be intimated in what G. Mensch described in 1871 as the autonomous collectivity of the string quartet, whose "four personalities move with and against one another . . . accompanying and complementing each other, but also challenging and fighting, asking and answering, affirming and denying, but always pursuing the same goal and remaining true to the alliance of protection and trust."[56] Herder assigned a similar quality to the "one and not one" of a choral group, whose voices "leave, search, follow, contradict, fight, strengthen, destroy each other, and awaken and animate and console and flatter and embrace each other again until they finally subside into one tone."[57] In these cases, collectivity produces expansiveness rather than anxiety—a listening "with" or "for." The result—then, as now—is a form of extended cognition generated by shared attention.

Ecstasy

Typical Enlightenment accounts of attention involved a rational response to environmental stimuli, whether voluntary or involuntary. For the French composer and music theorist Jean-Philippe Rameau, for example, careful attention enabled a listener to (voluntarily) discern overtones within a given sonority, while the philosopher and musician Jean-Jacques Rousseau held that our minds involuntarily responded to musical works by rewarding certain features with more or less focus.[58] Yet toward the end of the eighteenth century, reports of new, pathologized modes of attending begin to emerge, most distinctly in accounts of ecstatic, trance-like states experienced in the Parisian séances held by German physician Franz Mesmer. Mesmer claimed to have identified a unique energetic force present in all living things, which he termed "animal magnetism" (*magnétisme animal*). This force, he maintained, could be harnessed to induce action at a distance between people, as well as to generate trance states and even medical cures in certain, sensitive individuals.

A telling encounter between Enlightenment understandings of attentiveness and new medicalized forms of trance can be found in the 1784 investigation of Mesmer by a royal commission headed by Benjamin Franklin. Charged with explaining the phenomena reported by Mesmer's patients, the committee appealed to the faculty of attention. They remark:

> There is not an individual, in a state of the fullest health, who, if he paid a close attention to the point (*s'écouter attentivement*), would not be sensible to an infinity of interior motions and variations, either of a pain infinitely slight, or of heat in different parts of his body; these

> variations which exist at all times are independent of the magnetism. To turn and fix in this manner one's attention upon oneself, is not perhaps itself entirely without its effects. There is so intimate a connection, whatever be the vehicle of that connection, between the volitions of the soul and the motions of the body, that it is not easy to prescribe limits to the influence of attention, which appears to be nothing more than a train of volitions, directed, constantly and without interruption, to the same object. When we recollect that the arm is moved by the will as it pleases, how can we be certain, that the attention being fixed upon some interior part of our frame, may not excite some slight emotion in it, direct the heat towards it, and so modify its actual situation as to produce in it new sensations?[59]

The turn of phrase "*s'écouter attentivement*" meaning, literally, "to listen to oneself attentively" renders links among listening, embodiment, and attention explicit. Here, according to Franklin, the individual "listens" to his subjective physical sensations, whether of pain or of warmth, rather than subsuming them into the low sensory hum of embodied existence. An unhealthy attention to repressed sensations—whether real or imaginary—was the explanation converged on in the report, which concluded that

> the commissioners, even those who have their nerves most irritable, having expressly turned their attention to other objects, and having armed themselves with that philosophic doubt which ought always to accompany enquiry, have felt none of those sensations, which were experienced by the three patients of the lower class; and they have a right to suspect that these sensations, supposing their reality, were the fruits of anticipated persuasion, and might be operated by the mere force of imagination.[60]

As this passage indicates, the commissioners fully attributed Mesmer's alleged powers of animal magnetism to the overactive imagination of subjects under the influence of suggestion. Here the voluntary, rational attention of educated upper-class men is presented as a foil to the involuntary, pathologized attention of lower-class patients.

Despite the commission's comprehensive refutation of Mesmer's doctrine of animal magnetism, however, his influence continued to expand throughout the nineteenth century. Indeed, the phenomenon of hypnosis (a term used here to indicate the broad range of long-nineteenth-century practices having to do with animal magnetism, suggestion, and spiritualism, all of which stem from Mesmer's practice) was foundational to the emergence of Romanticism, offering a robust framework within which nineteenth-century subjects could understand a range of ecstatic, mystical experiences, ranging from trance states of various kinds to purported communication with otherworldly spirits.

A key thread linking hypnotic practices throughout the century is the concept of nervous sensitivity, which was associated with a heightened ability to attend, whether to oneself (as in Franklin's account of self-audition), to others, or to the environment. Thus the sensitivity of hypnotic subjects compelled them to pay heed to physical cues and vibrations that eluded regular people, including the commands of magnetizers, who "directed" the animal magnetism coursing through living things according to their own will. Given that sensitivity was understood to arise from innate refinement of the nervous system, women—often marked as more physiologically delicate than men—were regarded as particularly vulnerable to mesmeric manipulation. Members of the upper class were also considered to have more rarefied physiologies than members of the working classes, and white Christian Europeans were deemed more sensitive than any other group. Yet this sensitivity was a double-edged sword: Its excess rendered women hysterical and left men feminized, with both populations then potentially more susceptible to the machinations of unscrupulous magnetizers.

As a result, while Mesmer's ideas attracted many men (notable devotees included Samuel Taylor Coleridge, Friedrich Schlegel, and Charles Dickens), the practice of hypnosis was, from the beginning, marked as other, alternative, and feminine. Upper-class women, confined to vanishingly few possible paths of personal fulfillment, often sought hypnotic treatments as therapy for various ailments, paying both to have their own attention manipulated and to secure the attention of a (male) magnetizer. Poorer, working-class women found in animal magnetism financial gain and some measure of fame as "subjects," performing trance states for public consumption.[61] Moreover, hypnosis was both populist and radical: Magnetizers (and later mediums) often hailed from humble backgrounds, and séances were spaces where inhibitions were loosened, allowing members of different classes and the opposite sex to act and interact in normally unthinkable ways.[62]

The highly concentrated attention displayed in mesmeric sessions involved a subject's evident ability to block out all other sensations and fully attend to a single command. As Raz argues, the granting or withholding of attention within some hypnotic contexts can be understood as a form of rebellion against the status quo, a refusal to invest, a demand to be understood on one's own, marginalized terms.[63] The nineteenth-century literature is replete with accounts of women, feminized men, members of the lower social classes, and racialized subjects performing trance behaviors as a form of implicit resistance, even as the reporting witness, often a physician, ascribes these behaviors to intrinsic biological or psychological difference.[64] Music, which had been integral to mesmerism from the outset, played a central role both in hypnosis and in its critique. Mesmer himself had famously employed the soft strains of the glass armonica (an instrument ironically invented by Benjamin Franklin) as well as keyboard instruments to enhance his séances; the glass armonica, in particular, would later come to be associated with unsavory mesmeric activities. More broadly, the self-abandon and intense absorption so characteristic of (sonically induced) hypnotic trances would reappear in Romantic accounts of musical listening across the following several decades. One notable example is Heinrich Heine's report of the effect of Franz Liszt's concerts on Parisian audiences in 1844:

> The electric action of a daemoniac nature on a closely pressed multitude, the contagious power of the extase, and perhaps a magnetism in music itself, which is a spiritual malady which vibrates in most of us,—all these phenomena never struck me so significantly or so painfully as in this concert of Liszt. A physician whose speciality is the disorders of women, and with whom I conversed as to the magic which our Liszt exercises on his public, smiled mysteriously, and told many things of magnetism, galvanism, electricity, of contagion in an over-heated hall, in which are a vast number of wax-candles, and as many perfumed, perspiring mortals, of histrionic epilepsies or stage-fever, of the phenomena of tickling, of musical cantharides [Spanish fly], and other ticklish subjects, which have, I believe, relation to the mysteries of the Bona Dea [the ancient Roman goddess of fertility].[65]

Heine here invokes the trope of hypnotic trance to understand the ecstatic power Liszt appeared to exert on his audience. His explanation depends primarily on an unnamed medical authority who elides scientific causes such as electricity and heat along with esoteric and vaguely erotic references.[66]

As the century progressed, hypnosis became racialized and syncretic. Russ Castronovo has shown how mesmeric discourses from the 1830s on drew comparisons between the slave and the somnambulist, the master and the magnetizer, while also synthesizing European notions of animal magnetism with practices drawn from African and Caribbean trancing traditions.[67] Spiritualism, an American movement that blended European mesmerism with Swedenborgian mysticism, wishful thinking, and Yankee showmanship, traveled back across the Atlantic to exert a powerful hold on European, and particularly Victorian, culture. French travelers encountering the multisensory rites of mystical Islamic sects, notably the Aissawa Brotherhood in Algeria, also attempted to understand these practices through hypnotic and religious-ecstatic lenses, as **Céline Frigau Manning** explores.

The power and threat of trance states became a prominent focus of modernist psychology, including in the work of Hippolyte Taine, Jean-Martin Charcot, and Théodule Ribot, all of whom meditated at length on the nature of ecstatic (autohypnotic) processes. For both Taine and Ribot, ecstasy was securely pathological, characterized by a fixation of attention so complete that the normal fluctuations of consciousness were arrested and, along with them, rational thought and volition itself. In his 1883 *Maladies de la volonté*, Ribot underscores the susceptibility of women to such states and, even more robustly, their association with non-European traditions: "The religious and philosophical literature of the Orient, of India in particular, abounds in documents from which it has been possible to gather a sort of working manual for the attainment of ecstasy."[68] Restricted or repetitive behaviors are key, including the instruction "to remain motionless, to gaze fixedly at the sky, [or] a luminous object," or "to repeat continually the monosyllable *Om (Brahm)*."[69] Sound—in the form of rhythmical chant, devotional verses, or song—is central, not just in Asian traditions but also in mystical Christian histories.

Ribot dwells on descriptions of religious ecstasy left by the sixteenth-century nun and religious reformer Saint Teresa of Avila. In his *La psychologie de l'attention* (1888), he notes that St. Teresa describes the stages of ecstasy as stations of prayer in which the focus becomes incrementally ossified and totalized. The early stations involve audible recitation, which progresses to inward, unspoken supplication, then to the "prayer of quietude" (*oraison de quiétude*) in which the soul no longer recites but instead listens, as if entranced by its own echo. From there, the "prayer of union" (*oraison d'union*) brings the first sense of divine presence, followed by "rapture" (*oraison de ravissement*)

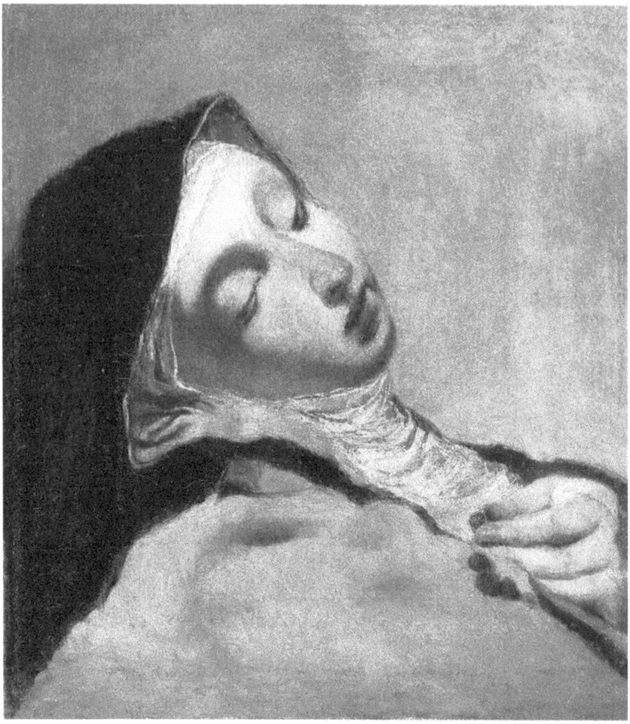

Figure 0.2. Ernst Josephson, "St. Teresa in Extasy." Before 1906. Oil on canvas, 37.5 cm x 33.5 cm. Nationalmuseum Sweden, artwork ID 19340. PD-US. https://collection.nationalmuseum.se/en/collection/item/19340/.

in which feeling and thought begin to fade, and finally "the flight of the spirit" (*vol de l'esprit*) involving "the very abolition of consciousness" (*l'abolition même de la conscience*), a dissolution of selfhood into an "excess of [divine] unity" (*excès de unité*).[70] (See Figure 0.2 for one of several nineteenth-century depictions of the ecstatic moment.)

At this point, attention is no longer an act of will but an eradication of it—a cessation of the fundamental motor activities of the brain such that neither choice nor action is possible. Already in *Les maladies de la volonté* (1883), Ribot had quoted Saint Teresa's account of the process by which "all its [the soul's] powers lose their natural activity; they have no knowledge of their operations. . . . As for the understanding, if it understands, it is by a mode which remains unknown to it, and it can comprehend nothing of what it understands."[71] The volitional power of attention is drained away in a process of mental evacuation eliminating rational forms of knowing and therefore the capacity for intellectual and aesthetic creation. Such a cessation was dangerous, according to Ribot, and he is careful to separate it from the deep forms of concentration associated with male genius—the "intellectual rapture" (*ravissement intellectuel*) of figures including Archimedes, Newton, and Pascal.[72]

But not all thinkers were so hostile to ecstatic attentional modes. Henri Bergson, in his 1889 *Essai sur les données immédiates de la conscience* (translated in English as *Time and Free Will*), positioned stilling of the volition as a prerequisite for the apprehension of beauty and grace. Among the primary benefits of art was its capacity to facilitate this form of evacuated openness, to "put to sleep the active or rather resistance powers of our personality, and thus to bring us into a state of perfect responsiveness."[73] Music was especially apt in this regard, since its "rhythm and measure" suspended the normal flow of sensations and ideas, "causing our attention to swing to and fro between fixed points," thus eliminating the natural attentional flux defining human consciousness.[74] The result was "a spiritualized version of the process commonly used to induce the state of hypnosis," a self-forgetfulness in which thought and will were attenuated.[75]

Twenty years later, the philosopher Joseph Segond, in his *La prière: étude de psychologie religieuse* (1911), extended this claim, arguing that the rhythmic regularity of music could absorb the attention utterly, producing a state of passivity conducive to prayer. As **Alexandra Kieffer** notes, Segond regarded such passivity not as metaphysical (in the manner of Saint Teresa) but purely physiological, reorienting mystical notions of ecstasy toward late-century materialist psychology. Rhythmic music, like the physical rocking it encouraged, brought about "regularized and lulled consciousness."[76] Temporarily quieting Hanslickian forms of focus—mental alertness to music's structural and aesthetic features—it encouraged instead a form of embodied absorption (entrainment) essential for contemplative practice.

Resistance to the forms of mental diffusion/dissolution embraced by Bergson and Segond was robust, including in critical responses to music thought to induce such a state. Debussy's compositions, for instance, when associated with Bergson's hypnotic ideal, were charged with emasculating, irrational, and even "narcotic" effects, the language of critics echoing quite clearly both the pathological terminology of Ribot and the moral charges levied by Europeans against non-Christian forms of ecstatic ritual.[77] In the early twentieth century, it was (perhaps unsurprisingly) a woman who offered the most compelling defense of ecstatic attention. Simone Weil (1909–1943), among the few published female philosophers of the period, synthesized mystical and aesthetic strands of ecstatic discourse, arguing in her posthumously published *La pesanteur et la grâce* (1947) that "extreme attention is what constitutes the creative faculty in man."[78] Such attention declined to explain, analyze, or grasp; in place of the intellectual striving associated with concentration, it aimed for nonvolitional plenitude, consisting of "that attention which is so full that the 'I' disappears."[79] For Weil, this self-negation was the purest form of prayer as well as the cornerstone of imaginative work. "The amount of creative genius in any period," she argued, "is strictly in proportion to the amount of extreme attention and thus of authentic religion at that period."[80] Her conception of creative focus as self-emptying resonated not only with Bergson and Segond, but, as she acknowledged, with aspects of both Hindu and Buddhist philosophy, whose devotional traditions she regarded as compatible with Christianity.[81]

Weil's work comes at the very end of our period, marking a point of culmination in the rich strand of scientific and mystical writing on hypnosis and ecstasy. Through the late nineteenth and early twentieth centuries this discourse was markedly French, blending materialist and experimental psychology with philosophy, sociology, and theology. In some respects, it contrasted with the tradition of empirical psychology unfolding in German centers during the same period, and yet points of intersection between the two existed. At the furthest reaches of empirically discernable attention, as **David Trippett** argues, lay a form of imaginative apprehension that seemed to open, if not onto a metaphysical realm, then onto an otherwise inaccessible domain of the human psyche. Bergson's conception of the ecstatic state as one in which the faint harmonics of a vibrating string were set in motion in the mind of the praying individual bears some resemblance to the Helmholtzian notion of distant overtones audible only to a mind endowed with extraordinary attentional virtuosity.[82] For some French theorists, these were echoes of divine revelation. In Germany, such sounds were more ambiguous, suggesting an unquantifiable mode of perception—hints of what Trippett terms an "auditory unconscious."[83]

Looming over German psychophysics in the final decades of the nineteenth century, the idea of unconscious perception interfaced with the emerging and, at times, quasi-mystical practices of psychoanalysis. Famously, Freud described the importance of a loose, undirected attention—a form of all-embracing awareness capable of intuiting the faint vibrations of the deep psyche. His student Theodor Reik (1888–1969) extended this idea, embracing a form of absorption that allowed for apprehension of the "most delicate communications . . . the secret messages that go from one unconscious to another."[84] Listening to such messages was not quite the same as hearing the voices of God, yet both involved a state of "free-floating attention" (in Reik's words) quite distinct from spotlighted concentration.[85] Nongrasping and anti-analytical, this inner focus required a quieting (in some accounts, hypnotizing) of the active intellectual powers. For psychologists, psychoanalysts, and mystics alike this state was regarded as a gateway to revelation—a psychology at once threatening, exotic, expansive, and annihilating.

Measurement

The long nineteenth century's fascination with attention is intimately linked to the emergence of new technologies of measurement. Instruments including clocks, metronomes, sonometers, resonators, plethysmographs, and kymographs played integral roles in the development of new theories of music and of perception alike. Speculations about the role of attention in human auditory cognition could be occasionally found in music theory treatises in the eighteenth century; by the second half of the nineteenth century, investigations of these topics were pursued at the most prominent psychological laboratories on both sides of the Atlantic. The disciplinary boundary between music, acoustics, and psychology at the time was highly porous; thus, we find key insights into the role of attention in structuring music first articulated by music theorists as introspective observations, and subsequently empirically verified in lab settings. Later in the century, these ideas would migrate from the lab to inform musical practice; as Alexander Bonus has compellingly argued, "the precise mechanical rhythm that most modern musicians practice, modern pedagogues teach, and modern audiences hear today" was forged in scientific laboratories, rather than on the concert stage.[86]

For one important case study, let us turn to what today is known as "subjective rhythmicization," that is, our perception of an undifferentiated series of clicks as a series composed of a "tick" followed by a "tock" (or by two or three "tocks," if we choose to perceive a triple or quadruple meter, respectively). The earliest account of this phenomenon appears in the *Essay towards a Rational System of Music* (1770) by English music theorist John Holden. Holden reports that

> if a drummer were to give a number of quick, but equal timed beats, each with just an equal force, a person attending to them would naturally parcel them by 4 and 4 together, by giving a greater regard to every 4th beat; by which regard these beats would to him acquire a kind of emphasis or accent, as if the drummer did really strike them harder than the others. . . . But the case would be the very same with a hearer, if an inanimate machine were made to beat the drum.[87]

Holden here ascribes the accents we perceive in a series of undifferentiated beats solely to our "regard," that is, to the attention we pay to these particular beats. (Attention, it should be noted, looms large in Holden's work, where it is also used to explain the experience of hearing tonal ambiguity.)[88] A similar awareness regarding the role of our minds in perceiving and constructing meter was independently articulated six years later by the German music theorist Johann Philipp Kirnberger in the second volume of *Die Kunst des reinen Satzes in der Musik* (1776).[89] This insight constituted a major break with earlier accounts of meter, which tended to ascribe the differences in the saliency of beats to acoustical differences arising from bodily movement relating to performance practice.

Holden's offhand mention of a drum-beating "inanimate machine" illustrates the aforementioned link between emerging conceptions of temporal attention and technology. Pocket watches (which Holden praises as useful for internalizing meter), chronometers (invented by Huygens in 1656 and subsequently adapted for musical use, notably by Jean-Étienne Despréaux and Gottfried Weber), and the metronome (patented in 1815 by Maelzel) swiftly revolutionized the way in which musical meter was understood. Prior to the metronome, time signatures had indicated tempo and genre/style of performance to a certain extent; these indications became far less codified by the middle of the nineteenth century, given that the desired tempo could be indicated by measuring clicks in absolute time.

The metronome was widely adopted by European performers and pedagogues in the first half of the nineteenth century, to a mixed reception.[90] Endorsed by composers including Cherubini, Pleyel, and Beethoven, some music theorists rejected the device as overly precise and hence a potential obstacle to expressivity.[91] However, a view that rhythm should in some way be "metronomic" was by no means widespread; Adolph Bernhard Marx's view of musical time perception, for example, prioritized the dimensions of duration and accent, regardless of whether the periodical succession of the events they were applied to was equal:

> I dwell upon a moment so long as it engages my interest and attention, or until I am drawn away by another of greater force of attraction; I hurry from moment to moment when a lively sensation impels me onward, and I expend a greater amount of force and energy upon those moments which are to me the most important.[92]

This vivid description of the experience of listening to music in time as dependent on attention, attraction, energy, and force would set the stage for later theories of musical forces, ranging from Ernst Kurth's approach to form as composed of various types of waves to the experience of time as a wave proposed by Victor Zuckerkandl in the 1960s.[93] Marx's (metaphorical) application of these metaphysically charged terms further reveals that both music theorists and psychologists were thinking about rhythm and meter with reference to concepts of force and motion.

Musicians were, of course, not the only ones interested in metronomes. By the 1850s, the metronome was employed by Dutch physiologist Franciscus Cornelius Donders as a tool of temporal measurement in his physiological experiments.[94] At the dawn of modern empirical psychology a decade or so later, the metronome became a staple of the laboratory, facilitating groundbreaking work on the nature of attention and distraction in the hands of the renowned German psychologist Wilhelm Wundt and his students. From the 1860s on, Wundt used the metronome in a series of important experiments intended to ascertain the duration and capacity of human consciousness. Acutely aware of the technology of his day, however, he first noted the absolute importance of finding a metronome adequately suited for scientific research: "If it happens that we can at will hear into the beats of the metronome an ascending or a descending beat, i.e. we can hear one and the same beat now emphasized and now unemphasized, then we may regard the instrument as suitable for all the psychological experiments to be described in the following pages."[95] Wundt here describes the phenomenon of subjective rhythmicization, and suggests his readers perform a version of Holden's thought experiment with an "inanimate machine" in order to test the accuracy of the scientific equipment required to probe consciousness.

By the 1890s, the phenomenon of subjective rhythmicization had been thoroughly experimentally studied in labs in Europe and in the United

States.[96] The landmark work in this regard is Thaddeus Bolton's article on "Rhythm," published in *The American Journal of Psychology* in 1894. Bolton, whose advisor, G. Stanley Hall, had studied with Wundt in Leipzig, explains subjective rhythmicization as follows:

> This rhythmical grouping [is] a series of efforts to attend to the sound. The grouping results from a sequence of acts of attention. When the attention is directed to the sensation, it lays hold upon the first impression with great force and makes it the sole object of consciousness. If this were the only sound, the attention would turn to something else, but as succeeding impressions follow before the first wave of attention has subsided, they are seized upon with less force than the first impression, and are subordinated to it in different degrees according to the strength of the apperceptive act. Subsequent waves of attention follow the same process as long as the will directs the attention to the phenomenon. The attention accommodates itself to a certain number of impressions, which fall easily within the period of a wave, providing there is no objective difference in the impressions. If there is a regularly recurrent difference, this becomes the signal for a new act of attention, providing only that the span does not exceed or fall much under the normal period of a wave. If this recurrent difference follows at too great intervals, the attention breaks up the span in two portions, the one more emphatic than the other.[97]

Bolton's reference here to the attention wave has to do with insights into dynamic attention, which were coming out of European laboratories and reverberating across the Atlantic (Bolton's experimental setup is shown in Figure 0.3). The notion of the attention wave linked our mental experience to our nervous physiology, explaining various thresholds and transitions in alertness as arising from the behavior of physiological processes. In his contribution to this volume, **D. Graham Burnett** surveys the rise of the attention wave from Viktor Urbantschitsch's physiological experiments in 1870s Vienna through research in Wundt's lab in Leipzig to its ultimate flowering in the kymograph experiments conducted by Edward B. Titchener, Wundt's student, at Cornell University circa 1905.

Attention also played a crucial role in the investigation of pitch perception, which similarly moved between the worlds of music theory and laboratory

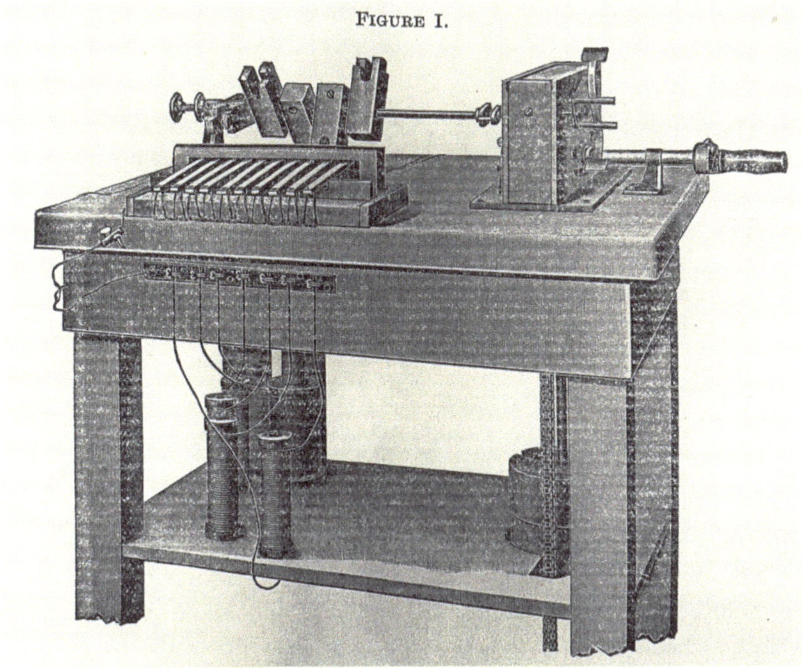

Figure 0.3. Bolton's experimental setup with a chronograph and telephone. Thaddeus L. Bolton, "Rhythm," *The American Journal of Psychology* 6.2 (1894).

psychology. As early as 1726, Rameau described the role of attention in distinguishing partials in his *Nouveau système*, observing that the overtones

> may be distinguished again in cymbals, in the lower sounds of a trumpet, in bells, etc., if one gives all his attention to the undulations that create a sort of murmur in the air just after the predominant sound of the instrument proper to this effect. It is also essential not to be occupied with this dominating sound and not to be distracted by any foreign noise.[98]

The notion that paying closer attention would enable perception of hitherto unrecognized sounds links Rameau's report with Helmholtz's empirical investigations undertaken with the aid of harmoniums and resonators, as Benjamin Steege has shown.[99] Helmholtz, for example, observes that in order to recognize overtones, the ear needs the assistance of special performance techniques, including touching a piano key, singing various vowels, or listening with resonators:

> On applying to the ear the resonator corresponding to any given upper partial of the compound *c*, such as *g′*, this *g′* is rendered much more powerful when *c* is sounded . . . the increase of the loudness of *g′* caused by the resonator may be used to direct the attention of the ear to the tone it is required to distinguish. On gradually removing the resonator from the ear, the force of *g′* will decrease. But the attention once directed to it by this means, remains more readily fixed upon it, and the observer continues to hear this tone in the natural and unchanged compound tone of the given note, even with his unassisted ear. The sole office of the resonators in this case is to direct the attention of the ear to the required tone.[100]

Here, the resonators assist the ear in singling out part of the compound tone via amplification. With practice, Helmholtz continues, one can eventually largely dispense with the resonators, though not without the "certain amount of undisturbed concentration [that] is always necessary for analysing musical tones by the ear alone."[101]

The experience of attentively listening to—and identifying—overtones that illustrated Helmholtz's resonance theory of cochlear mechanics (as presented in *Die Lehre von den Tonempfindungen als physiologische Grundlage für die Theorie der Musik*, 1863), reappears in subsequent investigations of the nature of attention by prominent psychologists including Georg Elias Müller, Wundt, William James, and Titchener.[102] Wundt, for example, observes that if we attend to a specific overtone within a sound,

> we always find that in this case we first recall the memory image of the tone to be heard and only then hear it out of the whole sound. . . . We hear a sound in which we suspect a certain overtone due to certain associations; only then do we recall the image of the tone-to-be-heard in our memory [*das Erinnerungsbild des zu hörenden Tones zurückruft*] and then immediately notice it from the sound we hear.[103]

That is to say, listeners can perceive overtones if they know what to listen for; if this stimulus is not provided separately (by means of priming or a resonator), they must mentally prepare to make the identification by means of their prior knowledge, which is to say, they must be prepared to attend to *that* particular stimulus. Moreover, this process does not occur instantaneously, but rather in time. Thus:

> Every idea requires a certain amount of time to rise to the focus of consciousness. During this time we always find within ourselves the peculiar feeling of paying attention. This feeling is all the more vivid the more the focus of consciousness is concentrated, and in this case it tends to continue even after the idea is completely clear to consciousness. In its preparatory stage, it is most clearly perceived in the state of reflection or expectation for an expected impression . . . We experience attention as something that comes from ourselves, even if we clearly imagine the impression that draws our attention to itself as an external one.[104]

William James invoked both Helmholtz's overtone experiments and Wundt's response to emphasize the importance of priming in our process of attention, noting that "where the impression to be caught is very weak, the way not to miss it is to sharpen our attention for it by preliminary contact with it in a stronger form."[105] The larger point, for James, however, is that perception depends on mental receptivity in the form of attention. Thus:

> Whilst the object excites [a brain cell] from without, other brain-cells, or perhaps spiritual forces, arouse it from within. The latter influence is the "adaptation of the attention." The plenary energy of the brain-cell demands the co-operation of both factors: not when merely present, but when both present and attended to, is the object fully perceived. . . . Organic adjustment, then, and ideational preparation or *preperception* are concerned in all attentive acts.[106]

Key to James's (and Wundt's) notion of attentional priming are the mechanisms of memory and prediction, both of which rely on experience and therefore enculturation. We attend most closely, and with the finest focus, to what we expect and know. Stimuli—from individual sounds to musical systems—outside our epistemic framework easily escape our notice or simply register as uncomputable. This is among the points made by **Jacob Olley**, whose exploration of "calculative listening" in Ottoman contexts underscores the degree to which perception of empirical data (including information revealed by devices like resonators and sonometers) was inflected by local cultural practices and values. In the late nineteenth century, he shows, researchers seeking to develop an empirical theory of Ottoman music used newly developed instruments such as the sonometer in order to quantify the experience of non-European audition and, in so doing, draw it into "modern" and "rational" discourse—render it legible. They thus laid the groundwork for an alternative, scientifically grounded attentive practice for Ottoman musics, including the meaningful "preperception" touted by James.

(Re)orienting and preparing the attention was not just an intellectual activity but a physical practice; indeed, for thinkers including James and, before him, Alexander Bain, Carl Georg Lange, and James Sully, attention was written in and on the body, reflected in material alterations of the nervous and muscular systems. These thinkers, James explains, regarded "muscular adjustment . . . [as] the essence of the attentive process throughout."[107] They devoted considerable time to studying physical attentional cues including inclination of the body, turning of the head, or swiveling of the eyes, often identifying such actions as products of a Darwinian evolution of physical attention.[108] It is unsurprising, therefore, to find these ideas informing music pedagogy of the same period. Once attention was conceived as a physical act, the bodily technique of musical performance ceased to be regarded as separate from psychic intention, or simply as its outcome or demonstration; instead, it became synonymous with alertness and even the first site of its generation. That is to say, rather than mental attention preceding or providing a preparation for physical performance, performance itself produced attention—a player could experience musical focus as an embodied phenomenon, now quantifiable in haptic (or more broadly somatosensory) terms.

For the pianist and composer Marie Jaëll, as Michael Weinstein-Reiman observes, "a pianist's attention to haptic and aural impulses [was part of] an elusive, 'simultaneous and successive' process that collates the pianist's tangible sensation of the keyboard and the ineffable mental impressions conjured by sound."[109] According to the piano pedagogue Tobias Matthey, "the sensation of key-resistance and the sensation of pedal resistance, these are both but part and parcel of that single thought and purpose, summed up as 'performing-attention'—and that means: *attention to Music*—THROUGH OUR INSTRUMENT."[110] The influence of motor theories of attention on the development of piano pedagogy is the subject of **Youn Kim**'s contribution to this volume, which explores how musicians applied scientific ideas about attention in relation to muscular action and motor innervation to improve specific aspects of pianistic technique. Musical performance thus became a site of experimental psychology, and the piano itself an instrument central to the honing and tracking of states of attention.

Legacies

> To pay attention / this is our endless and proper work.
>
> —Mary Oliver

The work of auditory attention—its marshaling, molding, measuring, and reimagining—continued unabated through the twentieth century, occupying an ever-wider contingent of researchers and artists around the globe. Many of the characteristic attentional orientations of the "long" nineteenth century persisted (albeit in modulated form), although these were juxtaposed with, and in some cases challenged by, new forms of alertness shaped by shifting media landscapes. The rise of broadcasting, recording, and amplification technologies, for example, generated noisy, multichannel auditory environments in which techniques of spotlighted attention became key.[111] The 1950s saw the identification of the "cocktail party effect," a term coined by British psychologist Colin Cherry, which referred initially to the ability of air traffic controllers to successfully attend to a single strand of the many simultaneous conversations with pilots coming in over their headphones.[112] (This problem continues to be investigated today as part of what psychologist Albert Bregman defined in the 1970s as auditory scene analysis.[113])

Ubiquitous sound emanating from radios and gramophones in commercial, domestic, and industrial settings necessitated a different kind of selectivity predicated on passivity—a low-level cognitive engagement with music that left attentional resources available for other auditory activities. This mode was prefigured, even inculcated, by musical cultures of the early twentieth century. To take one particularly telling example: In German-speaking centers, the 1920s saw the establishment of the genre of *U-Musik*, short for *Unterhaltungsmusik* (entertainment music), originally associated with light concerts put on by orchestras during the summer months at resorts. Unlike serious music (*ernste Musik*, or *E-Musik*), which demanded a listener's full concentration, *U-Musik* purportedly required only superficial focus, leaving auditors free to engage in conversation, or *Unterhaltung* (the word means both entertainment and conversation in German).[114] The main purpose of the distinction was to parcel out royalties between what was essentially pop music and art music—but it also entailed a psychic renegotiation resulting in dual-channel forms of ambient and spotlighted attentiveness among auditors.[115]

Passive forms of focus—those associated with boredom, dreaming, and sleep—emerged as central to experimental musical currents across the Atlantic, particularly among mid-century composers working with tape and related recording technologies. Exploration of loose forms of awareness including those associated with dream states was among the explicit goals of Pauline Oliveros's *Sonic Meditations*.[116] And drastically extended attentive exercises, often incorporating unchanging or slowly transforming musical material, shaped the music of Terry Riley, Julius Eastman, Steve Reich, and John Cage (among others). Sometimes linked to non-Western meditative practices, these experiments constituted new forms of ecstatic focus exploring the boundaries between the soporific and the revelatory. As John Cage put it, drawing on Zen philosophy, "If something is boring after two minutes, try it for four. If still boring, then eight. Then sixteen. Then thirty-two. Eventually one discovers that it is not boring at all."[117]

Within the lineage of the European concert tradition, on the other hand, analytical forms of concentration continued to be valorized, separating *Kenner* from *Liebhaber* ever more starkly. In his infamous essay "The Composer as Specialist," also known as "Who Cares If You Listen?," Milton Babbitt asserted that lay audiences could not help but be "bored and puzzled by what [they are] unable to understand."[118] A related observation was made by Michel Foucault in dialogue with Pierre Boulez a few decades later, namely that contemporary concert music required "a very difficult attention, one which is in contradiction to the familiarities woven by hearing of [older] classical music."[119] As late as the 1990s, composers were still wrestling with the dilemma of complexity versus cognizability, with the composer Fred Lerdahl turning to findings in psychology to argue that Elliot Carter's superposed musical meters were simply ungraspable, given that these contrasting lines made impossible demands upon the attention.[120]

The sense of musical listening as a testing ground for attention—a site for the practice, modulation, management, or measurement of alertness—was hardly new. Twentieth-century conversations along these lines were in many regards extensions of nineteenth-century debates around the virtue of concentrated listening, the ecstatic potential of highly focalized audition, and the role of musical focus in the evolution of specialized or extended attentional capacities. If some rhetorics and practices of auditory focus shifted through the mid-late twentieth century, or were newly demoted/valorized, others remained oddly unchanging. Most obvious among these was the practice of concentrated Hanslickian audition, which continued to be tacitly prescribed in recital hall settings by canonic repertoire. Indeed, paraphrasing Lydia Goehr, we might conceive of the experience of twentieth- (and twenty-first-) century concert listening as involving an imaginary museum of musical attentions, no less than one of musical works.[121]

As this essay has argued, attentive modalities are neither hard wired nor default, but are instead the product of political and social enculturation, technological conditioning, and scientific engineering. Especially central to their development have been techniques of listening, which emerged across the nineteenth and early twentieth centuries as inextricable from ideologies of focus/unfocus. Audition was predetermined by attention, even while attention was itself fundamentally shaped by audition. This feedback loop made listening the unstable measure of both concentrative acuity and—in some accounts—volitional selfhood. To explore this phenomenon, we need a stronger sense of both attention's sonic pasts and audition's attentional investments. Here, and in the chapters to follow, we take a step in this direction.

Notes

1. Augustine's *De musica*, trans. Robert Catesby Taliaferro, in *The Fathers of the Church, a New Translation: Writings of Saint Augustine*, vol. 2 (New York: Fathers of the Church, 1947), 345–46.

2. Gottfried Wilhelm Leibniz, *Nouveaux Essais Sur L'Entendement Humain* § 4. *Philalethes* (Amsterdam and Leipzig: Jean Schreuder, 1765), 90. The essays were written in 1704 but first published in 1765.

3. Thomas Reid, *Essays on the Active Powers of Man* (Edinburgh: John Bell, 1788), 79.

4. James Sully, *Sensation and Intuition: Studies in Psychology and Aesthetics* (London: Henry S. King & Co, 1874), 62.

5. "So sind wir auch fähig, unter einer Menge zugleich wirkender musikalischer Instrumente ein einzelnes und oft das schwächste mit Aufmerksamkeit zu verfolgen, während die Töne der anderen Instrumente des Orchesters nur dunkle Empfindungen in uns erregen." Johannes Müller, *Handbuch der Physiologie des Menschen für Vorlesungen*, Volume 1, Part 2 (Coblenz: J. Hölscher, 1834), 827.

6. This etymology is unpacked by Mladen Dolar in *A Voice and Nothing More* (Cambridge, MA: MIT Press, 2006), 75–76. Of particular interest is the Old English "*hlysnan*" ("hear, attend to, obey"); and Middle English "*listenen*" ("listen," "pay attention to"); see the entry for "listen" in the Oxford English Dictionary: https://www.oed.com.

7. John Mason Good, *System of Nosology with a Corrected and Simplified Nomenclature* (London: Bensley and Sons, 1817), 937.

8. The literature on attention is too voluminous to cite in full. Studies that have provided an important backdrop for this essay include Yves Citton, *Pour une écologie de l'attention* (Paris: Éditions du Seuil, 2014); Thomas H. Davenport and John C. Beck, *The Attention Economy: Understanding the New Currency of Business* (Boston: Harvard Business School Press, 2002); Bernard Stiegler, *Prendre soin: De la jeunesse et des générations* (Paris: Flammarion, 2008); Tim Wu, *The Attention Merchants: The Epic Scramble to Get Inside Our Heads* (New York: Vintage, 2017); Jonathan Crary, *Suspensions of Perception: Attention, Spectacle, and Modern Culture* (Cambridge, MA: MIT Press, 2001); Lucy Alford, *Forms of Poetic Attention* (New York: Columbia University Press, 2020); Nicholas Dames, *The Physiology of the Novel: Reading, Neural Science, and the Form of Victorian Fiction* (Oxford: Oxford University Press, 2007); David Marno, *Death Be Not Proud: The Art of Holy Attention* (Chicago: University of Chicago Press, 2016); Paul North, *The Problem of Distraction* (Stanford, CA: Stanford University Press, 2012); Natalie M. Phillips, *Distraction: Problems of Attention in Eighteenth-Century Literature* (Baltimore: Johns Hopkins University Press, 2016); Aleida Assmann and Jan Assmann, eds. *Aufmerksamkeiten. Archäologie der literarischen Kommunikation 7* (Munich: Fink, 2001); Lorraine Daston, *Eine kurze Geschichte der wissenschaftlichen Aufmerksamkeit* (Munich: Carl Friedrich von Siemens Stiftung, 2001); Michael Hagner, "Toward a History of Attention in Culture and Science," *MLN* 118, no. 3 (2003): 670–87; Johannes Wankhammer, *Creatures of Attention: Aesthetics and the Subject Before Kant* (Ithaca, NY: Cornell University Press, 2024).

9. A compact overview of these theories is given in Christopher Mole, "Attention," *The Stanford Encyclopedia of Philosophy* (Fall 2024 Edition), ed. Edward N. Zalta and Uri Nodelman, https://plato.stanford.edu/archives/fall2024/entries/attention/.

10. Crary, *Suspensions of Perception*, 24, 45.

11. "Diese Möglichkeit, gewissen Sinnesempfindungenden Eintritt in's Bewußtseyn zu verwehren, sie gleichsam zu annulliren, ist für das Gehör insofern von besondrer Wichtigkeit als wir nicht im Stande sind, unser Ohr in ähnlicher Art, wie das Auge, beliebig zu schließen und die Einwirkungen der Außenwelt von ihm abzuhalten. Darum vielleicht die fast noch größere Wichtigkeit, welche die Aufmerksamkeit im Gebiete der Schallempfindungen behauptet." Hermann Ulrici, *Gott und der Mensch*, vol. I: *Leib und Seele: Grundzüge einer Psychologie des Menschen*, 2nd ed. (Leipzig: T. D. Weigel, 1874), 289.

12. "So ist es denn natürlich, dass die Aufmerksamkeit, von welcher alle Geistesthätigkeit zuletzt abhängt, gerade in Gehörurteilen ihr Wesen und Wirken vorzüglich reich entfaltet und dass die Theorie der Aufmerksamkeit sich hier am besten entwickeln lässt." Carl Stumpf, *Tonpsychologie*, vol. 1 (Leipzig: Hirzel, 1883), 68.

13. "die wahre [Art des Genußes] . . . besteht in der aufmerksamsten Beobachtung der Töne und ihrer Fortschreitung; in der völligen Hingebung der Seele in diesen fortreißenden Strom von Empfindungen; in der Entfernung und Abgezogenheit von jedem störenden Gedanken und von allen fremdartigen sinnlichen Eindrücken." Heinrich Wilhelm Wackenroder, "Letter of May 5, 1792," in *Brief an Ludwig Tieck* 4, ed. Carl Holtei (Breslau: Eduard Trewendt, 1864), 173; "[eine] ruhige und gesammelte Aufmerksamkeit." Hermann von Helmholtz, *Die Lehre von den Tonempfindungen als physiologische Grundlage für die Theorie der Musik*, 4th ed. (Braunschweig: F. Vieweg und Sohn, 1877), 111; Salomon Jadassohn, *Manual of Harmony*, translated by Paul Tarek and H. B. Pasmore (Leipzig: Breitkopf and Härtel, 1884), 201.

14. "Die Aufmerksamkeit ist gleichsam die Hand, mit welcher die Seele in ihrem Grunde eine dunkle Vorstellung ergreift, sie in die Höhe hebt und dadurch an das Tageslicht bringt." Georg Friedrich Meier, *Metaphysik* (Halle: Johann Justinus Gebauer, 1757), 73.

15. Quoted in Margaret Koehler, *Poetry of Attention in the Eighteenth Century* (New York: Palgrave Macmillan, 2012), 1–60, at 6, 32.

16. James Beattie, *Dissertations Moral and Critical* (London: W. Strahan and T. Cadell, 1783), 12. Dugald Stewart, *Elements of the Philosophy of the Human Mind*, vol. 1 (Edinburgh: A. Strahan and T. Cadell, 1792), 108; see also Lorraine Daston, "Attention and the Values of Nature in the Enlightenment," in Lorraine Daston and Fernando Vidal, eds. *The Moral Authority of Nature* (Chicago: University of Chicago Press, 2004), 100–26.

17. Thomas Reid, *An Inquiry into the Human Mind*, 3rd ed. (London: T. Cadell, 1769), 228; Stewart, *Elements*, 130.

18. Matthew Riley, *Musical Listening in the German Enlightenment: Attention, Wonder and Astonishment.* (Abingdon: Routledge, 2017); see also W. Dean Sutcliffe, *Instrumental Music in an Age of Sociability: Haydn, Mozart and Friends* (Cambridge: Cambridge University Press, 2020), 346–70.

19. Walter Young, "An Essay on Rythmical [*sic*] Measures," *Transactions of the Royal Society of Edinburgh* 2 (1790), 95. On Young and his circle see Carmel Raz, *Hearing with the Mind: Proto-Cognitive Music Theory in the Scottish Enlightenment* (New York: Oxford University Press, 2025).

20. Nicholas Mathew, "Interesting Haydn: On Attention's Materials," *Journal of the American Musicological Society* 71, no. 3 (2018): 655–701.

21. "Die holden Sirenenstimmen deiner in bunter Mannigfaltigkeit prangenden Sätze locken mich immer tiefer und tiefer hinein." E. T. A. Hoffmann, "Kreisleriana," in Carl Georg von Maaßen, ed., *E. T. A. Hoffmanns Sämtliche Werke; historisch-kritische Ausgabe*, vol. 1 (Munich: G. Müller, 1908), 61.

22. On romantic idealism and musical concentration, see Mark Evan Bonds, *Music as Thought: Listening to the Symphony in the Age of Beethoven* (Princeton, NJ: Princeton University Press, 2006); On conceptions of organicism and deep knowledge, see Holly Watkins, *Metaphors of Depth in German Musical Thought from E.T.A. Hoffmann to Arnold Schoenberg* (Cambridge: Cambridge University Press, 2011); On attention and "sonically moving forms" in Hanslick and A. B. Marx, see Alexandra Hui, *The Psychophysical Ear: Music Experiments, Experimental Sounds*, 1840–1910 (Cambridge, MA: MIT Press, 2013), 23–54.

23. Theodor Adorno, "Anweisungen zum Hören neuer Musik," translated in Caroline Duttlinger, *Attention and Distraction in Modern German Literature, Thought, and Culture* (Oxford: Oxford University Press, 2022), 366.

24. Both in German-speaking lands and across the Atlantic, concentration was figured as an avenue to metaphysical uplift as well as a conquering/controlling power. See, for example, Edward Brooks, *Mental Science and Methods of Mental Culture: Designed for the Use of Normal Schools, Academies, and Private Students Preparing to be Teachers* (New York: Normal Publishing Company, 1882), 77: "The concentration of the mind in attention is like the massing of troops in a battle. A fortification is stormed and taken by the concentration of a battalion. . . . Attention may be regarded as the commander-in-chief of our mental powers. If we wish to conquer truth, to capture and hold new territory, or to occupy and appropriate that which is already taken, we must concentrate our mental forces, and, like a successful general, determine to fight it out on that line."

25. Natalie M. Phillips, "Fixated Attention: The Gothic Pathology of Single-Minded Focus," in *Distraction: Problems of Attention in Eighteenth-Century Literature*, Chap. 4, 132–73, at 140.

26. Francesca Brittan, "Berlioz and the Pathological Fantastic: Melancholy, Monomania, and Romantic Autobiography," *19th-Century Music* 29, no. 3 (2006): 211–39.

27. Stewart discusses this phenomenon in a literary context. See Garrett Stewart, *Novel Violence: A Narratography of Victorian Fiction* (Chicago: University of Chicago Press, 2009), 61–89.

28. "La distraction . . . c'est l'opposé de la stupidité, qui reste sur une même idée." Denis Diderot, "Distraction (*Morale*)," in Denis Diderot and Jean Le Rond d'Alembert, eds. *Encyclopédie, ou, Dictionnaire raisonné des sciences, des arts et des métiers* 4 (Paris: Briasson, David, Le Breton, & Durand, 1754), 1061.

29. Schiller's remarks on cognitive loosening and play weave through his epistolary treatise. See *On the Aesthetic Education of Man, in a Series of Letters* (1795), ed. and trans. Elizabeth M. Wilkinson and L. A. Willoughby (Oxford: Clarendon Press, 1967). Quotations given here are taken from letter 6 (at pp. 209, 215); letter 27 (at p. 43); letter 43 (at p. 103). Richard Adelman elaborates on Schiller's conceptions of cognitive passivity and creativity in *Idleness, Contemplation and the Aesthetic, 1750–1830* (Cambridge: Cambridge University Press, 2011), 52–67.

30. Friedrich Schlegel, "Idyll of Idleness" ("Idylle über den Müßiggang"), trans. in Peter Firchow, *Friedrich Schlegel's "Lucinde" and the Fragments* (Minneapolis, MI: University of Minnesota Press, 1971), 63–68.

31. Crary, *Suspensions of Perception*, 77.

32. Francesca Brittan, "Idle Schubert," in "Colloquy: Attention, Anxiety, and Audition's Histories," convened by Francesca Brittan and Carmel Raz, *Journal of the American Musicological Society* 72, no. 2 (2019): 557–63. See also Scott Burnham, "Schubert and the Sound of Memory," *Musical Quarterly* 84, no. 4 (2000): 655–63.

33. According to Rudolph Hermann Lotze, "He who tunes the strings of a piano and gives the most intense attention to his subject retains only the minimum of self-consciousness; he who is absorbed in a mathematical problem has scarcely more. But he who attentively considers a decision to be made should at least bring a specific recollection of his personality to this reflection. Both the self-conscious immersion in a single idea and the unconnected flight of many ideas are only compatible with the healthy pursuit of mental life when they occur temporarily; continual distraction as well as a prolonged narrowing of the flow of thoughts, are, as we will later recognise, the first stages of mental disorder." ("Wer die Saiten eines Clavieres stimmt, hat bei der angestrengtesten Aufmerksamkeit auf seinen Gegenstand ein Minimum des Selbstbewusstseins; wer versunken in seine Gedanken ein mathematisches Problem verfolgt, besitzt dessen kaum mehr; wer mit Aufmerksamkeit dagegen einen zu wählenden Entschluss überlegt, soll wenigstens zugleich eine bestimmte Erinnerung seiner Persönlichkeit zu dieser Reflexion hinzubringen. Sowohl das selbstbewusstlose Versenken in eine einzige Vorstellung, als die unverbundene Flucht vieler sind Zustände, die nur, wo sie momentan sich einstellen, mit der gesunden Bestimmung des geistigen Lebens vereinbar sind; eine dauernde Zerstreuung sowohl, als eine dauernde Verengung des Gedankenlaufs werden wir dagegen später als Anfangspunkte der Seelenstörungen kennen lernen.") Rudolph Hermann Lotze, *Medizinische Psychologie oder Physiologie der Seele* (Leipzig: Weidmann, 1852), 506–507.

34. On cyclic attention in literary culture, see Nicholas Dames, *The Physiology of the Novel*, esp. chap. 2. Many mid-century novels were published in serial installments, which itself suggested intermittent periods of attentiveness rather than an elongated effort; see Rob Allen,

"'Pause You Who Read This': Disruption and the Victorian Serial Novel," in *Serialization in Popular Culture*, edited by Rob Allen and Thijs van den Berg (New York: Routledge, 2014): 35–49.

35. Théodule Ribot, *Le psychologie de l'attention* (Paris: Alcan, 1889), 3; James, *The Principles of Psychology*, vol. 1 (New York: Henry Holt and Company: 1890), 420.

36. Benjamin Steege, "Joining In or Letting Go?" in "Colloquy: Attention, Anxiety, and Audition's Histories," 569–74.

37. For Morse's own account of this experiment, see *Samuel F. B. Morse: His Letters and Journals*, vol. 2, ed. Edward Lind Morse (Boston and New York: Houghton Mifflin Company, 1914), 74–76. An early press account was published by Frank Jewett Mather Jr., "Morse and the Telegraph: The Great Portrait Painter and Inventor Who Will Forever Be Memorialized in a Series of Dots and Dashes," reprinted in *The Nation*, July 8, 2009. https://www.thenation.com/article/archive/morse-and-telegraph/, accessed November 2025.

38. Citton, *The Ecology of Attention*, 18.

39. Ibid., 17–18, 31.

40. Gustave Le Bon's *Psychologie des Foules* (Paris: Alcan, 1895) paved the way for Gabriel Tarde, *L'opinion et la foule* (Paris: Alcan, 1901), which introduced "psychologie collective ou psychologie sociale" as an emerging area of study (see the unpaginated *Avant-Propos*).

41. Tarde, *L'opinion et la foule*, 40. In reference to "La foule": "Leur attention—et aussi bien leur inattention—est toujours plus forte et plus persévérante que ne le serait celle de chacun des individus qui les composent, s'il était seul." Ibid., p. 54: "Le public: car le fait seul de leur attention concentrée sur un seul objet, en une sorte de monoïdéisme collectif, les rapproche de l'état de rêve ou d'hypnose."

42. Wider explorations of nineteenth-century communication networks from technological, sociological, and economic perspectives include Carolyn Marvin, *When Old Technologies Were New: Thinking About Electric Communication in the Late Nineteenth Century* (Oxford: Oxford University Press, 1988); Laura Otis, *Networking: Communicating with Bodies and Machines in the Nineteenth Century* (Ann Arbor, MI: University of Michigan Press, 2001); Kenneth Rogers, *The Attention Complex: Media, Archeology, Method* (New York: Palgrave Macmillan, 2014).

43. For more on technologies of auditory networking, see Jonathan Sterne, *The Audible Past: Cultural Origins of Sound Reproduction* (Durham, NC: Duke University Press, 2003); Brian Kane, *Sound Unseen: Acousmatic Sound in Theory and Practice* (Oxford: Oxford University Press, 2014); Simon J. Potter, *Broadcasting Empire: The BBC and the British World, 1922–1970* (Oxford: Oxford University Press, 2012); Tom Lewis, *Empire of the Air: The Men Who Made Radio* (Ithaca, NY: Cornell University Press, 2021).

44. Francesca Brittan, "The Neural Orchestra: Cognitive Instrumentalities," *Journal of the American Musicological Society* 77, no. 1 (2024): 1–63. Musical metaphors appear throughout Tarde's text, including references to the leader of the crowd as a "conductor" (*L'opinion et la foule*, 15).

45. On Cubí y Soler's orchestral brain, see Brittan, "The Neural Orchestra," 20–23. Quotations given here are taken from Cubí y Soler, *La frenolojía y sus glorias* (1852), translated by the author as *Leçons de phrénologie scientifique et pratique* (Paris: Baillière et fils, 1857); see vol. 2, 369: "la volonté ou harmonisativité . . . c'est un principe d'attention suprême et souveraine" and vol. 2, 464: "une télégraphie électrique nerveuse."

46. "L'attention multiple lui revient; il a l'œil partout; il indique d'un regard les entrées vocales et instrumentales, en haut, en bas, à droite, à gauche; il jette avec son bras droit de

terribles accords qui semblent éclater au loin comme d'harmonieux projectiles; puis il arrête, dans les points d'orgue, tout ce mouvement qu'il a communiqué; il enchaîne toutes les attentions; il suspend tous les bras, tous les souffles, écoute un instant le silence . . . et redonne plus ardente carrière au tourbillon qu'il a dompté." Hector Berlioz, *Mémoires de Hector Berlioz: comprenant ses voyages en Italie, en Allemagne, en Russie et en Angleterre, 1803–1865* (Paris: Michel Lévy frères, 1870), 254; Hector Berlioz, *Memoirs of Hector Berlioz: from 1803 to 1865, comprising his travels in Germany, Italy, Russia, and England*, trans. Rachel (Scott Russell) Holmes and Eleanor Holmes, revised by Ernest Newman (New York: Alfred A Knopf, 1932), 270.

47. See, for instance, the *Gazette musicale* (18 November 1855), which describes "Berlioz, avec ses nombreux ministres ou lieutenants, avec sa cour ou son armée de douze cent cinquante artistes"; and *Le Constitutionnel* (16 November 1855), where we hear an account of "Des orchestres de sept cents musiciens, des choeurs de six cents voix, d'innombrables phalanges de violons, d'altos, de harpes, de violoncelles, de contre-basses, des ouragans de cuivre à effrayer les archanges du jugement dernier, six généraux de division sous le commandement d'un chef suprême, transmettant ses ordres, ses mouvemens, sa mesure à l'aide d'un métronome électrique."

48. See Brittan, "The Electrician, the Magician, and the Nervous Conductor," *Nineteenth-Century Music Review* 19, no. 1 (2022): 31–53; Inge van Rij, "'A Living, Fleshly Bond': The Electric Telegraph, Musical Thought, and Embodiment," *19th-Century Music* 39, no. 2 (2015): 142–66. A wider view of telegraphic imperialism is offered by Daniel R. Headrick in *The Tools of Empire: Technology and European Imperialism in the Nineteenth Century* (Oxford: Oxford University Press, 1981); idem, *The Tentacles of Progress: Technology Transfer in the Age of Imperialism, 1850–1940* (Oxford: Oxford University Press, 1988).

49. See, as starting points, Erin Johnson-Williams, "The Examiner and the Evangelist: Authorities of Music and Empire, c. 1894," *Journal of the Royal Musical Association* 145, no. 2 (2020): 317–50; ibid., "Archiving the Audible Debris of Empire: On a Mission Between Africa and Britain," *Postcolonial Studies* 26, no. 3 (2023): 360–385; Roe-Min Kok, "Imperial Music Examinations in South Asia: Colonial Imaginaries, Postcolonial Realities," *Postcolonial Studies* 26, no. 3 (2023): 386–405.

50. Johnson-Williams, "The Examiner," 326. See also Kok, "Imperial Music Examinations," 395, where she cites British reports on the alleged lack of "discipline" among North Indian ABRSM examination candidates. The cognitive effects of imperial music pedagogy—and the forms of very real resistance to these infringements—need greater attention. The editors of this volume would like to thank Erin Johnson-Williams for fruitful discussions and email exchanges on this topic over the course of 2023.

51. See Dylan Robinson, *Hungry Listening: Resonant Theory for Indigenous Sound Studies* (Minneapolis, MI: University of Minnesota Press, 2020).

52. David Riesman, *The Lonely Crowd: A Study of the Changing American Character*, rev. ed. (New York: Doubleday, 1953), 41–42. Emphasis ours. Crary also discusses Riesman in *Suspensions of Perception*, 53.

53. Kane's essay borrows from and critiques Jonathan Sterne's notion of "audile techniques" as explored in *The Audible Past: Cultural Origins of Sound Reproduction* (Durham, NC: Duke University Press, 2003).

54. Tarde, *L'opinion et la foule*, 40: "la mutuelle contagion des sentiments parmi les individus rassemblés."

55. Ibid., 45: "toute dissidence s'efface momentanément dans la communion d'un même désir, le désir de se voir, de se coudoyer, de sympathiser"; ibid., 47: "Les bons effets des foules

d'amour et de joie se cachent dans les replis du coeur, où, longtemps après la fête, subsiste un surcroît de disposition sympathique et conciliante qui se traduit sous mille formes inaperçues dans les gestes, dans les paroles, dans les rapports de la vie journalière."

56. G. Mensch, *Ludwig van Beethoven. Ein musikalisches Charakterbild* (Leipzig: Leudart, 1871), 161: "Diese vier Persönlichkeiten bewegen sich in dem echten Beethovenschen Quartett mit einander und gegen einander, indem sie sich begleiten und ergänzen, aber auch herausfordern und bekämpfen, indem sie fragen und antworten, bejahen und verneinen, aber immer dasselbe Ziel verfolgen und dem Schutz- und Trutz-Bündniß treu bleiben."

57. Johann Gottfried von Herder, *Kalligone* vol. 2 (Leipzig: Hartnoch, 1800), 159: ". . . sie sind Eins und nicht Eins; sie verlassen, suchen, verfolgen, widersprechen, bekämpfen, verstärken, vernichten einander, und erwecken und beleben und trösten und schmeicheln und umarmen einander wieder, bis sie zuletzt in Einem Ton ersterben." As Ryan Minor points out, the individuated collectivity of choral singing, as imagined by Herder, was all too easily complicated (and co-opted) by German nationalist currents across the nineteenth century. See Minor, *Choral Fantasies: Music, Festivity, and Nationhood in Nineteenth-Century Germany* (Cambridge: Cambridge University Press, 2015), 8–32.

58. Jean-Philippe Rameau, *Nouveau système de musique théorique* (Paris: Ballard, 1726), 17–18; Jean-Jacques Rousseau, *Lettre sur la musique françoise*, 2nd ed. (Paris: Pissot, 1753), 35–45.

59. Benjamin Franklin, Majault, Le Roy, Sallin, Jean-Sylvain Bailly, d'Arcet, De Bory, Joseph-Ignace Guillotin & Antoine Lavoisier, *Report of Dr. Benjamin Franklin, and Other Commissioners: Charged by the King of France, with the Examination of the Animal Magnetism, as Now Practised at Paris. Translated from the French. With an Historical Introduction* (London: J. Johnson, 1785), 39–40.

60. Ibid., 53–54.

61. For discussions of the implications of gender and class in hypnotic subjecthood see Jan Goldstein, *Hysteria Complicated by Ecstasy: The Case of Nanette Leroux* (Princeton, NJ: Princeton University Press, 2011); and Céline Frigau Manning, *Ce que la musique fait à l'hypnose: Une relation spectaculaire au XIXe siècle* (Dijon: les presses du réel, 2021).

62. Essential treatments of mesmerism can be found in Robert Darnton, *Mesmerism and the End of the Enlightenment in France* (Cambridge, MA: Harvard University Press, 1968; Alison Winter, *Mesmerized: Powers of Mind in Victorian Britain* (Chicago: University of Chicago Press, 1998).

63. Carmel Raz, "Talking to the Hand: The 'Hysterical Epistemology' of the Migrating Sensorium," in "Colloquy: Attention, Anxiety, and Audition's Histories," 552–57.

64. The case of Dr. John Elliotson's years-long deception by Jane Okey is an illustrative and entertaining example of this; see Winter, *Mesmerized,* 79–108; see also Kélina Gotman, "Ghost Dancing: Excess, Waste and the American West," in *Choreomania: Dance and Disorder* (Oxford: Oxford University Press, 2017), 224–51.

65. Heinrich Heine, "Lutetia—Musikalische Saison von 1844—Anhang," Paris, 25 April 1844, trans. Charles Godfrey Leland, in *Heinrich Heine: The Works*, vol. 4: *The Salon* (London: William Heinemann, 1895), 418–19.

66. Hypnotic trances induced by unusual sounds are reported as late as the 1890s in the experiments of Jean-Martin Charcot at the Sâlpetriere, who experimented on unusually "sensitive" working-class women with exotic instruments such as gongs and tuning forks. See James Kennaway, "Musical Hypnosis: Sound and Selfhood from Mesmerism to Brainwashing," *Social*

History of Medicine 25, no. 2 (2012): 271–89; Carmel Raz, "Of Sound Minds and Tuning Forks: Neuroscience's Vibratory Histories," in *The Science-Music Borderlands: Reckoning with the Past and Imagining the Future*, ed. Elizabeth Hellmuth Margulis, Psyche Loui, & Deirdre Loughridge (Cambridge, MA: MIT Press, 2023), 115–29.

67. See Russ Castronovo, "The Antislavery Unconscious: Mesmerism, Vodun, and 'Equality,'" *The Mississippi Quarterly* 53, no. 1 (1999): 41–56.

68. "La littérature religieuse et philosophique de l'Orient, de l'Inde en particulier, abonde en documents dont on a pu extraire une sorte de manuel opératoire pour parvenir à l'extase." Théodule Ribot, *Les maladies de la volonté* (Paris: Librairie Germer Baillière, 1883), 124; Théodule Ribot, *The Diseases of the Will*, 4th enlarged English edition, trans. Merwin-Marie Snell (Chicago: The Open Court Publishing Company, 1915), 94.

69. "Se tenir immobile, regarder fixement le ciel, ou un objet lumineux . . . répéter continuellement le monosyllabe *Oum* (Brahm)," ibid., 124, trans. *The Diseases of the Will*, 95.

70. Théodule Ribot, *Psychologie de l'attention* (Paris: Alcan, 1889), 144–49; translation by J. Fitzgerald in *The Psychology of Attention*, trans. J. Fitzgerald (Chicago: The Open Court Publishing Company, 1890), 97–100.

71. "Toutes ses puissances perdent leur activité naturelle; elles n'ont aucune connaissance de leurs opérations. . . . Quant à l'entendement, s'il entend, c'est par un mode qui lui reste inconnu, et il ne peut comprendre rien de ce qu'il entend." Ribot, *Les maladies de la volonté*, 130; trans. *The Diseases of the Will*, 99.

72. Ibid., 140, trans. *The Diseases of the Will*, 89.

73. "l'objet de l'art est d'endormir les puissances actives ou plutôt résistantes de notre personnalité." Henri Bergson, *Essai sur les données immédiates de la conscience* (Paris: Alcan, 1889), 11; translated by F. L. Pogson as *Time and Free Will: An Essay on the Immediate Data of Consciousness* (London: George Allen & Company, 1913), 14.

74. "le rythme et la mesure suspendent la circulation normale de nos sensations et de nos idées en faisant osciller notre attention entre des points fixes." Bergson, *Essai*, 11, trans. Pogson, *Time and Free Will*, 14–15.

75. "une forme atténuée, raffinés et en quelque sorte spiritualisés, les procédés par lesquels on obtient ordinairement l'état d'hypnose." Bergson, *Essai*, 11, trans. Pogson, *Time and Free Will*, 14.

76. Joseph Segond, *La prière: étude de psychologie religieuse* (Paris: Félix Alcan, 1911), 69. Quoted and translated in Kieffer, "Mystics and Skeptics: Joseph Segond's Psychology of Prayer," this volume, 215.

77. As in Raphäel Cor, *Le cas Debussy* (Paris: Bibliothèque de temps présent, 1910). For wider context see Alexandra Kieffer, "Bergson, Debussy, and Early Twentieth-Century *Psychologie*," Colloquy, "Attention, Anxiety, and Audition's Histories," 563–69; and idem, *Debussy's Critics: Sound, Affect, and the Experience of Modernism* (Oxford: Oxford University Press, 2019).

78. ". . . attention extrême est ce qui constitue dans l'homme la faculté créatrice." Simone Weil, *La Pesanteur et la Grâce* (Paris: Plon, 1947), 119, translated as *Gravity and Grace* by Emma Crawford and Mario von der Ruhr (London and New York: Routledge, 2002), 117.

79. "L'attention seule, cette attention si pleine que le "je" disparaît, est requise de moi." Weil, *La Pesanteur*, 120, trans. in *Gravity and Grace*, 118.

80. "La quantité de génie créateur d'une époque est rigoureusement proportionnelle à la quantité d'attention extrême, donc de religion authentique à cette époque." Weil, *La Pesanteur*, 119, trans. *Gravity and Grace*, 117. Weil's interest in attention spans much of her work including her slightly earlier *Attente de Dieu* [*Waiting for God*], written 1942, published 1951.

81. On Weil's relationship with these traditions, see A. Rebecca Rozelle-Stone and Lucian Stone, *Simone Weil and Theology* (New York and London: Bloomsbury, 2013). For a recent philosophical exploration of attention as selflessness see Jonardon Ganeri, *Attention Not Self* (Oxford: Oxford University Press, 2017).

82. As Bergson wrote to Joseph Segond (18 January 1911): "You have submitted to penetrating analysis the fundamental state of a praying individual and the states that are like harmonics, those that, reinforcing it, give it its psychological timbre. You have studied the mystics as they should be studied, with sympathy, but also submitting them to penetrating critique." In Henri Bergson, "Henri Bergson (1859–1941) Lettres à Joseph Segond (1872–1954)," *Revue philosophique de la France et de l'étranger* 205, no. 4 (October–December 2015): 543–560, at 543.

83. Trippett, "Franz, Stumpf, Riemann, Kafka: Towards an Auditory Unconscious ca.1883," this volume.

84. Theodor Reik, *Listening with the Third Ear: The Inner Experience of a Psychoanalyst* (New York: Farrar, Straus, and Giroux, orig. pub. 1948, repr. 1998). See chap. XV, "The Third Ear," 144–56 (here at 145). The connection between these "unconscious messages" and theories of an elusive inner language is explored in Viktoria Tkaczyk, *Thinking with Sound: A New Program in the Sciences and Humanities Around 1900* (Chicago: University of Chicago Press, 2023), 15–54.

85. Reik, *Listening with the Third Ear*, chap. XVI, "Free-Floating Attention," 157–72.

86. Alexander Evan Bonus, "The Metronomic Performance Practice: A History of Rhythm, Metronomes, and the Mechanization of Musicality" (PhD diss., Case Western Reserve University, 2010), 225.

87. John Holden, *Essay towards a Rational System of Music* (Glasgow: Urie, 1770), 113.

88. For more on Holden's ideas about attention see Carmel Raz, *Hearing with the Mind.*

89. Johann Philipp Kirnberger, *Die Kunst des reinen Satzes in der Musik* II (Berlin and Königsberg: Decker und Hartung, 1776), 114–15.

90. See Roger Mathew Grant, *Beating Time and Measuring Music in the Early Modern Era* (Oxford: Oxford University Press, 2014), 200–07.

91. Prominent voices include Peter Lichtenthal, François Joseph Fétis, and Gottfried Weber, though the latter, as Grant notes, had other reasons for disparaging the metronome, which competed with his own chronometer. Ibid., 201–203.

92. Adolf B. Marx, *The Music of the Nineteenth Century and its Culture: Method of Musical Instruction*, trans. August Heinrich Wehrhan (London: Robert Cocks and Co., 1855), 32.

93. See, e.g., Ernst Kurth, *Bruckner* (Berlin, Max Hesses Verlag, 1925); Victor Zuckerkandl, *The Sense of Music* (Princeton: Princeton University Press, 1956).

94. See Henning Schmidgen, "The Donders Machine: Matter, Signs, and Time in a Physiological Experiment, ca. 1865," *Configurations* 13, vol. 2 (2005): 211–56.

95. Wilhelm M. Wundt, *An Introduction to Psychology*, trans. Rudolf Painter (New York: Macmillan, 1912), 4.

96. See William James, "The Perception of Time," *Journal of Speculative Philosophy* 20, no. 4 (1886): 374–407, at 380.

97. Thaddeus L. Bolton, "Rhythm," *American Journal of Psychology* 6.2 (1894), 211–12.

98. Jean-Philippe Rameau, *Nouveau système de musique théorique* (Paris: Ballard, 1726), iii, translated in B. Glenn Chandler, "Rameau's '*Nouveau systeme de musique theorique*': An Annotated Translation With Commentary" (PhD diss., Indiana University, 1975), 159; Jean-Philippe Rameau, *Nouveau système de musique théorique* (Paris: Ballard, 1726), iii: ". . . se distinguent encore dans les Tymbales, dans le Son le plus grave d'une Trompette, dans les Cloches, et cetera

pourvû qu'on donne pour lors toute son attention aux ondulations qui forment une espece de murmure dans l'Air, immediatement après le Son dominant de l'Instrument propre à cet effet, sans s'occuper pour lors de ce Son dominant, ny sans y être distrait par aucun bruit étranger." A similar observation can be found in Holden, *Essay,* 47.

99. See Benjamin Steege, *Helmholtz and the Modern Listener* (Cambridge: Cambridge University Press, 2012),80–122.

100. Hermann von Helmholtz, *On the Sensations of Tone as a Physiological Basis for the Theory of Music,* trans. Alexander J. Ellis (London: Longman, 1875), 82; Helmholtz, *Die Lehre von den Tonempfindungen,* 88: "Wenn man die Resonanzkugel an das Ohr setzt, welche irgend einem bestimmten Obertone, z. B. g' des Klanges c, entspricht, und dann den Klang c angiebt, so hört man das g' durch die Kugel um Vieles verstärkt. . . . wohl kann man diese Verstärkung durch die Kugel benutzen, um das Ohr auf den Ton, den es hören soll, aufmerksam zu machen. Wenn man nachher die Kugel allmälig wieder vom Ohre entfernt, so wird das g' schwächer; indessen die Aufmerksamkeit, welche einmal darauf gerichtet worden ist, bleibt nun leichter an diesen Ton gefesselt, und der Beobachter hört diesen Ton nun auch in dem natürlichen unveränderten Klange der angegebenen Note mit nicht unterstütztem Ohre. Es soll also die Resonanzkugel hierbei nur dazu dienen, das Ohr aufmerksam zu machen auf den Ton, den es hören soll."

101. "eine gewisse ungestörte Concentration der Aufmerksamkeit [ist] nöthig, um die Analyse der Klänge durch das Ohr allein auszuführen." Helmholtz, *Die Lehre von den Tonempfindungen,* 89; Helmholtz, *On the Sensations of Tone,* 82.

102. On Müller's critique of Helmholtz's overtone experiments see Steege, *Helmholtz and the Modern Listener,* 108–10.

103. "So hören wir aus einem Zusammenklang einen vorher für sich angegebenen Ton besonders deutlich. . . . Dies führt uns unmittelbar auf die inneren Bedingungen der Aufmerksamkeit. Gehen wir von der zuletzt besprochenen Beobachtung aus, so kann das geübte Ohr einen schwachen Theilton eines Klanges bekanntlich auch dann wahrnehmen, wenn derselbe ihm nicht zuvor als gesonderter Eindruck gegeben wurde. Bei näherer Beobachtung findet man aber stets, dass man sich in diesem Fall zunächst das Erinnerungsbild des zu hörenden Tones zurückruft und ihn dann erst aus dem ganzen Klang heraushört." Wilhelm M. Wundt, *Grundzüge der physiologischen Psychologie* (Leipzig: Engelmann, 1874), 720–21.

104. "Jede Vorstellung bedarf einer gewissen Zeit, um zum Blickpunkt des Bewusstseins hindurchzudringen. Während dessen finden wir stets in uns das eigenthümliche Gefühl des Aufmerkens. Dasselbe ist um so lebhafter, je mehr der Blickpunkt des Bewusstseins sich concentrirt, und es pflegt in diesem Falle noch fortzudauern, auch nachdem die Vorstellung vollkommen klar vor dem Bewusstsein steht. In seinem vorbereitenden Stadium wird es am deutlichsten im Zustande des Besinnens oder der Spannung auf einen erwarteten Eindruck erfasst. . . . Wir empfinden das Aufmerken als etwas das von uns selbst ausgeht, wenn wir auch deutlich den Eindruck, der unsere Aufmersamkeit auf sich lenkt, als einen äusseren vorstellen. Hierdurch tritt dieser Zustand in die nächste Verbindung mit jenem, welcher der willkürlichen Bewegung vorausgeht." Ibid., 721.

105. William James, *The Principles of Psychology* 1 (New York: Henry Holt & Co., 1893), 439.

106. Ibid., 441, 444.

107. Ibid., 444.

108. See, for instance, James Sully, *Sensation and Intuition: Studies in Psychology and Aesthetics* (London: Henry S. King & Co, 1874); Sully, *Outlines of Psychology, with Special Reference to the Theory of Education* (Longsman, Green and Company, 1885).

109. Michael Weinstein-Reiman, "Printing Piano Pedagogy: Experimental Psychology and Marie Jaëll's Theory of Touch," *Nineteenth-Century Music Review* 18, no. 3 (2021), 427. Jaëll's haptic approach to pianistic attention is framed in her treatise *La musique et la psychologie* (Paris: Félix Alcan, 1896).

110. Tobias Matthay, *Musical Interpretation: Its Laws and Principles, and Their Application in Teaching and Performing* (Boston: Boston Music Company, 1913), 147. Original italics and small caps.

111. Media transformations during the early-mid twentieth century, and resultant shifts in auditory culture, are dealt with in foundational sound studies literature including Jonathan Sterne, *The Sound Studies Reader* (Abingdon: Routledge, 2012); Trevor Pinch and Karin Bijsterveld, eds., *The Oxford Handbook of Sound Studies* (Oxford and New York: Oxford University Press, 2012); and Michael Bull, Les Back, and David Howes, eds., *The Auditory Culture Reader* (London: Bloomsbury, 2015).

112. E. Colin Cherry, "Some Experiments on the Recognition of Speech, with One and with Two Ears," *The Journal of the Acoustical Society of America* 25, no. 5 (1953): 975–79.

113. Albert S. Bregman, *Auditory Scene Analysis: The Perceptual Organization of Sound* (Cambridge, MA: MIT Press, 1990). For a comprehensive account of the history of binaural listening, starting in the 1820s, see Gascia Ouzonian, *Stereophonica: Sound and Space in Science, Technology, and the Arts* (Cambridge, MA: MIT Press, 2021); see also Viktoria Tkaczyk, *Thinking with Sound.*

114. On the problematic distinction between *E-Musik* and *U-Musik* see Frieder W. Bergner, "Das E und das U in der Musik," *Glarean Magazin* (10 August 2007), http://glareanverlag.wordpress.com/category/u-und-e-musik/, accessed November 2025.

115. This capacity for music to siphon off some (but not all) cognitive energy became key to the development of "productivity" soundscapes including Muzak, which functioned as a mild distraction from repetitive or dehumanizing labor; see Allen Anderson, "Neo-Muzak and the Business of Mood," *Critical Inquiry* 41, no. 4 (2015): 811–40. On attentional dynamics of passive or "ubiquitous" listening, see Anahid Kassabian, *Ubiquitous Listening: Affect, Attention, and Distributed Subjectivity* (University of California Press, 2013).

116. Pauline Oliveros, *Deep Listening: A Composer's Sound Practice* (Lincoln, NE: Deep Listening Publications, 2005).

117. John Cage, *Silence: Lectures and Writings* (Wesleyan University Press, 1961), 93.

118. Milton Babbitt, "The Composer as Specialist," in *The Collected Essays of Milton Babbitt*, ed. Stephen Peles et al. (Princeton: Princeton University Press, 2003), 51. The essay was first published under the title "Who Cares If You Listen," in *High Fidelity* 8, no. 2 (February 1958): 38–40, 126–27; this title was the editors', not Babbitt's.

119. Michel Foucault, Pierre Boulez, and John Rahn, "Contemporary Music and the Public," *Perspectives of New Music* 24, no. 1 (1985), 11.

120. As Lerdahl explains, "Musical cognition has a bias for hierarchical organization. Simultaneous tempos instead produce independent organizations competing for attention. The difficulties in attending to more than one such organization are well known (Cherry 1953) and are ameliorated only if the vertical correspondences are coherent (Sloboda 1985), a condition that only partly obtains in this case." Fred Lerdahl, "Cognitive Constraints on Compositional Systems," *Contemporary Music Review* 6, no. 2 (1992), 106.

121. Lydia Goehr, *The Imaginary Museum of Musical Works: An Essay in the Philosophy of Music* (Oxford: Clarendon Press, 1992).

PART I

Concentration

CHAPTER 1

Colonial Concentration: Attention in Bengali Music Scholarship in 1885

Richard David Williams

What constituted "attention" in musical contexts in northern India at the end of the nineteenth century? In the context of British colonialism, how far were ideas about attention impacted by Western concepts and evolving conversations in Europe about concentration, focus, and distraction? As the contributions in this collection demonstrate, ideas around attention and distraction evolved over the long nineteenth century in conversation with new social contexts for music making, new theories of auditory perception and cognition, burgeoning discourses around psychology and the work of the mind, and the commodification of attention. However, although recent scholarship has nuanced our sense of these developments in Euro-American historical contexts, it remains unclear how far concepts like "attention" were significant to actors invested in music in other parts of the world, and whether ideas and practices cultivated in the West became global in their own time.

Clearly, these are large questions that resist easy answers. In South Asian history more broadly, there is an ongoing debate about colonial knowledge production in the nineteenth century: Were contemporary concepts and categories (from religions to the caste system) the imports and products of colonialism, or were they adaptations of earlier, precolonial ideas that had transitioned in response to colonial forms of knowledge and global conversations?[1] Colonialism took different forms in different regions, intellectual disciplines, and arenas of debate, and the responses of Indian intellectuals could vary enormously: Some continued to work within the frameworks provided by older, precolonial systems, while others renounced them as antiquated and adopted new approaches from Europe.

This essay concentrates on musical scholarship written in Bengali, focusing on one individual—Krishnadhan Bandyopadhyay (1846–1904)—who was part of an extremely dynamic and varied music scene in colonial Calcutta.[2] Bandyopadhyay, I suggest, was atypical in the way he thought about musical attention and in the extent of his subscription to ideas found in European studies. Nonetheless, his research sheds light on how Indian music scholars accessed information learned from English-language sources, selectively co-opted certain ideas, and brought them into conversation with materials in Sanskrit and Bengali. His work further indicates how intellectuals creatively negotiated multiple epistemologies of music and drew on European concerns with the nature of listening and attention to frame their own manifestos for the future of the practice and appreciation of music in South Asia. While not entirely in tune with debates and conversations in Europe, these manifestos drew upon a shared, circulating conceptual vocabulary for thinking about musical experience. At the same time, when this vocabulary was presented in Bengali, the local, vernacular terms of debate were themselves products of a longer colonial engagement between European and South Asian languages and knowledge systems. To make sense of this engagement, I begin by introducing the terms that eighteenth-century music scholars writing in Bengali used when they considered how the listening subject attends to music. I then turn to the broader conceptual vocabulary for attention and concentration available in colonial north India and explore the connected history of European and Indian litterateurs. I follow these terms through to Bandyopadhyay's work in the late nineteenth century and consider the place of attention in his vision for the future of Indian music.

Attending to Sound Through Visualization

Although Krishnadhan Bandyopadhyay was an idiosyncratic thinker, his work emerged out of a longer history of South Asian scholarship on art music. His generation of scholars prepared innovative studies on music, written in Bengali, which proliferated from the late 1860s onward.[3] Bengali intellectuals had been engaging with the study of music long before this, but the activity of the mid-nineteenth century represented a new development, reflecting larger changes in the language and medium of scholarship.[4] In the early modern period, South Asian musicology[5] had largely been carried out in Sanskrit (*saṅgīta-śāstra*) and Persian (*'ilm-i-mūsīqī*), though vernacular scholarship gradually took over from Sanskrit, especially in the eighteenth century.[6] A number of intellectuals from eastern India, operating out of aristocratic courts and religious institutions, developed poetic treatises on music theory in Bengali. These works followed the older models set by Sanskrit literature but were also informed by more recent trends in vernacular scholarship, especially the focus on the visualization (*dhyāna*) of modal entities, the *rāgas*.[7] Sanskrit music scholars had explained that a *rāga* could be understood in terms of its musical structure, its affective connotations, and through iconography, which could be illustrated through a painted image or a poem. The word *dhyāna* refers in particular to a practice of concentrated meditation, in which the practitioner assembles an image or icon in their mind in order to focus and integrate the object of concentration into their own internal environment.[8] Early examples of *dhyāna* visualization-concentration for *rāgas* certainly suggest an affinity with ritual meditation on deities.[9] Over the seventeenth and eighteenth centuries, chapters on *dhyāna* in music treatises moved into new directions, especially in vernacular languages such as Classical Hindi (Brajbhasha), in which the poets brought music theory into dialogue with "mannerist" (*rīti*) courtly poetry, which provided a nuanced vocabulary for discussing emotion, sensation, and pleasure through literary archetypes.[10] According to this aesthetic system, to attend to music was to construct and contemplate its imagined body: a woman waiting for her lover, a sage chanting by his shrine.[11]

To give one example, in 1717, Phajil Nasir Muhammad composed a *Rāganāmā* ("Book of Rāga") for Wahid Muhammad Caudhuri, the landlord of Sultanpur (Chittagong). The work opens with praise poems (*hāmd* and *nā't*) to Allah and the Prophet, then eulogizes the patron and his kingdom. Phajil Nasir contextualized and localized his work to the particular space of the Sultanpur court, describing how his lord had requested him to make Sanskrit musical lore (primarily drawn from the *Saṅgītadāmodara*, fifteenth to mid-sixteenth century) available to readers in Bengali.[12] Following the standard convention of the genre, the visualizations of the *rāgas* were arranged in a series, that is, selected and strung together to make up a "garland" (*mālā*). For his *rāga-mālā* series, Phajil Nazir cited a description in Sanskrit, then expounded upon it in Bengali in two ways: in a *paẏār* couplet, and then in *tripadī* triplets, as with the *rāga* called Mālava:[13]

deha-kānti śukla barṇa
sucāru kuṇḍala karṇa
mahāmatta mati atiśaẏ
pariẏā bicitra mālā
prabeśanta gīti śālā
pradoṣa samaẏa suniścaẏ
uttama basana pari
maneta kautuka kari
priẏā padmamukha cumbe ghana
nṛtyagīta mana sukhī
saṅhati saroja mukhī
saba nārī madanamohana
preme magna haiẏā cita
gāhasta susvara gīta
priẏā saba dekhiẏā susāja
daśa daṇḍa dibā ādi
rātri daśa daṇḍābadhi
gāha-e mālaba rāga-rāja

His radiant body, white in color,
charming earrings in his ears,
his thoughts excited in excess.
Wearing a colorful garland,
entering the music hall,
resolute at evening time,
Wearing the finest clothes,
feeling playful in his heart,
deep in a kiss with his lotus-faced love,
Delighted by music and dance,
alongside her lily face,
enchanting to all women.
His mind becomes steeped in love,
they sing a melodious song,
his beloved is finely adorned all over.
From the first *daṇḍa* of the day
to the tenth *daṇḍa* of night
Mālava, the king of *rāga*, sings.[14]

Following a poetic approach to musical scholarship, the gendered archetypes of the *rāga* series are brimming with emotions, immersed in love, spectacle, and the pleasures of music and dance. At least some of these descriptions were designed to be performed as songs—certain verses are marked as a refrain ("*dhru*." short for *dhrubā*)—which indicates how the *rāga* series could be brought to life, enacting a mental vision of musical pleasure through the interplay of sound and verbal, evocative imagery.

However, by the nineteenth century, courtly musicologists of this kind were largely forgotten. They were surpassed instead by a new class of gentlemen (*bhadralok*) scholars in Calcutta, the new colonial metropolis, who only selectively affiliated themselves with the older generations of intellectuals. This selectivity had lasting implications for the future of musical thought, as subsequent generations of Bengali writers located themselves between the precolonial musical conversations that interested them and colonial forms of knowledge production and dissemination. As I have argued elsewhere, a turning point was the 1818 publication of the *Saṅgītataraṅga* ("Wave of Music"), the first printed Bengali book on music theory, which was a digest of older ideas in Persian and Sanskrit.[15] Associated with the academy at Fort William, the book's author partially addressed a non-Indian readership and availed himself of the new possibilities posed by print technology. This book became an influential reference work for later generations of scholars, especially in the wake of a colonial language policy in 1837 whereby Persian ceased to carry official functions: Increasingly less familiar with the language, Bengali scholars also became increasingly disengaged from the vast corpus of Persian musicological thought—and by extension, the work of Muslim intellectuals—drawing what they needed from the Bengali summary and then reorienting themselves toward novel approaches and questions.

These breaks with convention led to a new landscape of musical publishing and the proliferation of several different types of music book: songbooks (often with introductions to the fundamentals of music); instrumental manuals; treatises on the history and theory of music; and antiquarian studies and editions of Sanskrit texts. While my focus here is on Bengali, we should remember that these works were part of a larger, multilingual conversation about the state and future of music studies in India, and that certain authors were directly influenced by concurrent trends in other languages, such as Hindi, Marathi, and Urdu.

This new generation of authors was interested both in the practice of music and in writing useful books, reflecting a larger sea change in access to musical training, which distanced their outlook from the likes of Phajil Nasir in the eighteenth century and called for a different kind of musical study. Nasir's royal patron was expected to appreciate musical aesthetics but uninterested in singing or performing himself: After all, music making was the work of professional, hereditary musicians, who were service providers to their patrons. However, in early nineteenth-century Calcutta, men from the elite classes were beginning to sing and play and to develop reputations as celebrated musicians and composers themselves.[16] By the 1860s, this had become increasingly typical, and collectives formed to appreciate, study, and perform music together. Of these, the Bengal Music School, established in 1871 by S. M. Tagore (1840–1914) and Kshetramohan Goswami (1813–1893), and another society, the Bhavanipur Saṅgīt Sammilanī, were especially influential. Writers in these circles engaged with developments in European scholarship to varying degrees; for example, there were debates over whether notation (including staff notation) was appropriate for Indian music, or how far the harmonium should be integrated into art music.[17] Certain authors, including S. M. Tagore and his colleague, Nabinchandra Datta, explicitly discussed ideas they found in European works on music in parallel to developing a neo-traditionalist approach to Indian music, one that framed innovative ideas as ancient and Sanskritic, often skipping backward in time over their immediate predecessors.[18]

One striking trend in this this period is the lack of interest in *rāga* visualization: While *rāga-mālā* had been a chief concern in writings on music in the eighteenth century, in the nineteenth it was increasingly framed as a curiosity or a form of poetry, but not as a significant theme for musicology. Thus, S. M. Tagore incorporated *rāga* visualizations in one work dedicated to the Prince of Wales (1877) and called them "meditations," but framed them as a decorative curiosity rather than as an essential way of concentrating on music: "To impart picturesqueness to the treatise, as well as to clothe it with orientalism, he has also given their emblematic representations."[19] I suggest the old form of musical concentration epitomized in *dhyāna* was no longer fit for purpose: Musical authors were no longer patron-listeners but performer-scholars, and they required a new vocabulary and technique for attending to music. While experiments with notation—both European and Indian—was a partial solution, some scholars also sought a new language for concentration and attention.

Defining "Attention" in Colonial India

Any attempt to find the equivalent of the word "attention" in multilingual north India indicates the variety of concepts and intellectual histories relating to consciousness that were already on the table before colonial interventions. In addition to the aforementioned notion of *dhyāna*, we can consider the Urdu word *tawajjŏh* (توجه), which has the sense of turning to face something or someone. In particular, the word was theorized in Sufi practices as the disciple offering his countenance to the face of the master and holding his image in his mind at all times, shutting out worldly distractions and absorbing the master's image to perceive the beauty of Allah therein.[20] The same sense occurs in other terms, such as to twist or turn (*multafit* ملتفت) to something to regard it attentively.[21]

Bengal was one of the first regions to be systematically and deeply colonized by the British, and Bengali intellectuals are renowned for their creative responses to the transformative influx of ideas from Europe. This influx can be traced in terms of a conceptual history of attention. In the earliest European vocabulary of the Bengali language, Fr. Manoel da Assumpçam's *Vocabulario em Idioma Bengalla e Portuguez* (1743), "Attençaõ" is listed as "Monsuba" (i.e., *mansubā*, a "fixed plan; contrivance; pretext")[22] and "Achenotá v. chit" (*ācenatā bā cit*),[23] where *cit* refers to the mind ("I think" is *cit kari*).[24] These were, perhaps, not the most obvious choices, and later authors developed their own set of equivalences. For example, in 1849, one Krishnamohan Bandyopadhyay had translated *The Improvement of the Mind* (1741) by Isaac Watts into Bengali as part of his *Encyclopedia Bengalensis* translation project. His abridged version of this exhortation to study culminated in the chapter "Of fixing the attention," or "*manaḥsthir karibār biṣaý*" (literally, "the subject of making the mind still"). Bandyopadhyay elaborated a range of synonyms for Watts's terms: *manaḥsthir* (mind-stilling) for "a steady fixation of thought"; *manaḥsaṅyogitā* (mind-joining) for "attention"; and *parīkṣā* (examination) for "inspection."[25]

While my focus here is on Bengal, the situation in Calcutta can be seen as part of a larger set of developments in response to European influence. From the eighteenth century, conversations in Europe around auditory attention and distraction were informed by emergent trends in literature, medicine, theology, economics, and technology.[26] Some of these trends, including the nineteenth-century fascination with mesmerism, circulated in colonial India where they were selectively integrated with earlier, indigenous systems. For example, Muhammad Taqi Sayyid's study of mesmerism in Urdu, *Tāsir al-anẓār* ("The Impression of the Gaze," 1877) was marketed as a compendium of specifically European talismanic devices (*ṯilism-i-farang*) yet embedded these Western ideas in a larger discussion involving Islamic medicine (*al-Ṯibb*).[27] In Hindi, Baldev Prasad Misra's *Mṛtak-milap: yā Pretnagar. Spirit Land* ("Meeting the Dead: or Ghost City. Spirit Land," 1897) examined how European spiritualism (both directly transliterated and translated as *spiricuyālizam* and *pret-tattva*) and mesmerism (*mesmerizam*) departed from Indian theories, including yoga, sorcery (*ojhā bidyā*), and tantra. The introduction to mesmerism defines a fit and able mind:

> Only those who conduct themselves well, those with the faculty of an intense remembrance and belief in God, endowed with a sharp consciousness (*tej buddhi*), a powerful brain (*sabal mastiṣk*), courage and forbearance and character, compassion, feeling, steadfastness and self-control can study this art (*bidyā*).[28]

This book outlined different techniques of mediumship and seances, preserving key phrases and terms in the original English. While the specific terms in this passage were hardly new to Hindi intellectuals, here they were deployed in new and self-consciously Occidentalist directions.

Krishnadhan Bandyopadhyay's Manifesto for Attentive Listening

Against this larger context, similar engagements and appropriations of European ideas can be seen emerging in Bengali musical scholarship. Krishnadhan Bandyopadhyay was one of the most original Bengali musicologists of the nineteenth century. Yet, as Sagnik Atarthi reminds us, his influence was curtailed because of his limited personal resources, especially compared to some of his peers and rivals in Calcutta's musical forums, who were able to set up their own schools and cultivate larger followings.[29]

Krishnadhan had a varied training: a Western-style education at Hindu College, an early career in Calcutta's musical theater scene as an actor-singer, followed by a period spent studying sitar in Gwalior with a hereditary professional master (*ustād*), Ahmed Khan. A voracious reader and prolific scholar,

Bandyopadhyay wrote a Bengali history of China in 1866, but also devoted himself to preparing new resources for music students. Over the 1860s–1870s, he published several music books about Indian art music presented with staff notation. Despite the criticism of scholars like S. M. Tagore who were busy promoting their own, "national" systems of inscription, Bandyopadhyay was a vocal champion of Western notation.[30] He also deployed concepts and arguments derived from his research on European music to make new inroads into the theory and practice of Indian music. This approach was spelled out in his *Saṅgīt Śikṣā* ("Music Instruction," 1868). This *Comprehensive Self-Instructor for the Setar, Esraj, Violin, Flute, and Harmonium; and also for Singing* was highly pragmatic in its arrangement of lessons and compositions: Bandyopadhyay retained the conventional vocabulary for Indian music (using terms in Hindustani, Sanskrit, and Bengali), but also provided a glossary of his neologisms that were entirely new to Indian musical discourse, including "Chromatic" (*acalasvārik*), "Minor" (*gauṇa*), "Enharmonic" (*mādhurācalik*), and "Acoustics" (*śabdabijñān*).[31]

Seventeen years later, Bandyopadhyay published his magnum opus, the *Gītasūtrasār* ("Quintessence of Music," 1885–1886). Here he abandoned the long-established Indian conventions for discussing sound and music and started from a new set of first principles, opening his book with Darwinian evolution, the biological development of the throat and voice, and the prehistoric role of sound in social communication.[32]

The *Gītasūtrasār* can be read as a manifesto: Bandyopadhyay exhorted his readers to pay music what he considered an appropriate level of attention. He attempted to provide a new model for musical scholarship and critical engagement with sound, since he thought the prevailing mode of discussing music in Indian languages was inadequate. He believed this deficiency was ingrained in the work of both ancient and contemporary musicologists, the performance practices of living musicians, and in popular listening habits:

> Many people have the impression that music which induces drowsiness in its listeners is, to their minds, the best; this is a terrible mistake. Listening to monotonous music lacking variety will bring on sleepiness. Listening to music that does something new and novel each and every moment, with *tāns* [fast phrases], and revelations of new and novel affects (*nūtun nūtun raser abirbhāb haẏ*), even sleepy

> and spiritless individuals are stirred up and come alive. Listening to the tambura or sitar for five or seven minutes, many people, especially children, fall asleep: this has been observed frequently. The reason is monotonous sound. . . .
>
> . . . The primary purpose of musical study should not just be to achieve the satisfaction of listeners. What should be done in music is whatever achieves mental satisfaction (*mānasik sukh*). Changes in mental mood (*cittabikār*), the elevation of the mind's passions (*maner ābeg*), and the captivation of the listener (*śrotāke mugdha karā-i*) is the finest work of music. While it is a challenging enterprise to give an account of how mental moods change and so on by means of arranging musical notes, this indeed is the legitimate work of music. Owing to the miserable state of our national (*jātīýa*) preferences and tastes, as with everything, [music] is not looking well.[33]

How far was Bandyopadhyay's dim view of Indian musical scholarship a product of Orientalist criticism? His model for critical engagement was certainly shaped by European studies of European music (below): while he had also read English studies of Indian music (e.g., William Jones and Augustus Willard), which informed his accounts of musical forms and their histories, he was less interested in these Orientalist works about India and more indebted to books and articles on vocal production, music education, and the scientific basis of sound. Bandyopadhyay was less concerned with musicology's pasts than its present and possible futures. While his Bengali contemporaries typically focused on the decline of music (especially in the hands of Muslim musicians) and focused their energies on ancient (and therefore purer and more authentic) Sanskrit music treatises, Bandyopadhyay regarded change as inevitable and considered the ancient authorities redundant for modern times. Since music—like everything else, he argued—had evolved over the centuries, it was futile to try to compare the music of his day to the music described in older texts.[34] Indeed, he explicitly called his former colleague Kshetramohan Goswami's attempt to devise modern practices based on ancient theories illogical.[35]

Although Bandyopadhyay engaged extensively with centuries-old music treatises, he read these Sanskrit works as historical texts rather than as tools for understanding the music of his day. He contrasted his own worldview and epistemology, which was proudly shaped by modern theories like evolution, to the antique modes of attending to and studying music: "All those treatise writers were poets; to them the entire subject of music was included within the poetic imagination."[36] To highlight the point, he turned to the *Saṅgīta-dāmodara*, a Sanskrit treatise from the fifteenth–sixteenth century, a once highly influential source for music scholars in eastern India:

> It is said in the *Saṅgīta-dāmodara* that when the air in the navel is hiccupped, and that within the throat and head is struck, a particular sound is produced, and since it is possessed of both fragrance (*gandhayukta*) and purity its name is *gāndhār*. If a certain note is pleasing it is very fragrant; some notes have a light fragrance, others have none; in our time, people do not have the faculty to make sense of this. That real [or material, *bāstabik*] notes should have fragrances is an impossible business . . . we are only introduced to this kind of marvelous imagination; but not to anything real.[37]

Dismissing earlier modes of canonical writings on music (*saṅgīta śāstra*) as poetry rather than musicology proper resonated with other critics of the late nineteenth century, including Bharatendu Harischandra (1850–1885), who made similar pronouncements in Hindi and argued against "blind tradition."[38] However, in Bengali circles the dominant trend involved reclaiming musical information from ancient texts and allocating a place for it in modern practice, something that Bandyopadhyay explicitly berated. Bandyopadhyay criticized his contemporaries, especially S. M. Tagore and Kshetramohan Goswami, for pursuing this line of inquiry and also critiqued earlier seminal works on music theory, such as the aforementioned *Saṅgītataraṅga* (1818), which was too conventional and uncritical for his liking.

Moreover, this particular example testifies to a friction between modes of musical perception: dissecting the experience of sound through a poetic palette—in this case, fragrance—was part of a larger epistemology of musical pleasure, and attentive multimodality, cultivated by early modern intellectuals. Bandyopadhyay urges his readers to reject this mode of experience in favor of what he called the "real" (*āsal*) and the "material" (*bāstabik*).

How did Bandyopadhyay account for the neglect of "real" music? He traced the problem back to the culture of musical professionals, which had permeated society at large and impacted the attention of individual listeners. In line with many other Hindu musicologists of his time, he dismissed hereditary Muslim masters (*ustāds*) as illiterate and suggested that because they were not erudite scholars they did not give their consideration (*anudhāban*) to the internal, affective dimensions of sound:

> *Rāgas* and *rāginīs*, whether certain rhythms (*tāla*) and tempos (*laẏa*) are "impure" or not—since *ustāds* are only concentrating (*manoyogī hauẏāte*) on this subject, the inner meaning of Hindustani music has disappeared.[39]

On a first reading, this claim sounds very familiar: Many writers in this period were actively accusing Muslims of being the inept custodians of musical culture.[40] Certainly, the myth of the "illiterate *ustād*" has been debunked in recent scholarship, including studies by Katherine Schofield and Max Katz on musicians who wrote music treatises in Persian and Urdu over the Mughal and colonial periods.[41] However, it is true that in Bandyopadhyay's own time, *ustāds* primarily contributed to the arena of Bengali music book production as consultants rather than authors.[42] Given that he also criticized the intellectual projects of elite Hindu writers, such as S. M. Tagore, Bandyopadhyay's scathing judgments evidently crossed communal boundaries. Like Orientalist writers, he characterized Muslim rulers by their love of pleasure (*bilās priẏatā*), but rather than developing this into the increasingly standard colonial account of degeneracy and decadence, he merely commented that they especially enjoyed music that cultivated the aesthetic of grief (*śok*).[43] Likewise, while other authors were hesitant to discuss certain genres like *ṭhumrī* without denouncing the immorality of courtesans and their Muslim accompanists, Bandyopadhyay had nothing but praise for *ṭhumrī* and its performers.[44] Given that Bandyopadhyay himself had chosen to study sitar with an *ustād*, Ahmed Khan, I suggest that his criticism was less concerned with denouncing Muslim involvement in the arts and more about his personal frustrations with the current terms of debate in musical scholarship. In particular, he was annoyed that musicians persisted in traditional frameworks for conceptualizing and composing musical sound:

> While superstition (*kusaṅskār*) and conservatism (*goṅrāmī*) are the king of the gods when it comes to progress in any given subject, in music it is even more so. Resistant to progress, adhering to

> superstition, the creativity of accomplished people in composing new sounds can never come into bloom.[45]

The possibilities of art music, he argued, had been narrowly defined and curtailed by a lack of forward-thinking vision.

Inspiration from Europe

Despondent about the state of musical creativity in India, Bandyopadhyay looked for inspiration elsewhere, especially from Europe. In doing so, he aligned himself with other Bengali intellectuals who were inspired by European precedents.[46] Bandyopadhyay articulates his own Occidentalism most dramatically when he describes his ideal musical genius, a poet of sound (*svar-kabi*), and disparagingly claims that even Tansen, the most celebrated court musician in the history of Hindustani music, would not qualify, since he did not pioneer composition beyond the hegemony of canonical *rāga*-based music.[47] While he does not name anyone instead, he offers a model for an artist explicitly based on European concepts:

> The fine English author Carlyle, said of the famous German poet and philosopher Goethe that: "He sees through every pore." Accordingly, the one I call the truly best singer should not only see but also feel in his every pore. At present, is there not a singer of this kind in our country?[48]

Bandyopadhyay here appears quintessentially cosmopolitan, invoking Thomas Carlyle's characterization of Goethe to formulate his aspiration for Indian music. At the same time, the global flow of ideas here was less than straightforward. It was not Thomas Carlyle who said this (in the English original: "for he seems to see out of every pore of his skin"), but rather Ralph Waldo Emerson: The two writers had been juxtaposed in a piece by Richard Holt Hutton, who quoted Emerson in order to demonstrate how "Carlyle worships Goethe blindly, but Emerson discerns the very core of the poet."[49] Hutton's own phrasing was a little confusing, and the Welsh singer and scientist Margaret Watts Hughes (1842–1907) misattributed the quotation as coming from Carlyle when she included it in her article, "A Glimpse into the Art of Singing," published in 1875. There we also find her commenting on the essential requirements of a good singer:

> Carlyle, in speaking of the great philosopher Goethe, says: "He sees through every pore of his skin." We can say of the truly great vocal artist, that he not only sees, but also feels, through every pore of his skin.[50]

Bandyopadhyay closely translated her phrasing into Bengali and then redirected it toward his larger argument about the poverty of music education in India. Indeed, Bandyopadhyay clearly used her article extensively, as many of her insights—including birdsong, pronunciation, and the techniques associated with her vocal teacher, Manuel Garcia—are closely followed, though not directly credited, in the the *Gītasūtrasār*.

Another significant source for Bandyopadhyay was John Curwen's *Singing for Schools and Congregations: A Grammar of Vocal Music* (1843). Bandyopadhyay seems to have relied on the third edition (1852), as he duplicated Curwen's table of "The Alphabet of Music," which prioritizes the "mental effects" of the seven notes.[51] This table (Figure 1.1) strongly appealed to Bandyopadhyay's fascination with the emotional architecture of music and his desire to scientifically deduce, dissect, and document the properties of sound and their relationship to the mind of the listener. He does not credit Curwen directly

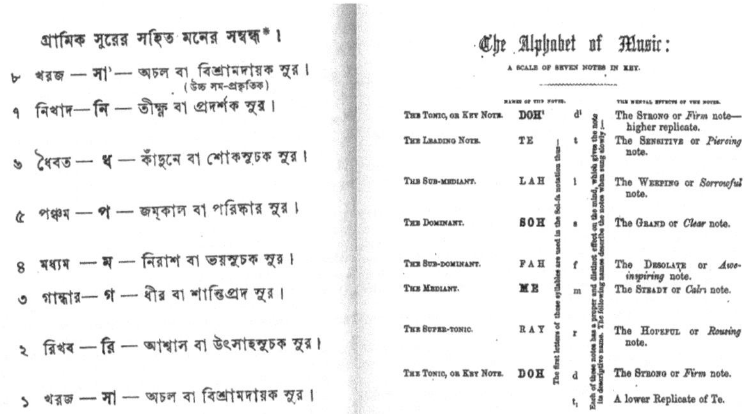

গ্রামিক সুরের সহিত মনের সম্বন্ধ* ।

৮ খরজ — সা' — অচল বা বিশ্রামদায়ক সুর ।
(উচ্চ সম-প্রকৃতিক)
৭ নিখাদ — নি — তীক্ষ্ণ বা প্রদর্শক সুর ।
৬ ধৈবত — ধ — কাঁদুনে বা শোকসূচক সুর ।
৫ পঞ্চম — প — জমকাল বা পরিষ্কার সুর ।
৪ মধ্যম — ম — নিরাশ বা ভয়সূচক সুর ।
৩ গান্ধার — গ — ধীর বা শান্তিপ্রদ সুর ।
২ রিখব — রি — আশ্বাস বা উৎসাহসূচক সুর ।
১ খরজ — সা — অচল বা বিশ্রামদায়ক সুর ।

The Alphabet of Music:

A SCALE OF SEVEN NOTES IN KEY.

	Names of the Notes.		The Mental Effects of the Notes.
The Tonic, or Key Note.	DOH'	d'	The Strong or *Firm* note—higher replicate.
The Leading Note.	TE	t	The Sensitive or *Piercing* note.
The Sub-mediant.	LAH	l	The Weeping or *Sorrowful* note.
The Dominant.	SOH	s	The Grand or *Clear* note.
The Sub-dominant.	FAH	f	The Desolate or *Awe-inspiring* note.
The Mediant.	ME	m	The Steady or *Calm* note.
The Super-tonic.	RAY	r	The Hopeful or *Rousing* note.
The Tonic, or Key Note.	DOH	d	The Strong or *Firm* note.
		t_1	A lower Replicate of Te.

The first letters of these syllables are used in the Sol-fa notation thus—

Each of these notes has a proper and distinct effect on the mind, which gives the note its descriptive name. The following names describe the notes when sung slowly:—

Figure 1.1. "The Alphabet of Music." Right: Curwen's original (1852). Left: Bandyopadhyay's adaptation (1885–86).

with this scheme but includes him in a footnote alongside several other "celebrated European musicologists": M. Jeu de Berneval, Dr. [John Wall] Callcott (author of the *Grammar of Music*), and Dr. R. J. Bryce, all of whom he probably read about in Curwen. These naming practices are intended to make Bandyopadhyay appear extremely well read and intimately engaged in conversations happening in Europe, as well as to highlight the variety of Western thinkers he admired and emulated, compared to the Indian musicologists he criticized and condemned.

While he followed the formula and related most of the mental effects directly, there are some differences in the properties of each note too: Doh, "The Strong or Firm note" becomes Sā, "The Firm or Restful note" (*acal bā biśrāmdāẏak sur*); likewise, Te, "The Sensitive or Piercing note" becomes Ni, "The Piercing or Revealing note" (*tīkṣṇ bā pradarśak sur*), and so on.

The affective properties of the scale were a long-established topic in canonical Indian musicology: Notes were said to have emerged from the sounds of specific wildlife and had emotional connotations, but perhaps Bandyopadhyay saw this as another expression of a poetic imagination and archaic orthodoxy. Instead, he preferred to think in terms of biology and psychology: Humans associated tones and notes with emotional states that were interpreted through the work of the mind. He invokes classical *rasa* theory (which examines how affect is generated in an aesthetic experience) but only superficially.[52] Instead, he emphasizes how emotional journeys are generated by the arrangement of the notes and their distinctive mental effects, which can then be mixed and contrasted like colors:[53]

> Now it will be demonstrated by what means various kinds of affect are introduced through meaningless, inexpressible notes. It is the relationship of the mind (*maner sambandha-i*) to the series of tones within the tones of the scale that is its root. If that relationship is not there, it is not possible for changes in mental mood and so on (*citta-bikārādi*) to manifest.[54]

He then explains this further by discussing specific melodic phrases or sequences, the interplay of expectation and excitement between notes, and how this might be explored in composition. This is emblematic of his larger endeavor to dissect, document, and demonstrate the inner workings of sound, while appealing to a physical universe of biological responses and resonating matter.

Conclusions

Bandyopadhyay saw the theoretical underpinnings of the emotional perception of music as fundamental: He maintained that because *ustāds* had neglected the art of singing pursuant to affective experience (*rasānusāre gān gāoẏā*),[55] they were also careless with their pronunciation of lyrics and articulation of musical structures and were undermining their audiences' engagement with musical meaning.

This emphasis on the mental stimulation of the listener, the emotional journey of the mind, and what Bandyopadhyay saw as the neglect in concentration and attention by lazy musicians and their sleepy audiences gestures to a larger divergence from older modes of understanding aesthetic emotion. As we have seen, before the nineteenth century, the key concept in musical reception theory was *dhyāna*, the remembrance or visualization of a musical entity. This visualization was tied to a specific iconography (as in *rāga-mālā*) and was interpreted through the vocabulary provided by classical *rasa* theory. This theory had originally been devised to dissect the emotions of a scene in a drama. In this system, the mind did not have a monopoly over emotions, and affective meaning was distributed across several components: the larger categorical emotions (*sthāyi-bhāva*) of the characters, their auxiliary feelings (*vyābhicāri-bhāva*), their embodied responses (*anubhāva*), and the stimulants provided by their surroundings (*uddīpana-vibhāva*).[56] Emotional states (*bhāva*) were cultivated across these components and played out at the level of the body. Bandyopadhyay regarded this system as inadequate and the idea of specific iconographies as limiting, archaic, and superstitious. In his system, *rasa* was an affect or emotional mood stimulated in the mind by configurations of notes, rather than an aesthetic substance that could only be understood through Sanskrit taxonomies of stimulants and responses. In other words, musical meaning was now a property of the mind.

While other Bengali scholars were critical of their peers and were actively pursuing new ways to talk about music, Bandyopadhyay was unusual in the ferocity of his critique and the radicalism of his reimagining of Indian musicology. A social and intellectual outlier, his ideas did not have a dramatic impact on discourses in Calcutta and beyond: Indeed, arguably his entire generation was overshadowed by the giants of the early twentieth century, especially Vishnu Narayan Bhatkhande (1860–1936), and later forums for discussion, especially the All India Music Conferences.[57] While the *dhyāna* system in *rāga-mālā* never had a revival,[58] other aspects of classical aesthetics, especially *rasa*, continue to provide a vocabulary for contemplating music and articulating musical experience—though this too has been deeply critiqued, including by the philosopher Daya Krishna, who called *rasa* "The Bane of Indian Aesthetics" and stressed that music required a less logocentric discourse, that is, one that was not adapted from literary theory.[59]

Bandyopadhyay certainly would have agreed, and over a hundred years earlier had written his music treatise as a manifesto for an evolution in musical discourse: He had argued that Indian musicians and scholars had neglected the work of the mind, that they were imprecise and blindly following tradition rather than engaging with modern, scientific paradigms. He was unabashedly in favor of adopting European ideas and practices, and he clearly saw himself as part of a global musical intelligentsia, yet also firmly at odds with his immediate neighbors in India. However, as Sagnik Atarthi and Charles Capwell have pointed out, while Bandyopadhyay had allies in Bengal, he ultimately struggled to compete with rival—and more influential—thinkers like S. M. Tagore and his followers.[60] Indeed, Bandyopadhyay's manifesto largely fell on deaf ears, as his demand for a modern, Occidental attentiveness in music ultimately did not gain significant traction, which itself raises yet further questions about how far concentration was colonized.

If we consider the larger picture and longer history behind Bandyopadhyay's work, three colonial-era developments emerge as especially influential. First, the imposition of British society's value system instigated a reckoning over the social nature of musical listening, one that foregrounded the idea of music as leisure and entertainment rather than an ethical, social, and political practice, as had been previously theorized through the codes of the Mughal salon assemblage (*meḥfil*).[61] As the conceptualization of what music was evolved, a space for amateur musicians from the patron classes grew: Musical practice was no longer imagined as the exclusive property of hereditary artists cum service providers. The language of articulating these conceptual and social changes was also evolving, with the political decline of Persian, the disenchantment with courtly forms of the vernaculars, and a steady preference for prosaic, accessible, and pragmatic forms of Bengali, Hindi, and so on. Taken together, these three developments—in society, in the performers, and in the language—cultivated a demand for new scholarly resources and literary tools for amateur musicians who wanted a systematic, theoretically informed approach to musical sound that was aligned with the Western understanding of music, one that was independent of the oral systems of hereditary masters and presented in a modern and critical vernacular register. From the perspective of attention, older modes of entrained listening and contemplation, from *dhyāna* visualization to *rasa* theory, were questioned if not judged as wholly redundant. Instead, from the 1860s in particular, notation, analytical writing, and studies that prioritized the specifics of musical form over poetic aesthetics and embodied modes of focus became increasingly desirable.

At the same time, the case study of Bandyopadhyay's work indicates how these factors did not result in a total sea change in musical attention in colonial India. Transformations in musical research certainly raised new questions and possibilities (often learned from European models), but these were not resolved with a standard, shared model of how to attend to music. There was a disconnect between these impressive tomes and lived listening practices, which often continued to uphold the aesthetic values of earlier times.[62] Western ideas were invoked by some thinkers, but again, when we look at Bandyopadhyay's actual sources, these were circulating in a piecemeal, fragmented manner. These ideas did not always find an eager audience either: Against the background of a nascent Indian nationalism and the related trend of Sanskritic neo-traditionalism in musical scholarship, appealing to European ideals did not become a standard practice, and the intrusion of Western musical concepts continued to be debated. Nonetheless, even if Bandyopadhyay's manifesto for critical attention did not transform the entire landscape of Indian art music, his work is a testament to how the field of musical scholarship had been substantially transformed, as the conventions and conceptual vocabularies for musical experience had been blown wide open and could now encompass new possibilities, propelled by expanding, modern, and colonial vocabularies for attention and concentration.

Notes

1. Key reference points in this debate include Nicholas B. Dirks, *Castes of Mind: Colonialism and the Making of Modern India* (Delhi: Permanent Black, 2002); C. A. Bayly, *Empire and Information: Intelligence Gathering and Social Communication in India, 1780–1870* (Cambridge: Cambridge University Press, 1999).

2. Richard David Williams, *The Scattered Court: Hindustani Music in Colonial Bengal* (Chicago: Chicago University Press, 2023).

3. Richard David Williams, "Music, Lyrics, and the Bengali Book: Hindustani Musicology in Calcutta, 1818–1905," *Music and Letters* 97, no. 3 (2016): 465–495.

4. For the significant developments in Bengali literary culture in this period, see Sudiptu Kaviraj, "The Two Histories of Literary Culture in Bengal," in *Literary Cultures in History:*

Reconstructions from South Asia, ed. Sheldon Pollock (Berkeley and Los Angeles: University of California Press, 2003), 503–566.

5. While Anglophone scholarship conventionally employs "musicology" as a translation of the nineteenth-century concept *Musikwissenschaft*, in this article the term refers to the systematic and canonical epistemology of music that developed in the South Asian context.

6. Emmie te Nijenhuis and Françoise "Nalini" Delvoye, "Sanskrit and Indo-Persian Literature on Music," in *Hindustani Music: Thirteenth to Twentieth Centuries*, ed. Joep Bor, Françoise "Nalini" Delvoye, Jane Harvey, and Emmie te Nijenhuis (New Delhi: Manohar, 2010), 35–64; Vishnu Narayan Bhatkhande, *Music Systems in India: A Comparative Study of Some of the Leading Music Systems of the 15th, 16th, 17th & 18th Centuries* (Delhi: S. Lal, 1984); Katherine Butler Schofield, *Music and Musicians in Late Mughal India: Histories of the Ephemeral, 1748–1858* (Cambridge: Cambridge University Press, 2024).

7. Ahmad Sharif, *Madhyayuger Rāg-Tālnāmā* (Dhaka: Bangla Academy, 1967); Richard David Williams "Singing in Tune with God: Bengali *Vaiṣṇava* Musical Scholarship in the Eighteenth Century," *South Asian History and Culture* 13, no. 2 (2022): 180–196.

8. Jinah Kim, *Garland of Visions: Color, Tantra, and a Material History of Indian Painting* (Oakland: University of California Press, 2021); Richard David Williams, "Playing the Spinal Chord: Tantric Musicology and Bengali Songs in the Nineteenth Century," *The Journal of Hindu Studies* 12 (2019): 319–338.

9. Sudhākalaśa, *Saṅgītopaniṣatsāroddhāraḥ: A Fourteenth-Century Text on Music from Western India*, trans. Allyn Miner (New Delhi: Indira Gandhi National Centre for the Arts and Motilal Banarsidass, 1998).

10. Richard David Williams, "Reflecting in the Vernacular: Translation and Transmission in Seventeenth- and Eighteenth-Century North India," *Comparative Studies of South Asia, Africa and the Middle East* 39, no. 1 (2019): 96–110.

11. This term also resonated with a body of theory developed by Hindi-speaking religious communities, who outlined the weakness of the embodied mind, which is assaulted and misled by the senses into thinking that the world of perception is real. These communities stressed that authentic listening and attending to the words of the guru offered a way out of this metaphysical trap. See Monika Horstmann and Dalpat Singh Rajpurohit, *In the Shrine of the Heart: Sants of Rajasthan from the Sixteenth and Seventeenth Centuries* (Heidelberg: Heidelberg Asian Studies Publishing, 2023).

12. Thibaut d'Hubert, *In the Shade of the Golden Palace: Ālāol and Middle Bengali Poetics in Arakan* (New York: Oxford University Press, 2018), 287; Emmie te Nijenhuis, "The Sanskrit '*Dhyānas*' of Johnson Album 35 and the *Rāga* Descriptions of Subhaṅkara's '*Saṃgītadāmodara*,'" *Journal of the Royal Asiatic Society of Great Britain and Ireland* 1 (1971): 52–58.

13. Sharif, *Madhyayuger*, 49–50.

14. A *daṇḍa* is a unit of time, approximately 24 minutes.

15. Williams, "Music, Lyrics," 468–470.

16. Williams, *Scattered Court*, 156–91; Justin Scarimbolo, "Brahmans Beyond Nationalism, Muslims Beyond Dominance: A Hidden History of North Indian Classical Music's Hinduization" (PhD diss., University of California, Santa Barbara, 2014).

17. Charles Capwell, "Musical Life in Nineteenth-Century Calcutta as a Component in the History of a Secondary Urban Center," *Asian Music* 18, no. 1 (1986): 139–163.

18. John Napier, "The Svarmaṇḍal and Its 'Ancestors': From Organological to Aesthetic Continuity," *The Galpin Society Journal* 58 (2005): 124–131, 225.

19. Sourindro Mohan Tagore, *Six Principal Rágas, with a Brief View of Hindu Music* (Calcutta: Calcutta Central Press, 1877), i.

20. Gerhard Böwering, "ENSĀN-E KĀMEL," *Encyclopædia Iranica*, Vol. VIII, Fasc. 5 (1998), 457–461.

21. Indeed, this resonates with the origins of the English word "attention," in the Latin *attendere*, "to stretch to."

22. Manuel da Assumpção, *Manoel da Assumpçam's Bengali Grammar. Facsimile Reprint of the Original Portuguese, with Bengali Translation and Selections from his Bengali-Portuguese Vocabulary*, trans. Suniti Kumar Chatterji and Priyaranjan Sen (Calcutta: Calcutta University Press, 1931), 88. English definitions from Sukumar Sen, *An Etymological Dictionary of Bengali, c. 1000–1800 A.D.*, Vol. II (1971): 737.

23. Assumpção, *Manoel*, 42

24. Other relevant Bengali terms include: *ābagāha-* (to plunge into, to pay attention); *gaur* (consideration, from Arabic *ghaur*, with sense of lowering into something, to go deep); *cita kari* (paying attention). Sen, *Etymological*, Vol. I: 48, 243, 277.

25. Krishnamohan Bandyopadhyay, *Bidyākalpadruma: Arthāt Bibidha Bidyā Biṣaẏak Racnā. Cittotkarṣa Bidhān* (Calcutta: R. C. Lepage and Co. and P. S. D'Rozario and Co., 1849), 139.

26. Francesca Brittan and Carmel Raz, eds., "Colloquy: Attention, Anxiety, and Audition's Histories," *Journal of the American Musicological Society* 72, no. 2 (2019): 541–580.

27. Muhammad Taqi Sayyid, *Tāsir al-anẓār* (Kanpur: Naval Kishor, 1877).

28. Baldev Prasad Misra, *Mṛtak-milap: yā Pretnagar. Spirit Land.* (Lucknow: Jain Press, 1897), 192.

29. Sagnik Atarthi, "Whither Musicology? Amateur Musicologists and Music Writing in Bengal," *Ethnomusicology Forum* 26, no. 2 (2017): 247–268.

30. Atarthi, *Whither*; Charles Capwell, "Marginality and Musicology in Nineteenth-Century Calcutta: The Case of Sourindro Mohun Tagore," in *Comparative Musicology and Anthropology of Music: Essays on the History of Ethnomusicology*, ed. Bruno Nettl and Philip V. Bohlman (Chicago: University of Chicago Press, 1991), 228–243.

31. Krishnadhan Bandyopadhyay, *Saṅgīt Śikṣā* (Calcutta: P. C. Doss, Day and Co., 1868).

32. Richard David Williams, "Epistemological *Jugalbandī*: Sound, Science, and the Supernatural in Colonial North India," in *The Acoustics of Empire: Sound, Power, and Media in the Long Nineteenth Century*, ed. Peter McMurray and Priyasha Mukhopadhyay (Oxford: Oxford University Press, 2024), 134–155.

33. Krishnadhan Bandyopadhyay, *Gītasūtrasāra* (Koch Bihar: Rajakiya Press, 1885), 97–98.

34. Bandyopadhyay, *Gītasūtrasāra*, 105.

35. Bandyopadhyay, *Gītasūtrasāra*, 107.

36. Bandyopadhyay, *Gītasūtrasāra*, 109.

37. Bandyopadhyay, *Gītasūtrasāra*, 109.

38. Essay first published in Hindi in 1875. See Bharatendu Harischandra and Kesarinarayan Shukla (eds.), *Bhāratendu ke Nibandha* (Varanasi: Nandakisora, 1967), 211–220.

39. Bandyopadhyay, *Gītasūtrasāra*, 97.

40. Lakshmi Subramanian, "The Master, Muse and the Nation: The New Cultural Project and the Reification of Colonial Modernity in India," *South Asia: Journal of South Asian Studies* 23, no. 2 (2000): 1–32.

41. Schofield, *Music*; Max Katz, *Lineage of Loss: Counternarratives of North Indian Music* (Middletown: Wesleyan University Press, 2017).

42. An exception to the rule is Asadullah "Kaukab" Khan's (ca. 1850–1919) introduction to music in Bengali, formatted as a question and answer session with one Bengali disciple. See A. K. Kaukab, *Saṅgīt Paricaẏ* (Calcutta: P. C. Mookerjee, 1915).

43. Bandyopadhyay, *Gītasūtrasāra*, 99. On the rhetoric of decadence in this period, see Williams, *Scattered Court*, 64–95.

44. Bandyopadhyay, *Gītasūtrasāra*, 86–87. On the evolving discourse around this genre, see Lalita du Perron, *Hindi Poetry in a Musical Genre: Ṭhumrī Lyrics* (Abingdon: Routledge, 2007).

45. Bandyopadhyay, *Gītasūtrasāra*, 101.

46. In particular, Bandyopadhyay was influenced by the poet Michael Madhusudan Dutt (1824–1873), most famous for his epic poem, *The Slaying of Meghnad* (*Meghnādbadh Kābya*, 1861), an adaptation of the *Rāmayaṇa* modeled on *Paradise Lost*. The two had first crossed paths in 1859 when Bandyopadhyay had acted in one of Dutt's plays, and Bandyopadhyay had dedicated his music self-instructor, the *Saṅgīt Sikṣā*, to Dutt nine years later. See Atarthi, "Whither," 256.

47. Bandyopadhyay, *Gītasūtrasāra*, 101.

48. Bandyopadhyay, *Gītasūtrasāra*, 104.

49. Richard Holt Hutton, "Ralph Waldo Emerson," *The Spectator*, May 6, 1882, 590–591. Reproduced in Richard Holt Hutton, *Criticism on Contemporary Thought and Thinkers*, Vol. 1 (London: Macmillan and Co., 1894), 49.

50. Margaret Watts Hughes, "A Glimpse into the Art of Singing," *The Nineteenth Century: A Monthly Review* 13, no. 75 (1883): 862–871, here 868.

51. Bandyopadhyay, *Gītasūtrasāra*, 94; John Curwen, *Singing for Schools and Congregations: A Grammar of Vocal Music* (London: T. Ward and Co., 1852), i.

52. On *rasa* theory, see Sheldon Pollock, *A Rasa Reader: Classical Indian Aesthetics* (New York: Oxford University Press, 2016).

53. Bandyopadhyay, *Gītasūtrasāra*, 95–96.

54. Bandyopadhyay, *Gītasūtrasāra*, 93–94.

55. Bandyopadhyay, *Gītasūtrasāra*, 103.

56. Pollock, *Rasa*.

57. Janaki Bakhle, *Two Men and Music: Nationalism in the Making of an Indian Classical Tradition* (Ranikhet: Permanent Black, 2008).

58. Aspects of the *rāga-mālā* have certainly continued in popular discourse among musicians and audiences. See Laura Leante, "The Lotus and the King: Imagery, Gesture and Meaning in a Hindustani *Rāg*," *Ethnomusicology Forum* 18, no. 2 (2009): 185–206.

59. Daya Krishna, "Rasa—the Bane of Indian Aesthetics," *The Journal of Indian Council of Philosophical Research* 21, no. 3 (2004): 119–135.

60. Atarthi, "Whither"; Capwell, "Marginality."

61. Katherine Butler Brown, "If Music Be the Food of Love: Masculinity and Eroticism in the Mughal *Mehfil*," in *Love in South Asia: A Cultural History*, ed. Francesca Orsini (Cambridge: Cambridge University Press, 2006), 61–83.

62. Williams, *Scattered Court*.

CHAPTER 2

The Attentive Eye / The Attentive Ear: Learning (and Unlearning) Focus in the Enlightenment

Ewan Jones

We live, it's often asserted, in a distracted age. New digital technologies, the widespread prevalence of ADHD, a sharpening corporate focus upon the "attention economy": All these factors contribute to the sense of a pressing crisis for concentration.[1] Johann Hari's *Stolen Focus: Why You Can't Pay Attention—and How to Think Deeply Again* (2022) offers but one recent instance of a spate of popular works that would counteract informational overload.[2] Scholarship impels and responds to this social concern: A discursive space that we can loosely designate "attention studies" encompasses cutting-edge work in experimental psychology, computer science, neuroscience, philosophy of mind, and empirical aesthetics.[3]

Concurrent with such empirical work, a number of genealogies of attention have explored the historical and social circumstances that condition and inflect our intrinsic cognitive capacities. Jonathan Crary's *Suspensions of Attention: Attention, Spectacle, and Modern Culture* (1999) offers an influential account of how the developing nineteenth-century psychology of attention (and the depredations of a rapidly developing global capitalism) transfigure spectatorial focus in impressionist painting. David Marno, Natalie M. Phillips, and Lily Gurton-Wachter have explored how poetry inherits and transfigures cultural practices of attending.[4] Matthew Riley, Nicholas Matthew, and Carmel Raz are among those to have charted similar narratives for music theory and performance.[5]

Such work, however, has seldom inquired how the specific channel of sensory input itself changes the nature and modality of attention. How does the attentive eye differ from the attentive ear? The question holds cross-historical pertinence, for the overwhelming association of focus and absorption with specifically ocular forms of culture extends from the Enlightenment into contemporary cognitive psychology. Its persistence is the substance of this chapter, which advances three interlocking claims. First, it charts the extent to which visual art sought to represent and to prove Descartes's revolutionary conceptualization of attention, as a mental faculty that mediated cognition and affect. Second, it examines how a smaller number of Enlightenment thinkers tested this ocular emphasis by appealing to other senses—most obviously the haptic and the aural. For Henry Home, Lord Kames, poetry represented a privileged site of contention, given its constitutive appeal both to the eye (through imagery) and to the ear (through prosody). For Thomas Reid, meanwhile, both touch and sound facilitated access to nondiscursive experience, which might overcome the "habits of inattention" that invariably beset human subjects.

Finally, by distinguishing between these separate channels of sensory input, this chapter endeavors to demonstrate that the Enlightenment posited an attentive ear that differed in kind—and not just degree—from the attentive eye. More specifically, aurality enabled the recovery of a truly radical suggestion in Descartes's conception: that attention and distraction, so far from existing as opposed or mutually exclusive states, might instead exist in a dynamic and dialectical relation. However much a quasi-scientific mode of taxonomic Enlightenment portraiture would seek both to model and to engender attention, the competing stimuli of any visual array necessarily involve distraction for the subject who beholds for any length of time. The constitutively temporal nature of music only underscores this fact. Accordingly, I argue, the attentive ear represents a unique means of challenging both the visualist paradigm and the preference for undivided attention, which continue to dominate modern aesthetics and empirical psychology alike.

* * *

Cultural historians frequently agree that distraction is a constitutively modern problem—whatever "modernity" means. For Jonathan Crary, the nineteenth-century concern over attentional deficit, facilitated by the physio-psychological work of Wundt, Helmholtz, and others, signals the demise of the "transcendental" Kantian subject, who formerly and assuredly bound the world, via the synthetic unity of apperception. Cognitive focus, in this account, proved so little a problem for the eighteenth century that Anglophone authors from the period "had no need of an idea of attention (the word is of marginal significance in the work of Locke, Hume, and Berkeley if present at all)."[6]

Historians of the eighteenth century beg to differ. Natalie Phillips's *Distraction: Problems of Attention in Eighteenth Century Literature* demonstrates just how much Enlightenment thought *did* worry over cognitive bottlenecks. The mass circulation of the commercial novel, the rise of the leisured (and "distractable") reader, the growing circulation of exotic commodities: All these contribute to a newly dispersed subject. In Phillips's case, the prior moment of attentional plenitude was marked not by the Kantian self but the attentive and self-sufficient Cartesian *cogito*.[7] Now, we cultural historians are naturally bound to our periods: We have a strong and understandable urge to perceive our own area of specialization as the birthplace of whichever idea assumes newfound contemporary salience. But I do not believe that cultures were ever not distracted. As David Marno has persuasively demonstrated in a lengthy exploration of premodern and early modern cultures of Christian spiritual exercises, Western culture has from time immemorial betrayed anxiety over the perfectibility (or decay) of human attention.[8] While the current demands that digital culture places upon the modern subject are indeed importantly unprecedented, this cultural continuity enables the historical archive to offer surprising resources for the present.

To argue that point, let's start with the figure whom Phillips and others have often taken to incarnate the ideal of the properly attentive subject—Descartes. It certainly is true that often attention serves, in his writing, to procure clear and distinct knowledge: "So long as we attend to a truth which we perceive very clearly," he states in his reply to the seventh set of Objections to the *Meditations*, "we cannot doubt it."[9] I think, therefore I am; I *attend*, therefore I truly am.

Yet the later Descartes posits a more complex role for attention, which interestingly mediates between cognition and affective states such as wonder, curiosity, and obsession. *Les passions de l'âme* §53 introduces wonder (*admiration*) as "the first of all the passions," in that it marks our encounter with a surprising or novel object.[10] Yet the simple appearance of a novel object "does not make us retain it in our memory unless our idea of it is strengthened in our brain by some passion, or perhaps also by an application of our intellect as fixed by our will in a special state of attention and reflection" (§76).[11] As we mature, "we should try to free ourselves from this inclination [to wonder] as much as possible. For we may easily make good its absence through that special state of reflection and attention which our will can always impose upon our understanding when we judge the matter before us to be worth serious consideration."[12] Volitional attention, which had supplemented instinctive wonder, now replaces it entirely.

Yet Descartes immediately goes on to suggest that this "special state" is not always quite so simply under a subject's control:

> [Wonder] seems to diminish with use, for the more we encounter unusual things which we wonder at, the more we find ourselves accustomed to stop wondering at them and to regard any we subsequently come upon as common. Nevertheless, when it is excessive and makes us fix our attention solely on the first image of the objects before us without acquiring any further knowledge about them, it leaves behind a habit which makes the soul disposed to dwell in the same way on every other object coming before it which appears at all novel.[13]

The very same faculty of attention, which should curb wonder's excess, now only exacerbates it, devolving instead into monomania, the *idée fixe*. Where the wonderstruck child proves merely absent-minded, the adult suffers from an attentional hypertrophy, which eagerly "fixes" salient stimuli, to the detriment of other possible experience. Descartes thus postulates a remarkably dynamical relation between attention and its putative antonyms: Riveted focus itself generates inattention to the world.

Over the ensuing centuries, a dazzling array of visual art—painting, engraving, photography—would seek both to represent and to confirm the cognitive-affective taxonomy of *Les passions de l'âme*. The first and most significant of these adoptions is the *Conférence sur l'expression générale et particulière*, a series of monthly lectures delivered in 1690 by Charles Le Brun, whose status as chancellor of the Royal Academy of Painting and Sculpture gave him unrivaled control over the training and advancement of court painters. Le Brun's lectures were published eight years after their delivery, with a series of accompanying etchings and diagrams that seek to represent with mathematical precision the full Cartesian suite of passions (hate, desire, joy, sadness, fear, hope, despair).[14] Le Brun specifies these physical corollaries with a meticulousness that is frequently comical—attention itself being a case in point:

> Esteem can only represent itself through attention and by the movement of those parts of the body that seem to be fixed upon the object that causes this attention: for here the eyebrows will appear to advance over the eyes and to the side of the nose, while the other end (of the eyebrows) will appear slightly raised. The eye will be wide open and the pupil fixed, the veins and the muscles of the forehead will appear somewhat inflated, along with those around the eyes, the nostrils will seem to incline towards the lower part of the face, the cheeks will also be slightly lowered toward the jaws, the mouth slightly open, its sides pulling back and leaning down. The head will also appear to move forward and lean upon the object that causes the admiration.[15]

Le Brun's coupling of esteem and attention attempts to work through the problem that beset Descartes's discussion of wonder. "Esteem" is clearly a positive state, reserved for objects that merit such exalted regard. But this spiritual condition can only manifest itself through the mediation of attention, along with its attendant physiological corollaries. We thereby discover one of the defining characteristics of the attentive eye, as Enlightenment thought would develop it: *an isomorphism between the subject and object of vision.* The subject learns to attend by looking correctly; states of attention are themselves one of the first things that she learns to behold. It's surely not incidental that Gustave Audran's engraving of the attitude is promoted to the frontispiece of the 1760 English-language edition of the *Conférence*, where it operates as a preparatory visual imperative: attend (Figure 2.1). The isomorphism of the attentive eye persists all the way to Michael Fried's famous articulation of absorption, in which the art viewer concentrates upon subjects of concentration (the boy engrossed in his task of blowing bubbles).[16]

But as with Descartes before him, Le Brun struggles to regulate the relation between affect and appropriately channeled focus. If "esteem" can only represent itself through the cognitive state (and physical corollaries) of attention, this does not mean that every act of attention necessarily implies the benign moral value of esteem. Earlier versions of the *Conférence* vacillate between the cognate (but not identical terms) of *attention* and *admiration*. Take a look back at Audran's plate: Might we not speculate that the subject seems rather unhealthily riveted upon his unseen object?[17]

Figure 2.1. Frontispiece to an English edition of *Charles Le Brun's Expressions of the passions of the soul*, with Gustave Audran's engraving of attention, ca. 1760.

The fraught attempt to secure a "pure" form of attention continued into the eighteenth and nineteenth centuries. The French archaeologist Alexandre Lenoir's *Observations scientifiques et critiques sur le génie* (1821) follows Le Brun's example in describing attention as "a mental exertion [*tension d'esprit*] so concentrated upon an object that it cannot be deviated from it," yet which is imprecisely demarcated from the "love, or any other carnal sentiment" that would produce an *idée fixe*.[18] Lenoir dutifully recites the standard physiological indices: raised eyebrows, open mouth, focused pupil. Yet even Le Brun's own illustrations, he concludes, fall short: "The attention that he has drawn is not *simple attention*, by which I mean that state that observes in a calmness of spirit; his is combined with fear, because he has raised the eyebrow of the figure, opened his mouth and given him an expression of pain."[19] Johann Kaspar Lavater's remarkably successful *Physiognomische Fragmente zur Beförderung der Menschenkenntnis und Menschenliebe* (1775–78) demonstrate the ongoing struggle to distinguish good and bad forms of attentiveness within minute physiognomic changes. Attention here represents a supreme moral virtue: "The attentive," writes Lavater, "are capable of all of which man ought to be capable"; even so small a task as the "taking, holding, or returning a tea-cup"

Figure 2.2. Johann Casper Lavater, "A woman, the physiognomy of whom expresses attention excited by desire," 1794.

can reveal exquisite sensibility.[20] Yet none of Lavater's sketches clearly achieve the "*attention simple*" that Lenoir strove to attain: attention "excited by desire" clearly falls short (Figure 2.2); attention "without real interest" (Figure 2.3) suggests a vacancy of spirit.

This Cartesian prioritization of the ocular endures in a remarkable variety of settings, down to the present age. The nineteenth-century French electrophysiologist Duchenne de Boulogne, to take but one example, used new technologies so as to compel and capture attention—passing electric currents through his experimental subjects, so as to stimulate the raised eyebrows that for Le Brun corresponded so exactly to the relevant attitude, and then capturing such states through the recent innovation of photography. De Boulogne

Figure 2.3. Johann Casper Lavater, "Attention without (real) interest," 1794.

visualized affective states that remained remarkably unchanged from *Les Passions de l'âme*: fear, anger, joy, and, naturally, attention, which, as with Lavater, assumes "pure" and "lascivious" forms. One plate reveals a young girl's ocular muscles forced, by electric current, to contract, while a bright light forces her other eyebrow to lower for the sake of contrast (Figure 2.4).

The domination of the attentive eye similarly persists in the discursive field that most fully shapes the contemporary understanding of cognitive focus: empirical psychology. While theoretical postulates such as the "cocktail party" theory of attending do consider how subjects process parallel channels of aural stimuli, the overwhelming majority of empirical studies rely upon optical recognition tasks. The expedient is in large part practical: It simply

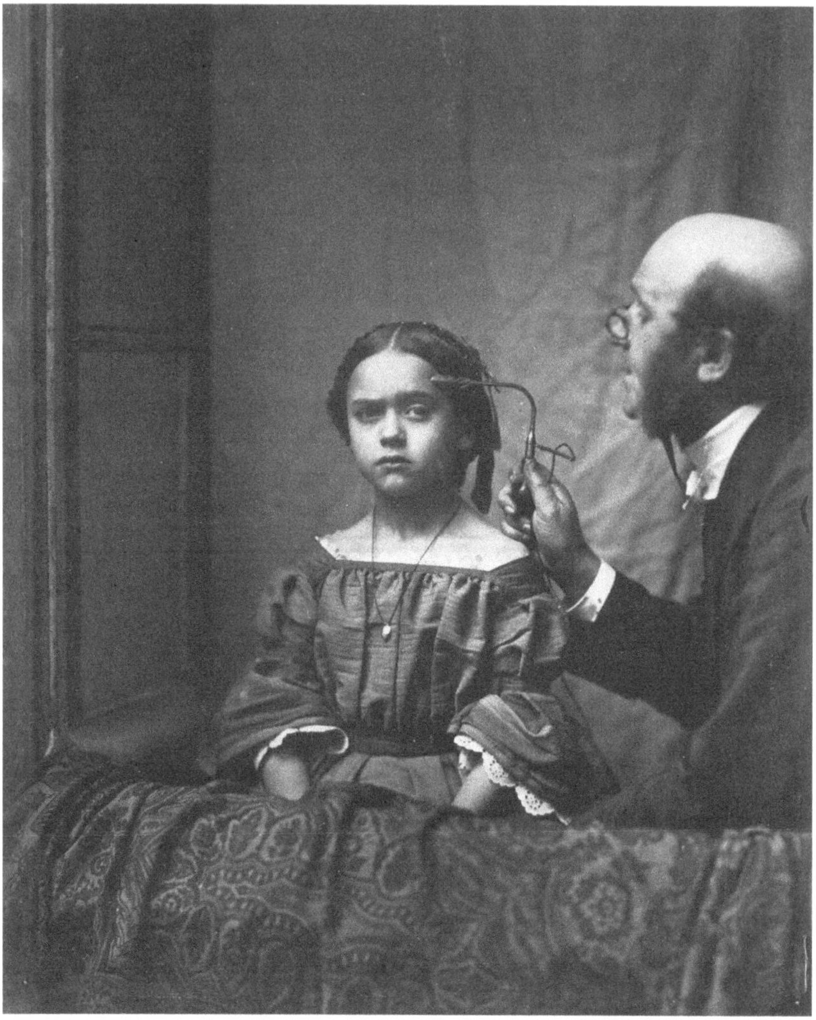

Figure 2.4. Guillaume-Benjamin-Amant Duchenne (de Boulogne), "The muscle of attention" (1854–56).

is easier to measure whether a subject does or does not detect a punctual stimulus within a given visual array than it is to probe the more complex and temporally embedded nature of auditory input.[21]

Yet there are strong reasons to believe that the optical paradigm that so dominates cognitive science—with its emphasis upon punctual stimuli and the binary yes/no approach to the perceptual capture of salient events—does not adequately describe even visual experience. Let me briefly give an example, before showing why such an approach proves still less adequate to the phenomenology of sound that justifies this collection of essays. A moment ago, I stressed the isomorphism of the attentive eye: The subject attends correctly to correct practice of attention. The engravings that accompany Le Brun's *Conférences*, like the photographs with which de Boulogne illuminates his electro-physiology, have something in common with the punctual stimuli of contemporary cognitive science, insofar as the trained observer can "correctly" perceive the state of attention in a snapshot glance. Yet visual experience often does not possess this quality: Salient stimuli unfold in time, make competing multifocal demands upon a beholder, exist on the fringe of consciousness.

To indicate what I mean, consider an artwork that Michael Fried takes to exemplify absorption: Carle van Loo's "The preaching of Saint Augustine before Valère" (Figure 2.5), which depicts a mainly attentive group that hearkens to the address of the eponymous orator.

Figure 2.5. Carle van Loo, "The preaching of St. Augustine before Valere," 1755.

An obvious exception to this general focus, as Fried points out, is the absent-minded boy at the right of the tableau, whose distraction, he claims, merely "throws into relief the intense absorption of everybody else." But does this summary capture a viewer's plausible phenomenology? The foregrounded prominence of Augustine (along with his pointing index finger) naturally also captures our preliminary focal attention, in keeping with the isomorphism described above. Yet it is not quite true that "everybody else" is focused: Note the woman in the very center of the tableau, who, in contrast to the companion to her side, casts her gaze not at Augustine, but down at the ground. She may even be asleep—or she may be in a reverie that enables deeper reflection upon the discourse. If we contrast her attitude to the boy, two very different forms of distraction arise: Where her eyes are closed and downcast, he gazes levelly out of the picture frame, as if to engage not simply "something that occupies him" so much as the beholder directly. It is as if he asks: You, who a moment ago was following Augustine's discourse properly, like all the rest—have you allowed yourself to become distracted? Yet perhaps, if you have, this is not so simply a bad thing. Perhaps—like the woman, who may slumber yet may also attend differently—this painting calls into question whether unifocal absorption really is the simple good that Fried claimed. Perhaps we might further contrast the unabashed boy to the scribe, whose dutiful transcription significantly points in the opposite direction from the orator's index figure. All this taken together, we might well start to wonder whether the amanuensis's absorption amounts to true attending: whether von Loos's painting does not rather set up a tension between living flesh and dead letter.

* * *

My brief reading of von Loos's painting teases out the difficulty that visual art attempts to encounter when representing and testing the Cartesian theory of attention. This difficulty is compounded, so soon as we consider the artwork not only as a simple object in space, to be taken in with a punctual glance, but as also subject to a temporal unfolding. In order for us to dismiss the boy as "distracted," we must first (ironically) have been distracted from the foregrounded saint to his unruly influence. What, then, if we switch from the attentive eye to the attentive ear? We might expect that music theory would find itself in a better position to explain the temporal nature of attention and to permit a more nuanced account of its relation to distraction than does the optical tradition that stretches from Le Brun to Fried. In fact, as several recent critics have pointed out, much relevant scholarship from the same period considers music from broadly the same rigidly Cartesian perspective that so dominated visual art. Roger Grant has recently demonstrated the extent to which a Cartesian taxonomy of passion proved one of the central planks in eighteenth-century musical theory (especially in the specific case of choral music). As with the examples of visual art that we considered above, music could represent a given affect, given its formal structure. Or—as Grant expounds at length—music could also engender a given state of passion more directly by operating upon the auditor's nervous system, itself conceived as a series of "tuned" sympathetic strings.[22]

Several good examples of what this taxonomic consideration of affect looks like in practice arise in Matthew Riley's study of German Enlightenment musical theory. Riley cites Johann Mattheson's first-person account of musical experience, whose separate elaborations (awe, wonder, fear, joy) are familiar from the preceding discussion:

> When I hear a solemn symphony in church, a sense of reverential awe falls over me. If an instrumental chorus joins in, this brings about an elevated sense of wonder within me. If the organ begins to storm and thunder, a divine fear arises in me. And if everything concludes with a joyful Halleluia, my heart leaps within my body.[23]

So far, the constitutive isomorphism that I identified with visual art holds: The affect of the auditor corresponds to (or is induced by) the formal properties of the artwork. Yet when we focus upon the question of attention, this equivalence vanishes. In contrast to visual art, which can seek directly to represent states of concentration or absorption, music has no obvious correlate. For what would be the aural equivalent to the raised eyebrows that for Le Brun signaled intense focus? That various answers could be essayed—harmonic resolution, motif, melodic surprise—itself suggests that no one constituent part of music can play the role.

It is perhaps for this reason that the anxiety over cognitive distraction, which even the visual appropriation of the Cartesian passions could not fully banish, emerges still more powerfully for the would-be attentive ear. The subject's struggle to maintain focus emerges in communal settings such as church services, as satirized in the second plate from William Hogarth's moral series of engravings, *Industry and Idleness* (1747), where the upstanding Francis Goodchild diligently singing with his master's daughter from a hymnal, not quite concealing a sleeping man with his frockcoat; and in the rapidly changing context of concert performance, through which the forms of dispersal tolerated during the Baroque period (where auditors were permitted to wander around, talk, and even eat during recitation) gave rise to the expectation of silent absorption. Hogarth again satirized the difficulty of undistracted performance with his 1741 etching *The Enraged Musician*, where a violinist is unable to proceed with his practice given the commotion outside his window, where drunkards, ballad-hawkers, and squawking parrots congregate.

Riley's study amply demonstrates the broader consequences of these changing attitudes to musical attention. Once again, the contrast with visual art proves instructive: Where Le Brun sought to train aspirant painters and illustrators through a series of mathematical, readily reproducible specifications, the question of expertise becomes both more important and more amorphous in a musical setting. "Attentive listening," Riley summarizes in reference to the music theory of Johann Nikolaus Forkel, "is now an attitude that the listener brings to the music, not something that flows from the 'give and take' between stimulus and response."[24] The capacity to attend correctly separates the connoisseur (*Kenner*) from the amateur (*Liebhaber*), the civilized man from the "savage."

In light of this growing focus upon expertise, the identification of "proper" attention with the formal features of a given artwork becomes still more difficult to secure than was the case for visual art. As Riley notes, C. P. E. Bach is frequently celebrated by connoisseurs as a composer who encouraged attentiveness; yet his frequent deviations from regular presentation of the *Hauptsatz*, in terms of phrase structure and hierarchy, produced "disorientating harmonic digressions, sudden pauses, apparently unmotivated contrasts."[25] Nicholas Matthew has similarly treated the "surprise chord" in the Andante movement of Haydn's "Surprise Symphony" as an instance of forceful cognitive capture. Yet such a conception presupposes a listener hitherto distracted: As Matthew notes, public opinion quickly understood the chord as "Haydn's bid to startle sleeping concertgoers or to scare inattentive ladies."[26] The critic whose concentration remains undisturbed misses the force of surprise.

What, then, would stand as the musical equivalent to the raised eyebrow with which visual art sought to train and represent attention? The problem is not a lack of structural features but rather a surfeit—surprise chords, dynamic contrasts, melodic variation, regular meter—which pushes subsequent analysis in mutually exclusive directions. Johann Georg Sulzer's *General Theory of the Fine Arts* (1771–74) serves as a case in point. The first two parts of this triumvirate of essays respectively consider the foundations of aesthetic unity (with particular concern for visual art) and the creative process of the artist. In both cases, attention is a desideratum: The painter can model variety in uniformity by presenting common physiognomies while altering posture and dress; the artist learns his or her craft by inducing a state of absorptive isolation.[27] When Sulzer moves to consider music in the third and final essay, however, his successive identification of attention with both rhythm and melody generates significant tension. The former demonstrates how repetition scaffolds cognitive focus: "There is nothing better for sustaining ourselves in the same emotion than a uniform, evenly measured motion, by which attention to a given object may be secured."[28] Melody, however, seems to enforce the diametrically opposite message: The goal of music, which is to maintain the listener in a given state of emotion, "is not possible by simply repeating the phrase, no matter how excellent it may be, since the repetition of anything is boring, and one's attention can quickly wane."[29]

It is not that these competing attentional directives could not possibly be reconciled. We can readily imagine an integrationist account that viewed metrical uniformity as preparing (or accompanying) melodic variation. Yet Sulzer's rather abstract and schematic treatment does not demonstrate how they might unite in actual experience, as, for example, common facial physiognomy and variety of posture obviously intersect within the picture plane. The remainder of this essay shifts its own focus to a different school of eighteenth-century aesthetics, which certainly *did* account for the dynamic, time-bound nature of aural experience. The Scottish Common Sense school of philosophy did not so extensively treat music as a separate domain as did their German near-contemporaries. Yet their broader consideration of aurality involved three crucial factors: first, a more sustained contrast of sensory experience that have hitherto been considered separately (the attentive eye and the attentive ear); second, the treatment of artistic modes (such as poetry) that combined representational and nonrepresentational aspects (imagery and prosody); and third, a more positive role for distraction than the simple negation of an absorption for which most aesthetic theory continues to press. In the process, they anticipate the more dynamic accounts of attention that emerge in the nascent psychology of Ribot and James, before being superseded by the later cognitive focus upon the punctual visual stimulus.

Both works situate this treatment of aurality within their wider struggle against what Reid called The Way of Ideas—the representational philosophies of Locke and Hume, which contend that we come to know the world only through the mental mediation of "Ideas" or "impressions." Lord Kames's *Elements of Criticism* (1762) has strong grounds for being considered the single most successful work of eighteenth-century aesthetics in the English-speaking world. This substantial two-volume work offers a hodgepodge of off-the-cuff theorizing and delineation of choice specimens from several artistic media. Kames intends his work to resist the "chimerical" nature of "ideal" philosophy—offering, in the process, what to my mind is the first significantly cognitive account of aesthetic experience. Throughout this endeavor, "the pleasures of the eye and the ear" are brought together and teased apart.[30] Kames is commonly read as part of the "triumph of associationism," according to which a broadly Lockean framework came to replace Francis Hutcheson's intrinsic conception of a "Common Sense" for beauty—with the consequence that statements of taste devolve into expressions of culturally relative values.[31] In fact, however, Kames extends Hutcheson's work into a remarkably sensitive account of the generative constraints—the bottlenecks and bandwidths—inherent to human cognition.

The contrast of sound and vision anchors this account from the start: "I shall begin," Kames declares, "with those beauties of language that arise from sound; after which will follow the beauties of language considered as significant: this order appears natural; for the sound of a word is attended to, before we consider its signification" (*EC* 4). Sonority, then, captures focus, and in a wide variety of ways: Kames successively enumerates the manner in which grammar (*EC* 28), sentence structure (*EC* 47), pause (*EC* 80), and rhyme (*EC* 109–110) all work to "rouse" the attention. A "uniform" prosody, such as he alleges French to be, proves "extremely fatiguing" (*EC* 90–91); the admixture of smoothness and harshness, peculiar to the English vernacular, counteracts this tendency (*EC* 10).

More remarkable still than this cognitive reconceptualization of what had previously been considered largely in evaluative terms (the alleged "superiority" of French or English), is the manner in which Kames sets it to work—in service of the evocation of lived temporal experience. *Elements of Criticism* is nearly unique among canonical eighteenth-century aesthetic texts for the richness of its proto-phenomenological description—for its conveyance of how it feels to read, in real time. Cognitive focus, on this account, is no simple ideal for which to strive; rather, it is as a dynamical state that paradoxically requires continual disruption in order to endure. The varied caesuras within the heroic couplets of Alexander Pope's *Rape of the Lock* offer a case in point.

Where much criticism praises Pope's epigrammatic concision, or putative concord of sound and sense, Kames emphasizes how the varying placement and effect of the medial caesuras interrupt the flow of verse. The pause that follows an accented syllable "occasions a sudden stop without preparation, which rouses the mind" (*EC* 97); a caesura after an unaccented syllable, by contrast, prepares the mind for rest, and as such is "calm and sweet" (*EC* 98). Kames's stipulations may sound programmatic; but what really matters is the way in which these separate medial caesuras are integrated within a broader experience of reading—where a mind craves not only variation, but also variation in variation itself.

Having defined and listed exempla from his taxonomy of pauses, Kames gives way to a sudden doubt: "Musing on the foregoing subject, I begin to doubt whether all this while I have not been in a reverie, and whether the scene before me, full of objects new and singular, be not mere fairy-land. Is there any truth in the appearance, or is it wholly a work of imagination?" (*EC* 102). The interjection proves ironic to the extent that it mirrors the "chimerical" idealism from which Kames wished to deliver himself. Yet this self-interruption performs precisely the work that *Elements of Criticism* had imputed to Pope's abrupt masculine caesura: The mind, having been drawn in, is roused to life by sudden self-suspension. It is for this reason that Kames immediately answers his skeptical self-query in the emphatic negative: "We cannot doubt of its reality." A generation previously, Francis Hutcheson had established uniformity within variety as *the* mark of beauty, identifying it most powerfully in geometrical objects (squares, equilateral triangles, rhombuses).[32] Kames now self-consciously cites "uniformity within variety," while converting it from a spatial into a temporal register: It marks the pressing need for a subject with finite cognitive resources to replenish attention through shifting its focus.

We can contrast Kames's "reverie" with a variety of cognitive cognates that this essay has previously thrown up: absorption, spotlighting, "*attention pure*." Kames, in contrast to Lavater and others, proposes a discontinuous mode of experience, where interruptive auditory stimulus corresponds to a subject who has taken conscious leave of herself. That Kames used "reverie" advisedly is shown in the frequency with which he returns to the term: "The reader's passions are never sensibly moved, till he be thrown into a kind of reverie; in which state, forgetting that he is reading, he conceives every incident as passing in his presence, precisely as if he were an eye-witness" (*EC* 68–69). We can differentiate this formulation sharply from Locke's own conception in *Essay Concerning Human Understanding*. "When *Ideas* float in our mind, without any reflection or regard of the Understanding," he states, "it is that which the French call reverie; our language has scarce a name for it."[33] Locke immediately and categorically opposes this state to attention.

Kames's reverie, by contrast, does not exclude cognitive focus. Indeed, it enables a subject to attend differently to a given object (Pope's varied couplets) than a purely volitional and unifocal absorption allows. Poetry is peculiarly able to discharge this work, to the extent that we shuttle continually between its signifying qualities (the paraphrasable statements that a Popean couplet advances) and rhythmical effects, which exist on both a larger and a smaller level, and which may not always be consciously perceived. But *Elements of Criticism* offers several other real-life examples of where sensory suspension sharpens perception, as "between sleeping and waking, when all the senses are bound and suspended, music is far sweeter than when one is fully waking" (*EC* 337n). In such cases, "no object in the field is seen distinctly, but that singly which fixes the attention: in a profound reverie that totally occupies the attention, we scarce see what is directly before us" (*EC* 338). (In an interesting aside, Kames explains the terror of the stillness of night as the impossibility of being distracted.)

These repeated references to reverie import into aesthetics something like what cognitive scientists now call "flow states," where the human subject can discharge intense cognitive focus in a quasi-automatic fashion.[34] This dynamic conception differentiates Kames's picture of the mind from the contemporaneous thought of Rousseau, which accepts Locke's characterization of the mode as unreflective—yet merely inverts the valuation, such that distraction trumps rational thought. *Elements of Criticism* rather suggests that such dynamic limit-states for attention themselves hold cognitive significance. To clarify what this may be, the final section of this essay turns to the related philosophy of Thomas Reid.

* * *

The *Elements of Criticism* tells us, with an acuity that rivals any other work in aesthetics from the eighteenth century, how it feels to attend to an artwork. Yet its bricolage-like nature impeded Kames from offering any sustained definition of attention, or broader consideration of its quotidian operation and etiology. Several thinkers associated with the Aberdeen and Glasgow schools of Common Sense probed such questions.[35] These included Thomas Reid, who extends three of the fundamental features that characterized my earlier account of Kames's aesthetics: an abiding interest in the differing attentive modalities of the sense organs, taken together and in isolation, with a particular concern for how aural and tactile stimuli differ from the visual; a concern to probe the difference between representational and nonrepresentational modes of thinking; and a dynamic conception of attention itself, which relies upon variation, interruption, or distraction in order to sustain itself.

"Let scholastic sophisters entangle themselves in their own cobwebs," Reid declares by way of prolegomenon to *An Inquiry into the Human Mind, on the Principles of Common Sense* (1764), "I am resolved to take my own existence, and existence of other things, upon trust; and to believe that snow is cold, and honey sweet, whatever they may say to the contrary."[36] "They," here, are the Lockes and Humes of the world, whose vocabulary of "ideas," "mental representations," and "schemata" he, like Kames, dismisses from the start. In order to demonstrate how we can take the world "on trust," Reid introduces a central distinction between two forms of processing sensory stimuli. "Artificial signs," on his account, name a kind of relation that proves merely conventional: The word *gold* bears an arbitrary relation to the valuable metal that it indicates. "Natural signs," by contrast, bind together sensory experience and cognitive content through "the original constitution of our nature" (*IHM* 42).

Facial expression offers a ready example of the sort of "natural sign" that Reid has in mind. Reid agrees with Le Brun that affects and dispositions convey themselves immediately and unreflectively to the mind of the observer. The *Inquiry* argues that

> the thoughts, purposes, and dispositions of the mind have their natural signs in the features of the face, the modulation of the voice, and the motion and attitude of the body: That, without a natural knowledge of the connexion between these signs and the things signified by them, language could never have been invented and established among men: and, That the fine arts are all founded upon this connexion, which we may call *the natural language of mankind.* (*IHM* 61)

Reid enumerates three examples of the natural signs: the visual ("features of the face"), the aural ("modulation of the voice"), and the kinesthetic ("motion and attitude of the body"). A recent resurgence of interest in Reid's philosophy of perception has tended—in keeping with my discussion above—to concentrate upon the first of this triumvirate.[37] They have good reason to do so, given the *Inquiry*'s pioneering formulation of the "geometry of visibles": a non-Euclidean description of spatial properties that significantly anticipated the later work of Gauss, Riemann, and others. Yet by focusing upon the comparatively overlooked sensory modalities of touch and sound, I aim to demonstrate that Reid produces perhaps the most significantly dynamic formulation of attention that the period has to offer.

In *Thomas Reid's Theory of Perception* (2007), Ryan Nichols understands the *Inquiry* to argue for a distinctive form of realism—according to which we perceive the world directly, without necessary recourse to ideas, impressions, or other mediatory apparatus provided by the Way of Ideas. Nor, Nichols contends, does perception need to depend upon the subjective "sensations" of primary qualities, which the *Inquiry* introduces as an alternative to such bridges. "Sensations," he summarizes in a chapter on the topic, "far from being necessarily perceived intermediaries, or the typical immediate objects of acts of perception, are not even contingent causal components in visual perception and are probably not even physically necessary corollaries to perception."[38] Nichols has good textual authority for concluding that such subjective qualia remain "intangible and uncertain."[39] For Reid himself elsewhere acknowledges that the very act of attending to sensations can itself lessen or even entirely efface conscious awareness:

> When the mind is agitated by any passion, as soon as we turn our attention from the object to the passion itself, the passion subsides or vanishes, and by that means escapes our enquiry. This, indeed, is common to almost every operation of mind: When it is exerted, we are conscious of it; but then we do not attend to the operation, but to its object. When the mind is drawn off from the object to attend to its own operation, that operation ceases, and escapes our notice.[40]

Reid's dynamical theory of attention—which operates in tension with conscious awareness, both of which compete for a subject's finite bandwidth—here appears negative. We perceive the world until the moment that we foreground that perception—whereupon we no longer do.

Elsewhere, however, Reid would consider this faculty (and its dynamic propensity) in a more productive manner. To do so, we need to turn—once again—from the visual modality, which anchors Nichols's analysis, so as to consider other forms of sense experience. Touch—which Reid offers as an exemplary natural sign (*IHM* 61)—offers a case in point. Tactile and visual forms of experience most obviously come into conjunction through Reid's contribution to the longstanding Molyneux Problem, which asked whether a blind subject, having had their sense restored, could perceive the newly available visual manifold in light of past experience. As Nichols summarizes, Reid held that a sophisticated mathematician such as Nicholas Saunderson (who was indeed blind) could be said to perceive the revealed world immediately through possessing geometric concepts. To this extent, concepts do not require sensations. Yet three-dimensional objects (cubes and spheres) prove different. "To acquire the concept of depth," Nichols acknowledges, "one must have tactile sensations."[41]

I think we can say more about the importance of tactile experience, in addition to its relation both to aurality and to attention in general. To do so, I want to consider at length a passage from the *Inquiry*'s chapter on that sense. Reid asks the reader to undertake one of his characteristic self-experiments: Lay your hand on the table and attempt to attend to the sensation without forming some idea of the object or its qualities. "It is one thing," he contends, "to have the sensation, and another to attend to it, and make it a distinct object of reflection. The first is very easy; the last, in most cases, extremely difficult." We routinely encounter such a sensation; "yet it is never attended to, but passes through the mind instantaneously" (*IHM* 57).

A moment ago, we saw how attention could exacerbate the fugitive nature of experience: Dwelling on a passion, the passion ceased to be. Here Reid suggests something subtly different: If quotidian experience proves elusive, this might well be because we have been attending in the wrong manner—moving too rapidly from a sensation to that which it customarily signifies. Such, at least, is the suggestion of a passage from the *Inquiry* that I cite at length for its relevance and beauty. Indeed, it offers an apt summary of my argument as a whole:

> May we not hence conclude, that the knowledge of the human faculties is but in its infancy? That we have not yet learned to attend to those operations of the mind, of which we are conscious every hour of our lives? That there are habits of inattention acquired very early, which are as hard to be overcome as other habits? For I think it is probable, that the novelty of this sensation will procure some attention to it in children at first; but being in nowise interesting in itself, as soon as it becomes familiar, it is overlooked, and the

> attention turned solely to that which it signifies. Thus, when one is learning a language, he attends to the sounds; but when he is master of it, he attends only to the sense of what he would express. If this is the case, we must become as little children again, if we will be philosophers: we must overcome this habit of inattention which has been gathering strength ever since we began to think; a habit, the usefulness of which, in common life, atones for the difficulty it creates to the philosopher, in discovering the first principles of the human mind. (*IHM* 59)

That human knowledge remains in its "infancy" proves ironic—to the extent that children are precisely what we have forgotten to be. "Habits of inattention" befall us all, precisely through excessive focalized attention to signification. Philosophy might seek to unlearn such habits. Like Kames before him, Reid seeks to deliver us back to a world where "the sound of a word is attended to, before we consider its signification."

Like Kames before him, moreover, Reid's emphasis upon prediscursive, nonsignifying, and preconscious mental operation leads him to supplement haptic experience with the attentive ear.[42] In the section immediately preceding its invitation to tactile self-experiment, the *Inquiry* extends the discussion of the "natural language" of "brutes," who communicate via vocal modulations, facial expressions, and gesture. Language, once recited or sung aloud, decays into silent reading. Aesthetics recapitulates at the social level what the inattentive subject experiences singly: "The arts themselves are merely the language of nature, which we brought into the world with us but have unlearned through disuse, and so find the greatest difficulty in getting it back" (*IHM* 54).

If such talk might superficially resemble Rousseau's noble savage, we should remember that Reid's "brutes," or children, or dumb individuals, or advanced Cambridge mathematicians, alike lay claim to knowledge. Their adoption and comprehension of "natural signs" does not mark the triumph of expressiveness over science, but a means of knowing the world directly. The *Inquiry* returns to tactile experience, to drive home the point:

> From the time that children begin to use their hands, nature directs them to handle every thing over and over, to look at it while they handle it, and to put it in various positions, and at various distances

> from the eye. We are apt to excuse this as a childish diversion, because they must be doing something, and have not reason to entertain themselves in a more manly way. But if we think more justly, we shall find that they are engaged in the most serious and important study. (*IHM* 241)

It's hard, in short, to think of another Enlightenment thinker who argues that a fetus can have sensations of time and space (*IHM* 71); or who grants authority to the midwives and nurses that he has consulted (*IHM* 130).

* * *

By supplementing his sustained consideration of visual experience with a broader treatment of touch and sound, my reading of Reid has reproduced, in microcosmic form, the thinking of this essay as a whole. In doing so, we catch sight of the most dynamic formulation of attention available to eighteenth-century thought, which harnesses the potential of Descartes's theorization of the affects. We have seen how that Cartesian taxonomy sponsored a series of attempts to police and train proper attention, which visual art itself—and music and poetry all the more—resisted. In the hands of Kames and Reid, over-fixated attention devolves into distraction, while reverie represents a means of thinking (or knowing) differently.

I believe that the replacement of mind-wandering and multifocal attention for the old humanistic value of absorption proves truer to the nature of aesthetic experience, in addition to representing a productive means of relating such experience to contemporary empirical research. Some recent studies challenge the reductive emphasis upon periodic stimuli that, as I noted above, continues to dominate cognitive science. One experiment contrasts the way in which "experts" and "amateurs" contemplate paintings: Eye-tracking devices show that the former's focus roves widely around the tableau, while the latter fixate upon a foregrounded object.[43] In that wandering eye, we locate again the "distracted" boy who gazes back at us from the tableau; the wavering attentive ear, carried downstream by Pope's couplet prosody; the eighteenth-century subject, who tries to learn to touch a table. We might redescribe the "amateur" as the modern subject who moves straight to signification. The "expert," in turn, names the body who has rediscovered how to be a child.

Notes

1. Timothy Wu made the phrase "attention economy" proverbial in his *The Attention Merchants: The Epic Scramble to Get Inside Our Heads* (New York: Alfred A. Knopf, 2016).

2. Johann Hari, *Stolen Focus: Why You Can't Pay Attention—and How to Think Deeply Again* (London: Bloomsbury, 2022).

3. Attention is by some measures *the* single most-researched topic in William P. Seeley, *Attentional Engines: A Perceptual Theory of the Arts* (Oxford: Oxford University Press, 2020).

4. Natalie Phillips, *Distraction: Problems of Attention in Eighteenth Century Literature* (Baltimore: Johns Hopkins University Press, 2016), 6.

5. Matthew Riley, *Musical Listening in the German Enlightenment: Attention, Wonder and Astonishment* (London: Routledge, 2004); Nicholas Matthew, "Interesting Haydn: On Attention's Materials," *Journal of the American Musicological Society*, 71, no. 3 (Fall 2018), 655–701; Carmel Raz, "'To 'Fill Up, Completely, the Whole Capacity of the Mind': Listening with Attention in Late Eighteenth-Century Scotland," *Music Theory Spectrum*, 44, no. 1 (2022), 141–54.

6. Jonathan Crary, *Suspensions of Perception: Attention, Spectacle, and Modern Culture* (Boston, Mass.: MIT Press, 2001), 18.

7. Phillips, *Distraction: Problems of Attention*, 6.

8. David Marno, *Death Be Not Proud: The Art of Holy Attention* (Chicago: University of Chicago Press, 2016).

9. René Descartes, "Reply to Objections," in *The Philosophical Writings of Descartes*, trans. John Cottingham, Robert Stoothoff, and Dugald Murdoch, 3 vols. (Cambridge: Cambridge University Press, 1988), I, 309.

10. René Descartes, "The Passions of the Soul," in *The Philosophical Writings of Descartes*, I, 325–404, at 350.

11. Ibid., 355.

12. Ibid.

13. Ibid.

14. The 1698 edition collated several lectures given by Le Brun, who had died eight years previously; it went through numerous editions with various engravings.

15. Claude Nivelon, *Vie de Charles Le Brun et description détaillée de ses ouvrages* (Geneva: Droz, 2004), 344; translation mine.

16. Michael Fried, *Theatricality and Absorption* (Chicago: University of Chicago Press, 1976).

17. Bernard Picart's etchings, based on Le Brun's sketches, identify the same facial gesture with "veneration." André Félibien, who as secretary to the Royal Academy was partly responsible for the confusions and inconsistencies in the publication of Le Brun's lectures, proposes the Greek attendant in Le Brun's historical painting *The Tent of Darius* as a model for attention; yet the figure in question more closely resembles Le Brun's profile of admiration. For more on the history of Félibien's involvement, see Stephanie Ross, "Painting the Passions: Charles LeBrun's *Conférence Sur L'Expression*," *Journal of the History of Ideas*, 45, no. 1 (January–March, 1984), 25–47.

18. Alexandre Lenoir, *Observations scientifiques et critiques sur le genie* (Paris: 1824), 124; translation mine.

19. Ibid.

20. John Kasper Lavater, *Essays on Physiognomy*, trans. Thomas Holcroft (London: William Tegg & Co., 1853), 162.

21. Mari Riess Jones's dynamic attending theory (DAT) remains a peripheral rival approach in this regard, with its attempt to incorporate the detection of punctual acoustic events into larger rhythmical hierarchies, which operate in the mind of a subject who nevertheless may not consciously perceive them. However intuitive such phrasings may feel to musical experts and amateurs alike, it is easy to see why their complex modeling is difficult to operationalize in a controlled empirical setting. Mari Riess Jones, *Time Will Tell: A Theory of Dynamic Attending* (Oxford: Oxford University Press, 2018).

22. Roger Grant, *Peculiar Attunements: How Affect Theory Turned Musical* (New York: Fordham, 2020).

23. Johann Mattheson, *Der vollkommene Capellmeister* (Hamburg: Christian Herold, 1739; repr. Kassel: Bärenreiter, 1954), 208–9; cited and translated in Bonds, "Aesthetics of Instrumental Music," 393; cited in Matthew Riley, *Musical Listening in the German Enlightenment: Attention, Wonder and Astonishment* (London: Routledge, 2004), 21.

24. Riley, *Musical Listening in the German Enlightenment*, 109.

25. Ibid., 151.

26. Nicholas Mathew, "Interesting Haydn: On Attention's Materials," 677.

27. Georg Sulzer and Heinrich Christoph Koch, *Aesthetics and the Art of Musical Composition in the German Enlightenment: Selected Writings of Johann Georg Sulzer and Heinrich Christoph Koch*, ed. Nancy Baker and Thomas Christensen (Cambridge: Cambridge University Press, 2009), 47, 58.

28. Ibid., 94.

29. Ibid., 101.

30. Henry Home, Lord Kames, *Elements of Criticism*, 2 vols. (Charles Ingham: Dublin, 1772), II, ii; subsequent references are given parenthetically within the text. While my reading concentrates upon sound, landscape gardening (*EC* 277–89) both reflects the contemporary discourse of taste and rises above it, through a discussion of how the "wandering path" through variegated scenery engages and interrupts cognition. (Kames holds Chinese gardens to be the most successful in this regard.)

31. Peter Kivy elides Kames with Archibold Alison's attack on Hutcheson. See Peter Kivy, *The Seventh Sense: Francis Hutcheson and Eighteenth-Century British Aesthetics* (Oxford: Clarendon, 1975), 203. Paul Guyer judiciously designates Kames's work as the most significant eighteenth-century attempt to reconcile Shaftesbury's cognitivism with Addison's "free play" (*A History of Modern Aesthetics*, 3 vols. (Cambridge: Cambridge University Press, 2014), I, 177–94).

32. Frances Hutcheson, "Treatise I: An Inquiry Concerning Beauty, Order, Harmony, Design," in *Philosophical Writings* (London: J. M. Dent, 1994): 7–44, at 15–16.

33. John Locke, *An Essay Concerning Human Understanding*, ed. Peter H. Nidditch (Oxford: Oxford University Press, 1999), 227.

34. See Mihaly Csikszentmihalyi, *Flow: The Psychology of Happiness* (London: Random House, 2013).

35. Alexander Gerard's *Essay on Taste* (London: 1759), for example, considers how the salience of an aesthetic object depends not only upon the cognitive affordances of the perceiving subject, but also the customary preferences of the culture in which she lives (111).

36. Thomas Reid, *An Inquiry into the Human Mind, on the Principles of Common Sense* (London: 1764), 17; further references follow parenthetically in the main body of the text.

37. While I thoroughly agree with Gideon Yaffe that Reid's novel theory of attention differs from the "spotlight" view of cognitive science, the pattern of thinking that I will outline differs

from his formulation of concerted conscious awareness over extended temporal envelopes. Gideon Yaffe, "Thomas Reid on Consciousness and Attention," *Canadian Journal of Philosophy* 39, no. 2 (June 2009), 165–194, at 183.

38. Ryan Nichols, *Thomas Reid's Theory of Perception* (Oxford: Oxford University Press, 2007), 159.

39. Ibid., 208.

40. Thomas Reid, *Essays on the Intellectual Powers of Man*, ed. Derek Brookes (Pennsylvania: Penn State Press, 2002), 61.

41. Nichols, *Thomas Reid's Theory of Perception*, 255.

42. For a broader consideration of Reid within Scottish Enlightenment musical theory, see Raz, "To 'Fill Up, Completely, the Whole Capacity of the Mind,'" especially 144.

43. Bence Nanay, "Aesthetic Attention," *Journal of Consciousness Studies* 22, nos. 5–6 (2015), 96–118, at 109. Nanay cites the experiment contained within Stine Vogt and Svein Magnussen, "Expertise in Pictorial Perception: Eye-Movement Patterns and Visual Memory in Artists and Laymen," *Perception* 36 (2007), 91–100.

CHAPTER 3

Attention and Value: Music Aesthetics as a "Science of Ought"

Benjamin Steege

Attention Implicates Valuation

Attention goes to what we value—which is to say, what we already value. Or so one might think. Is this not the basis on which one selects certain things and not others from out of the buzzing multitude? Yet at the same time, attending to something may uphold its given value, affirm it—and in that sense produce new value. What first draws attention is in turn invested with more robust presence, more robust significance, in the extreme arriving at a fetish object. "It is because millions of tourists have come to see the Mona Lisa that millions of tourists rush to see the Mona Lisa," writes Yves Citton, maintaining that "value-creation strongly depends on the way in which we distribute our attention."[1] In a similar vein, music scholars have often noted how acquired ways of listening, and their correlate acts of aesthetic description, deepen the grooves of perception by fixing and amplifying certain features over others.[2] It seems, then, that attention and valuation coincide, neither preceding the other, neither reducing to the other.

This chapter begins to consider the way the history of attention implicates the history of value, focusing on one extended moment in these interlocking developments. At the beginning of the period in question, in the 1880s, inevitably stands Friedrich Nietzsche, with his iconoclastic declaration of an imminent "*re*valuation" or "*trans*valuation of all values" (*Umwertung aller Werte*).[3] With this quotable but ambiguous slogan, Nietzsche both pointed to the startling fragility of what had long passed for authoritative moral and aesthetic guardrails, and also launched an effort to understand what might emerge as a basis for the establishment of new values. In the generations succeeding Nietzsche, though by no means universally under his direct influence, a minor industry of value theory took off on both sides of the Atlantic, among other places, marked by sharp discontinuity from ancient or classical conceptions.[4] By the end of this period, it had become mundane to find a broad range of intellectuals preoccupied with what the psychologist Wolfgang Köhler called *The Place of Value in a World of Facts* (1938).[5]

Intellectual history is the basic mode of this essay. But of course, intellectual history usually has a more presentist impetus. In this case, it is the ongoing liveness of anxiety about value in music scholarship of our own past quarter century. In relation to Western classical music in particular, to which North American liberal arts departments tethered themselves at the moment of their founding, proponents have long worried over an apparent decline in symbolic and cultural capital (not to mention financial capital), articulating their concerns in monographs such as Julian Johnson's *Who Needs Classical Music? Cultural Choice and Musical Value* (2002) and Lawrence Kramer's *Why Classical Music Still Matters* (2007).[6] In the case of popular music, by contrast, one might suppose that twentieth-century uncertainties about intellectual and cultural value would have been essentially laid to rest by now-classic work such as Simon Frith's *Performing Rites: On the Value of Popular Music* (1996) and the roughly contemporaneous "arrival" of popular music studies within the academy.[7] But another decade, and then another, passed with such worries popping up in monographs again and again, often in the form of either tacit or explicit concern that the comparatively robust market viability of popular dance music and song has continued to distract us, ironically, from perceiving their true worth.[8] An even more recent, and indeed ongoing, index of the insistence on the "mattering" of popular music is the (highly engaging) "Music Matters" series from University of Texas Press, which launched in 2018 with *Why the Beach Boys Matter* and *Why the Ramones Matter*, and has since seen parallel titles on (in order of appearance) Karen Carpenter, Lhasa de Sela, Labelle, Solange, Bushwick Bill, Marianne Faithfull, Patti Smith, Tammy Wynette, Mariah Carey, Willie Mae Thornton, and Sinéad O'Connor.[9]

We get it. Stuff matters! Indeed, lots of *different* stuff matters.[10] I will not venture a theory as to why value discourse has crescendoed so volubly over the past quarter century or so. (What comes to my mind is as likely to come to yours: scarcity, precarity, the crisis of democracy, the slow ebb of religion, a fragmented and saturated cultural marketplace, eroding confidence in the societal contributions of the proverbial "neoliberal academy," new moods toward "matter" and "mattering" more broadly, a *schadenfreudig* farewell to the perilous apathy of an aging Generation X for whom purportedly *nothing* once mattered, and so on.) This chapter has a much narrower goal, which is to highlight certain aspects of early value discourse that intriguingly do *not* figure so prominently in our most recent fretting about music's worth. In particular, it identifies a historical position that might be described as asserting the *obligatory* character of attention to music, manifested here in two overlapping but contrasting varieties on either side of the Atlantic. Rather than quickly dismissing this sense of "ought" as a suspicious ideological trap, the following discussion pauses to wonder if it might in fact harbor certain possibilities worth becoming sensitive to.

The two cases on the table are, first, the little-known political economist and music scholar Arthur Wolfgang Cohn (1894–1920)—little-known at least in part because he died at the age of twenty-six, nevertheless leaving behind a substantial and thoughtful enough body of work to merit a reading today;[11] and second, the far better-known philosopher and cultural critic Alain Locke (1885–1954), sometimes referred to as "Dean" of the Harlem Renaissance, largely for his edited anthology, *The New Negro* (1925), a collection of essays, poems, fiction, artworks, and cultural criticism that remains one of the most iconic cultural documents of its milieu.[12] Although their specific aims were far from aligned, Cohn and Locke maintained broadly similar thematic concerns.[13]

What provides the central motive for this essay is that they shared at a distance a distinctive twofold theoretical maneuver. First, both entered into early twentieth-century value discourse with the aim of overcoming a false choice, which they saw as a pernicious legacy of nineteenth-century philosophy and morals: a choice between, on the one hand, ascribing absolute and inherent value to things prior to any direct engagement with them, and, on the other hand, simply asserting that value is imported from the outside, arbitrarily projected onto otherwise neutral things according to the unaccountable desires or positions of individuals and institutions—in short, a choice between what they variously called objectivism and subjectivism, formalism and sensualism, realism and nominalism, among other binary formulations. And second, at the same time that they sought to set aside the malign ethics of evaluative absolutes, both Cohn and Locke nonetheless insisted on not just the possibility, but indeed the *obligation* to realize, to recognize, or to actualize the value of things, and specifically musical things—to participate in the coming to actuality of what, in the engagement with them, they are worth. This question of "should" or "ought"—Cohn refers to music aesthetics as needing to become a *Sollenswissenschaft*, a "science of ought"—is of historical interest precisely because it is so foreign to current ways of thinking and talking.[14]

What could it mean that these two figures, otherwise so different in orientation despite certain shared intellectual reference points, would both fixate on the *imperative* of value, and indeed more specifically, the imperative to cultivate a *feeling* of the value of musical things? Beyond theoretical concerns, moreover, what is at issue is a pragmatic orientation toward the material import of evaluation. Cohn and Locke are linked not only by a common investment in thematizing the theory of value, but also through parallel projects of world-building, which is to say, projects of care: the curation of a world to live in. The thematization of valuing in these cases, and the attentional practices that go along with valuing, are thus a matter of both *letting things show up*, and also of acknowledging a certain obligation or responsibility to that very event of appearance. And it is ultimately the character of that obligation to things that is of interest here, as an underappreciated problem in the study of attention.

There Is No Valueless World

One of the seminal articulations of the intuition of a value-laden world appeared between 1900 and 1920 with the emergence of the philosophical style that would become known as phenomenology. I hasten to insist that I am not interested in promoting an orthodox phenomenological approach in this context. Yet there are at least two good reasons to begin with this material. The first is that this was Cohn's own point of departure. Cohn was among the post–World War I generation of younger German scholars and students for whom the conceptual resources of phenomenological teaching were taken as a sign of the imminent renewal of scholarly and ethical ideals.[15] The second reason is that phenomenology put on especially vivid display the complexity of the relationship between the task of characterizing our attentiveness to things and the task of characterizing the way their value appears for us. Phenomenology was understood as offering a particular way of regarding the world, and while it did not in itself prescribe a specific attentiveness—after all, "attention" must surely name but *one* way of relating to things, whereas phenomenology was meant to enfold a virtually inexhaustible range of relational modes, including (as is far too often forgotten) nonperceptual ones—then it was certainly invested in thinking afresh through what attentiveness might be like in the first place. What is broadly of interest in the present context is phenomenology's tendency toward description that seeks to identify the manner in which things are given to us as we inhabit the world. In fact, it is the very possibility of value's being given in attention, of the possible *givenness* of value itself, that is the initial challenge entertained in the following reflections.[16]

The theme emerges nearly right away in a canonical passage early in Edmund Husserl's introduction of a broad program for the redescription of consciousness and its relation to the constitution of the world: his *Ideas for a Pure Phenomenology and Phenomenological Philosophy* of 1913. Husserl (largely, though by no means solely, responsible for shaping the early trajectory of the phenomenological movement as such) is not normally associated with Nietzsche's transvaluative project or its rhetoric.[17] Yet in an obliquely related way, he *also* called for a radical "transvaluation" of the world. In fact, the characteristic launching point of Husserlian thought—the so-called "suspension" or "parenthesizing" (*epochē*) of the world that is supposed to lead eventually to a clearer appreciation for just *how* the world is constituted—is explicitly an act of "transvaluing" for Husserl, an *Umwertung* that reads as a more or less direct, and rather surprising, echo of Nietzsche. Beginning the work of phenomenology means adopting a "peculiar manner of consciousness" that approaches the everyday "natural" attitude toward the world and "revalues" or "transvalues" it. "This revaluing (*Umwertung*) is a matter of our utter freedom and stands in contrast to every manner of taking up a position in thinking."[18] The prefix *um-* in Nietzsche's and Husserl's usage means neither a simple inversion of value from positive to negative (or vice versa), nor a regression to a previous state, nor any kind of simple change in the perceived "worth" of things—hence translators' irresolvable vacillation between the false equivalent "re-" and the etymologically remote "trans-"—but instead a radical putting in question of the status of worldly value and the act of valuing.

If Husserl presented his program as "transvaluative," this raises the question of how our reflective consideration of the world relates to the constitution of value within it. From a phenomenological perspective, the world as it presents itself to us just *is* already saturated by value. Value is, in a word, given. Yet an acknowledgment of this givenness is not the end, but instead the beginning of the task of thinking. "The world," as a foundational passage from Husserl goes, "is continuously 'on hand' for me, and I am myself a member of it." Yet being "on hand" (*vorhanden*) is not just a matter of brute availability, a neutral thingliness which only acquires meaning and value in some unnoticed, secondary stage of awareness. From the outset, what distinguished phenomenology from most available styles of thought was the conviction that worldly presence always already entailed value qualities, without which in effect there would be no "world" properly so called.

> This world is not thereby there for me as a mere *world of things* (*Sachenwelt*); instead, with the same immediacy it is there *as a world of values, a world of goods, a practical world* (*Wertewelt, Güterwelt, praktische Welt*). Without further ado, I find the things before me outfitted with the make-ups of things but just as much with valuable characteristics; I find them to be beautiful and ugly, pleasing and displeasing, agreeable and disagreeable, and the like. Things stand there immediately as objects of use, the 'table' with its 'books,' the 'drinking glass,' the 'vase,' the 'piano,' and so forth. These valuable characteristics and practical characteristics also belong *constitutively to the objects 'on hand' as such*, whether or not I turn to those characteristics and to the objects in general.[19]

The conviction that there simply *are* no things that are not also things of some value lies at the heart of the phenomenological project. And, of course, "value" here means more than simply greater or lesser worth (practical, or monetary, or whatever), but also all conceivable shadings of those qualities that might make something worthy in the broadest sense, or indeed *un*worthy, understood as its own kind of value. Yet what could it mean for "valuable characteristics"—"value characters" or *Wertcharakteren*—to belong "constitutively" to something? The language of "constitution" is misleading when taken out of its original expository context. It was meant to suggest neither that we mentally "construct" something in a certain way, nor that something is "built" out of certain physical or psychological elements, but rather just that something shows up in such and such a way in our manner of relating to it.[20] We are to infer that one may expect things to show up as valuable things, as a matter of how they are in the context of the world we share with them.

But, of course, Husserl does not think that it is not possible to make a meaningful distinction between something and its value. In fact, there is a subtle bifurcation here. And it is at this point that the action of attention comes into play more directly, in the context of a commentary on the particular act of noticing things:

> In the widest sense of the term, apprehending (*Erfassen*) coincides with attending-to-something (*Auf-etwas-achten*), noticing it (*Bemerken*), whether as a matter of being especially attentive to it (*aufmerksam sein*) or merely paying notice to it in passing (*nebenbei beachten*). . . . That is to say, turning toward something (even if it be in fiction) is eo ipso 'apprehension,' 'noticing.' In the act of evaluating, however, we are turned toward the value, in the act of joy we are turned toward what is enjoyable, in the act of love we are turned toward the beloved, in acting toward the action, without apprehending any of that. . . . We are turned to some subject matter in the course of valuing something, that coentails, to be sure, the apprehension of it; yet it is not the *mere* subject matter but instead the subject matter of *value* or [simply] the *value* . . . that is *the full intentional correlate of the act of valuing.*[21]

In paying attention to something, you may attend to something you hold valuable even as its value as such escapes your attention. Indeed, Husserl is at pains to say that we do *not* "attend" to value itself, only to the valued thing. At the same time, there is a distinctive, *non*-attentional "turning toward" value that nonetheless resembles attention. Valuing and attending are described here as parallel species of orientation, which share a kind of loose morphology without collapsing into one another. Or to refine things further, it would seem that valuing depends on a broader category or moment of attending, which gives the valuation its general feeling of directedness without at the same time bestowing on the valuing act the characteristic sense of clarity and apparent presence that attention seems to have.

In fact, for Husserl, the multivalent nature of attention to valuable things is a matter of a fundamental doubleness:

> In acts . . . of valuing, we thus have an *intentional object in a double sense*. We have to distinguish between the *mere 'subject matter'* (*die bloße 'Sache'*) and the *full intentional object* and, correspondingly, a *double intentio*, a twofold way of being turned toward something. If, in an act of valuing, we are oriented (*gerichtet*) to some subject

> matter, then the orientation to the subject matter is a case of paying attention to it, of apprehending it; yet we are also 'directed' at—only not in the manner of apprehending—the value.[22]

We are always in a mode of valuing, simply because things are laden with value, or because what it means for something to be a worldly thing at all, something belonging to our world, is already to have some value. Hence what it means for them to be present for us, to be given, is to be valued, even if we are not paying attention to them. When we do specially turn our attention to something, that attention does not necessarily and immediately import a *new* value to the thing, yet it nonetheless seems to cast a new shadow on, or evoke a different resonance in, the value that was already latently there. So while attending for Husserl, as in other accounts of attention, may bring about a moment of change in its object, rather than merely fixing or stabilizing it, it is nonetheless essential that attention submits to a kind of thingly or worldly priority that pushes back on it and exerts a subtle pressure against it.[23]

Arthur Wolfgang Cohn: *Musikwissenschaft* Is a *Sollenswissenschaft*

The way music, specifically, might exert pressure on us, and hence demand something of us, was the preoccupation of the twenty-five-year-old Cohn as he began to sketch out the program for a new orientation in music scholarship in the months before his death. The son of a classical philologist in Prussian Breslau (now Wrocław, Poland), Cohn was trained in law, musicology, and political economy, and had just attained the position of *Assistent* (roughly, assistant professor) in economics at the University of Frankfurt.[24] His work on music, then, was developed not so much from the perspective of a parochial disciplinary insider, but from that of a more general theorist of value seeking orientation within the brand-new legal and political landscape of the Weimar Republic. His vision, as abstract, incomplete, and aspirational as it would remain, can best be appreciated in three essays from 1919–20 in the *Zeitschrift für Musikwissenschaft*, where Cohn, drawing liberally on phenomenological literature, attempts to characterize, first, what it means to "know" music, and second, how (from his perhaps overly optimistic perspective) the spirit of German musicology was beginning to manifest an ethical "awakening" that would turn it away from the limiting choice between psychologism and positivism that had blocked its progress prior to the war.[25]

The shift in tone is unmistakable right away in the opening of the first of the essays, which features some basic yet idiosyncratic definitional work. "'Knowledge' means: the being-concerned of a knower (the subject) with a circumstance (an object). Or the other way round: the coming-into-effect of the object through the subject."[26] This is a surprising formulation indeed. "Knowledge" is simply "being-concerned" (*Befaßtsein*)? Now, "to come into effect" (*Zur-Geltung-kommen*)—or, rendering *Geltung* more literally, "to come into validity"—self-evidently belongs to the realm of value discourse. And the formulation is also couched here in terms that could only come from someone immersed in the language of contemporaneous phenomenology. The characterization of "knowing" as a two-poled dynamic in which, from the one direction, the thing "grips" (*befaßt*) the knower, while from the other direction, it is only through the knower that the thing acquires its gripping power (any effect, or *Geltung*) in the first place: This characterization roughly reformulates in more casual terms the basic phenomenological dynamic of "constitution" from Husserl. In fact, the entirety of Cohn's assessment of the state of German musicology circa 1920 rests on his Husserlian conviction that "what 'appears' (the phenomenon) is grasped as 'self-given.'"[27]

But if Husserl is, today, the most recognizable presence in Cohn's musical writings (the latter quoting extensively from passages discussed above, among others), another member of Husserl's early circle, the ethicist Max Scheler, plays an equally prominent role, accounting for Cohn's thematic drift not so much toward the analysis of consciousness per se and its relations to objects, as toward emotional life and the experience of value. Although the concern with "intentionality" is inextricably associated with the canonical Husserlian strand of phenomenology, in the same inaugural 1913 volume of the *Yearbook for Phenomenology and Phenomenological Research* in which Husserl's watershed work first appeared, Cohn had discovered equally pathbreaking texts by Scheler as well as by the aesthetician Moritz Geiger, neither of whom had much use for the jargon of intentionality, and instead typically describe the fundamental manner of engaging worldly phenomena as a disposition of *feeling*.[28] In Scheler's work, which became the premiere oeuvre of value-theorizing in 1920s Germany, value emerges for us *only* through (quasi-intentional) acts of feeling, as if values exist in the world as latencies that radiate when touched, or as calls that emanate to us from the world.

Value's "call" is exceptionally explicit in Scheler. No less explicit is the force with which it enjoins a response from us. A value, Scheler writes, is

> for '*me*' in the sense that there is an experienced *reference* to me which is contained [in] this good-in-itself, something that comes from this content, the experience of something *pointing a finger* at me, something that whispers, 'For you.' And this content places me in a *unique* position in the moral cosmos and obliges me with respect to actions, deeds, and works, which, when I imagine them, all call, 'I am for you' and 'You are for me.'[29]

Scheler's emphasis on an ethos of *Weltoffenheit*, or "openness to the world," meant a readiness to receive such calls and an acknowledgment that the individual person is in effect a medium for worldly value: A being who, in loving the world as it is given, is responsible for bringing the value of that world into the sharpest possible clarity, precisely *through* love.[30] Scheler, along with other phenomenologists in Husserl's early circle, would refer to a kind of relation to things he called *Wertnehmen*, modeled on but distinct from the commonplace *Wahrnehmen*, "perception," which had been the long-standing obsession of professional philosophers. If *Wahrnehmen* is typically rendered as "perceiving"—or, to capture the German more directly, "taking for *true*"—the untranslatable *Wertnehmen* (sometimes rendered unpalatably as "value-ceiving") may be glossed as "taking for *valuable*" or, more vividly, "apprehending *as value*."[31]

No less than Scheler's far grander (and far more realized) philosophical project, Cohn's incomplete music-critical project was to have hinged on such a modulation from *Wahrnehmen* to *Wertnehmen*. Yet, we must reiterate, for Cohn, to prioritize "valuing" was not to retreat to a position of subjectivism or perspectival aesthetic criticism. As far as he was concerned, the phenomenological priorness of value, its own peculiar "always already," remained indispensable. This mode of thought pointed toward a way of valuing musical things without saying either that certain musical phenomena are inherently, absolutely, or "objectively" valuable on their own terms, *or* that value is only a matter of what is projected onto them by skilled hermeneutic writers or by especially sensitive empathetic listeners. In fact, as in Scheler, perception—and here, attentive listening itself—is defetishized and even decentralized in Cohn's account, since what he envisions as a paradigmatic scholarly approach to music is a modality in which linguistic communication is used

			Erklärung	Bestimmung	
Apriorismus, Deduktion			Philosophie (Idealästhetik)		Allgemeine Musiklehre
	Intuition		Phänomenologie (Realästhetik)	Kritik	
Empirismus	Induktion	Sachkunde	Akustik	Technik	Besondere Musiklehre
		Menschenkunde	Anthropologie (Physio-, Psychologie)	Pädagogik (Gestaltungs-, Vortrags-, Deutungslehre)	
		Gemeinschaftskunde	Soziologie	Politik	
		Geschichtskunde	Geschichtsschreibung	Entwickelungslehre	

Figure 3.1. Table schematizing modes of contemporaneous music scholarship; Arthur Wolfgang Cohn, "Die Erkenntnis der Tonkunst" (1919).

to reach out toward the potentially valuable in music, without insisting on the predominance of "facts" and "concepts." Music is not to be approached in the guise of preexisting facts in the manner of empiricism, nor is it articulable only in terms of abstract concepts in the manner of idealism. Drawing on a formulation in Moritz Geiger's contribution to the phenomenologists' inaugural 1913 *Yearbook*, Cohn insists that proper phenomenological intuition "proceeds neither deductively nor inductively, pursues aesthetics neither from above nor from below."[32] (See Figure 3.1, a table in which Cohn positions "Phänomenologie" in line with the general category of "Intuition," as distinct from those approaches typical of aprioristic "Deduktion" above and those typical of empiricist "Induktion" below.)[33] That is, it does not argue for musical value from general principles any more than it builds up an appreciation for musical value from a series of discrete facts about musical things. Instead, musical knowing "finds" value in music, where Cohn's verb *vorfinden* (faintly echoing the Husserlian adjective *vorhanden*) suggests an action neither of mere "discovering" (*entdecken*) nor, even less, "inventing" (*erfinden*), but rather of coming to settle upon something that is in some sense prior to the act and yet also only first constituted within it.[34] It is "before" (*vor*) us both in the sense of preceding us and also of con-*front*ing us. As is characteristic of phenomenology as a whole, this line of thought is challenging: Value at once seems to precede the valuing and yet also to arise within the moment of the valuing act, and the trick lies in holding both of these elements in mind at once. The contribution of phenomenology on this score is to have shown that values are not simply "there," but that they *come* *forth*, they *show up*.[35] Cohn's adaptation of this insight only began to make this clear.

Musical knowledge, then, which we have already seen Cohn define as a matter of being "gripped" and "concerned," is, to be more precise, "just as much the esteeming of value (*Wertschätzung*) and summoning of value (*Wertschöpfung*) as it is research, wisdom, communication. Knowledge (*Erkenntnis*) in this sense, as a cultural realization of the idea of truth, is called *science* (*Wissenschaft*)."[36] The move from *Erkenntnis* to *Wissenschaft* is a move from an individual moment of cognition and feeling to a broader, communal enterprise of shared understanding. Knowing is to be thought of not just as a clever personal ability, but as a "praxis," a historically developed skill in the actualization of value. The "summoning" (*Schöpfung*) of value in this perspective is not a matter of fabricating something novel out of whole cloth, but instead of fulfilling of something latent.[37] Aesthetics, which Cohn takes to mean extant and ongoing work in the contemporaneous German music criticism and scholarship he is surveying, just *is* the fulfillment of the value of the music under study.[38] Writing, talking, teaching, reading—all manner of interpersonal linguistic communication: These forms, so often denigrated (or forgotten) as subsidiary or subservient to "actual" music-making, are instead to be celebrated for their crucial role in what Cohn calls *Musikpflege*, whose root is related to *Pflicht* ("duty" or "obligation") and hence has a more tendentious connotation than the usual English equivalents, "musical training" or "cultivation," might suggest. Indeed, Cohn is very explicit that the concept of musical "praxis" should be expanded to include not simply the production and performance of music, but also the many forms of activity with which performance is interwoven, so long as these are understood to traffic in the "valuable." *Musikpflege*, he writes, "implies nothing other than a musically valuable doing (*ein musikalisch wertvolles Tun*), whether this means a bringing-forth or receiving, a creating, reproducing, or recording."[39] From this perspective, what some might dismiss as "mere talk," for example, is not simply *about* value: it *is* the cultivation of value, its bringing-about. When Cohn conceives of an expanded cast stepping into the role of *Pfleger*—that is, those who are responsible for "caring for," "tending to," or indeed "*at*tending to" music—he also expands the community of those held responsible for the preservation and continuance of value-laden worldhood. And this, Cohn tells us, bears the weight of an imperative mood not only through implicit obligation (the *Pflicht* that belongs to the *Pfleger*), but also through a more explicit necessity that is inherent to his theory of value. Cohn borrows a line from Scheler's value ethics to the effect that an "ideal ought" leads to a worldly "demand," which "is *stressed* by the inner command of a knowing-oneself-to-be-obliged."[40]

Yet dangers lurk here. How are we to adjudicate the suggestion that we are in any way "obliged" to musical things? How do we know about this obligation? Or hear its call? Or learn what it requires of us? And here we run into a familiar trap in the flirtation with phenomenology: If you give givenness an inch, it will take a mile. So when Cohn winds up saying that what music aesthetics really needs to do is to declare itself a *Sollenswissenschaft*, "A Science of Ought," in which the actualization of value is the participation in an already prescribed value system, it is difficult to know how to judge the proposition other than as an arch-conservatism that seems to hold little contemporary appeal.[41] At best, we must attend to the value of music as its stewards; at worst, we are simply compelled to "observe" value in the dual sense Jonathan Crary wrought from this word: not *simply* "paying attention to" it but also "adhering to" its rule of conduct or disciplinary regimen.[42] And while this may have been all very well for Cohn's model in Scheler, for whom values are arrayed in a ranked series—with *sensory values* such as pleasure and pain ranked at bottom, followed by *vital values* such as health and well-being, followed by *spiritual* or *intellectual values* such as beauty and justice, followed at the highest level by *sacred values* or holiness itself—Cohn could by no means plausibly transfer such a hierarchy onto music aesthetics.[43] Indeed, in the three value-theoretical essays of 1919 to 1921, he appears to have been unable or unwilling even to venture any specific value predicates for the musical repertories he had in mind, so general and "theoretical" was his basic disposition toward the material.

And what is the status of the "transvaluation" that would seem to have motivated the ferment in value theory in the first place? Scheler was called the "Catholic Nietzsche" for his dogged pursuit of the spiritual and emotional reawakening of values that would overthrow the joint dominions of Protestantism and Kantian rationalism in European thought.[44] For Cohn, meanwhile, as for his phenomenologically minded contemporaries in music discourse, the demons to be cast out included naturalism, psychologism, and positivism—in short, the premature incursion of then current natural-scientific premises and attitudes toward a field of knowledge that seemed not yet to have fully exhausted its own local forms of knowing and valuing. Yet in contrast to the Nietzschean precedent, the gist of Cohn's valuation seems to reduce too easily to an acceptance of things as they are, rather than a clear affirmation of new value principles.

Alain Locke: Transvaluation Is Imperative

How might we stake a claim for the obligation to attend, in terms that do justice to the reality of historical change and at least the real possibility of transvaluation? Alain Locke's answer to this question was inflected by his thematic focus on an ongoing revaluation of a kind that was anything but theoretical—namely, what he referred to in 1925 as the "revaluation by white and black alike of the Negro."[45] The opening rhetorical gestures of "The New Negro," his own introductory title essay from the landmark art and literature anthology he edited, narrate a sequence of historical transformations, all of which are to be understood as instantiating this "revaluation": a "metamorphosis" of "stock figure" or "formula" into "human being," "myth" into "man," "shadow" into "personality"—in general, "historical fiction" into "realistic facing of facts."[46] This metamorphosis amounts to something more than the dismantling of stereotype (though it includes that). What Locke has in mind is the entrance into the space of public appearance of a cultural figure or persona that, it is argued implicitly and explicitly throughout, had not previously existed there.[47] This is, needless to say, in large part a function of the massive demographic fact of the Great Migration, which Locke memorably described as the "wash and rush of this human tide on the beach line of the northern city centers" and "a deliberate flight not only from countryside to city, but from medieval America to modern," a history that underlies the central editorial thesis of the volume as a whole: that "in the very process of being transplanted, the Negro is becoming transformed."[48] Locke's introduction to his volume presents matters as a great American mass *Bildungsroman*, an economic and political event from which he derived an optimistic celebration of cultural advancement that would prove not uncontroversial from the moment of its publication.[49]

But alongside the sunnily liberal narrative of an achievement of "cultural recognition," which would "in turn prove the key to that revaluation of the Negro which must precede or accompany any considerable further betterment of race relationships," there is another story here.[50] Although Locke was not by any means a card-carrying phenomenologist, the matter of the "revaluation" of a particular historical persona in his thinking is indissolubly connected with his keen attunement to historical transformations in the fluctuating appearance of art forms—the manner of their public showing-up and *demanding attention*—and indeed of one especially significant genre in particular: the African American spiritual, or "Sorrow Song," which had already loomed large in Black cultural theory at least since its prominent textual and rhetorical positioning within W. E. B. Du Bois's *The Souls of Black Folk* (1903).[51]

"Recall how suddenly," Locke observes, "the Negro spirituals *revealed themselves*; suppressed for generations under the stereotypes of Wesleyan hymn harmony, secretive, half-ashamed, until the courage of being natural *brought them out*—and *behold*, there was folk-music."[52] Locke, himself avowedly non-Christian (he was Bahá'í), deploys the mystical language of "revelation" advisedly. What is "Behold!," in rhetoric as much as in actual effect, if not a quasi-biblical enactment of the secular command to "pay attention"? Compare this with the social-psychological register in which the sociologist Charles S. Johnson, in his own contribution to *The New Negro*, assessed Black Americans' revaluation of the spiritual. Johnson emphasized instead the post–World War I shift from "an utterly conscious effort to forget the past" with its "memories of the painful degradation of slavery," and hence an effort to "rule out the Sorrow Songs as the product of ignorant slaves," toward the tendency now, "at a more comfortable distance," to "find a new beauty in these heritages, and new values in their own lives."[53]

In other words, the narrative emphasis in Johnson's account falls rather more on the cultural strategies individuals (albeit as a group) fully intentionally undertake to adapt their attitudes to changing historical conditions. Locke's view by no means rules out this more commonsense social perspective. But, as throughout his lifework as whole, his reflections also participate in a broader reflection on the constitution of value as such. There is a sense in which Locke sought not just to furnish an account of why the spirituals might have been—and in 1925 remained—historically valuable, or even to explain shifts of cultural value within Black communities, but also to provide a theoretical justification for the *necessity* of their transvaluation, "by white and black alike." This is a more ambitious task, and one that cannot be read directly from Locke's own contributions to *The New Negro*, but may be partially reconstructed in conjunction with his several value-theoretical publications.

With Cohn still in mind, it is hard not to be struck by the almost exact contemporaneity of a distinct but uncannily similar insistence on the place of an imperative mood in ethics, written by an American author who declared himself over and over to be not just a value pluralist, but an outright relativist.[54] Like Cohn, Locke was a precocious polymath. Educated at Oxford University, the University of Berlin, and Harvard University, where he wrote a

1917 dissertation in philosophy on "The Problem of Classification in Theory of Value," he published and spoke on value sporadically throughout his life.[55]

Yet Locke's counterexample compels us to consider the question as to how an obligation to attend and hence to actualize value might play out in an overtly anti-authoritarian mode. For him, the world is no less value-laden than it is for Husserl, Scheler, or Cohn: "Man does not, cannot, live in a valueless world," he insists.[56] And similarly, the problem of how to analyze the undeniable and insistent presence of value, without rendering it absolute, guides his approach: "To my thinking, the gravest problem of contemporary philosophy is how to ground some normative principle or criterion of objective validity for values without resort to dogmatism and absolutism on the intellectual plane, and without falling into their corollaries, on the plane of social behavior and action, of intolerance and mass coercion."[57] But in sharp contrast to Cohn and his influences, for Locke, values themselves are mutable and fungible, and he would consistently characterize his theoretical preferences as "functionalist" rather than "realist." A functionalist value theory, with its roots in American pragmatism after William James and John Dewey, regards values as arising within individuals' interaction with the world.[58] From this standpoint, values emerge as ethical norms that operate with all the force of imperatives. They are not simply "there" in the world at large, and neither are they arbitrarily projected onto otherwise neutral worldly affairs. Like the phenomenological approach, the functionalist one is neither objectivist nor subjectivist. Yet rather unlike the early stages of the phenomenological movement, this view tends to ascribe a dynamic and fluctuating role to values, and hence involves a certain historicist sensitivity to context.

Locke's 1935 essay, "Values and Imperatives," is of special interest, presenting the fullest published statement of his thinking in this area. Although the essay remains true to his dissertation's explicit strategy of "classification"—that is, elaborating an ornate system of axiological categories—its main thrust in general terms is clear and can be presented in three basic claims. First, the error of previous theories of value lies in their conflating after-the-fact *rationalizations* of value with the reality of valuing itself. Schematic formulations of judgment are what tend to produce absolutes, which are effectively detached from the circumstances in which they arose and thus threaten a certain violence to the texture of reality. For Locke (as for Cohn), the act of valuing something is less like formulating a judgment and more like feeling. The folly of philosophy, he thinks, had been to make "truth too exclusively a matter of the correct anticipation of experience, of the confirmation of fact" (and here he includes even James and Dewey). By contrast, since "truth may also sometimes be the sustaining of an attitude, the satisfaction of a way of feeling, the corroboration of a value," then value theory should involve itself with the exploration and understanding of those attitudes (and we should take careful note of Locke's crucial hedges: "too exclusively," "also sometimes," etc.).[59]

Second, acknowledging that the world in which valuing takes place is itself dynamic and fluctuating, value theory should concern itself not simply with relatively stable forms of value, but even more with the ways in which one can observe values *changing*. Yet, third, acknowledging such flux does not mean that we should thereby resign ourselves to "value-anarchism" or "anarchic relativism," but rather that we should in fact become all the more attuned to the possibility that we might feel ourselves obligated or compelled to respond to circumstances in certain ways, precisely *because* of the fluctuations in those circumstances.[60]

A distinctive and in this context particularly significant feature of Locke's value theory is the way it incorporates what we might call a translational dimension to value change. Confronted with something new and unfamiliar, whose value for us is at first indeterminate, confusing, or alienating, we tend to transpose onto it feelings associated with an existing value, deriving from an earlier or more traditional context. The convergence of the old value-feeling and the new state of affairs then results in an altogether new evaluative complex, whose affiliation with its precursor tends to be effaced, such that we do not typically notice the transference.

> The further we investigate, the more we discover that there is no fixity of content to values, and the more we are bound, then, to infer that their identity as groups must rest on other elements. We know that a *value-genre* often evades its definition and breaks through its logical barriers to include content not usually associated with it. The awe-inspiring scene becomes "*holy*," the logical proof, "*beautiful*," creative expression, a "*duty*," and in every case the appropriate new predicates follow the attitude and the attitude cancels out the traditionally appropriate predicates. . . . We are forced to conclude that the feeling-quality, irrespective of content, makes a value of a given kind, and that a transformation of the attitude effects a change of type in the value situation.[61]

MODAL QUALITY Form-Quality and Feeling-Reference	VALUE TYPE or Field	VALUE PREDICATES	VALUE POLARITY *Positive*	 *Negative*
EXALTATION: (Awe-Worship) a. Introverted: (Individualized): Inner Ecstasy b. Extroverted: (Socialized): Religious Zeal	*Religious*	Holy—Unholy Good—Evil	Holiness Salvation	Sin Damnation
TENSION: (Conflict-Choice) a. Inner Tension of "Conscience" b. Extrovert: Outer Tension of "Duty"	*Ethical* *Moral*	Good—Bad Right—Wrong	Conscience Right	Temptation Crime
ACCEPTANCE or AGREEMENT: (Curiosity—Intellectual Satisfaction) a. Inner Agreement in Thought	*Logical Truth*	True (Correct) and Incorrect	Consistency	Contradiction
b. Outer Agreement in Experience	*Scientific Truth*	True—False	Certainty	Error
REPOSE or EQUILIBRIUM a. Consummation in Contemplation b. Consummation in Creative Activity	*Aesthetic* *Artistic*	Beautiful—Ugly Fine—Unsatisfactory	Satisfaction Joy	Disgust Distress

a: Value: introverted type.
b: Value: extroverted type.

Figure 3.2. Chart from Alain Locke, "Values and Imperatives" (1935).

Locke went some way toward schematizing a typology of value-feelings in terms of their "modal quality" (see Figure 3.2). This would consist of, on the one hand, what he called "form-quality" (that is, a generalized affective structure belonging to one of the four categories, "exaltation," "tension," "acceptance or agreement," and "repose or equilibrium") and, on the other hand, "feeling-reference" (that is, the sense that feeling is oriented *toward* some domain, as captured here in respective variations on the Jungian binary of "introverted" and "extroverted"). These are further correlated with disparate but potentially overlapping fields of application, from religious to aesthetic.

In the flow of the text, Locke typically refers to these complex categories by the shorthand "form-feeling." And it is worth noting simply that, in this theoretical conception, "feeling" in fact *is* a "form": an organizing mode of our evaluative engagement with the world. On this score, Locke's approach is quite different from the way Husserl, Scheler, or Cohn saw values as phenomena that might be activated in the world at the moment of an intentional act of feeling. In the phenomenological account, the word "feeling" refers to an act whose quality is determined in the moment that it is oriented toward something, whereas in Locke's functionalist account, "feeling" refers to the formal structuredness a sense of value seems to have, even if it has been initially shaped in a context discontinuous from the one in which it is presently felt. If a logical proof appears "beautiful" to you, you are not about to claim that something in the proof inherently conforms to an aesthetic standard of beauty, only that there are certain qualities about it that you are prepared to appreciate in a manner that you have already developed in your experiences of the traditionally beautiful, and that this appreciation has a certain compelling character about it, even if nothing is thereby said about the actual validity of the proof.

Values here, as form-feelings, are thus dependent on contextual norms, yet nonetheless continue to exert pressure on us in a delicate balance of the contingent and the obligatory: "In de-throning our absolutes, we must take care not to exile our imperatives, for after all, we live by them. We must realize more fully that values create these imperatives as well as the more formally super-imposed absolutes, that norms control our behavior as well as guide our reasoning."[62]

True to the scholarly genre of his era, Locke's writing on these topics is usually abstract, presented without extended examples. So, then, returning to this section's starting point in the African American spiritual, where might we locate the pressure to attend to this repertoire—to respond to its historical imperative to "behold," as Locke wrote in *The New Negro*'s introduction? The answer lies in first recognizing the role that art forms themselves play in providing a space for transvaluation. Kobena Mercer writes that a crucial premise in Locke's writings on Afrodiasporic visual culture is that art should not be held merely to "reflect" or "express" identities, reducing it to a neutral conduit for the actual substrate of reality. Rather, "to say culture comes first in 'producing' the positions in which identities take shape is to argue—as Locke did—that culture is the realm in which value-laden racial attitudes can be shaken up and opened to change."[63] Locke's position has been repeatedly contested, and this is hardly the occasion to retell that history.[64] Yet there is no way to appreciate the imperative mood of his value theory without articulating this initial stance, according to which the space of aesthetic appearance is itself the central scene of negotiation over cultural value, even if not always in an explicit manner. On the one hand, we pay attention to music in the first place "because of the way it strives to be seen and heard,"[65] because it is in the nature of sounding artworks that they demand attention within a public sphere—Scheler's "'I am for you' and 'You are for me'"—however limited, damaged, or distorted that public space may be (and however imbricated with private or counterpublic spaces). But Locke's thinking on these matters suggests that there is an essential second moment, in which the initial value of music's sheer presence also enjoins a modulation of value-feeling, a reshaping of its "form."

Thus, on the one hand, Locke notes that in the period between roughly 1875 and 1925, when spirituals increasingly circulated within an expanding cultural marketplace in the form of concert-hall performances, sheet music arrangements, and finally records, this repertory "saved a folk art and preserved as no other medium could the folk temperament, and by maintaining them introduced the Negro to himself." But, on the other hand, Locke does not end with the affirmation of a preservationist; he does not argue for a value-stabilizing outcome to this historical development. Instead, he points first to the awkward, possibly even cringeworthy stages of "valuing" in which already available feelings were yanked from the shelf, as it were, and clung to in order to make sense of an art form that had only partial precedents in the public imagination and hence confounded immediate judgment: "Still the predominant values of this period in estimating the Spirituals were the sentimental, degenerating often into patronizing curiosity on the one side, and hectic exhibitionism on the other. Both races condescended to meet the mind of the Negro slave, and even while his moods were taking their hearts by storm, discounted the artistry of genius therein. It was only as the musical appreciation of the Spirituals grew that this interest changed and deepened."[66] Sentimentality is a "form," which may be learned and then adapted to unfamiliar things, especially things that do not otherwise offer us the kinds of affective affordances that aid interpretation and evaluation. Religious kitsch, the aesthetic commodity that wants to inspire feelings of piety despite the very fact of its commodified form, is a pertinent case in terms of its particular affective environment, and arguably traces its genealogy more or less to the beginnings of the time period in question. And as unappealing as you may find such responses—Locke himself is by no means celebrating them—his suggestion is that they can also function as cultural placeholders that make room for later developments in the possible forms of valuing available to us. Or here, rather, not simply make room for but, in "taking their hearts by storm," effectively *oblige*.

It is the art-for-art's-sake rhetoric, with its too optimistic hint of redemption and too explicit hint of transcendence, that usually puts people off Locke, then as now.[67] That is fair. But without getting diverted by a separate (and already long, far-flung, and fiercely contested) debate, we can still train our focus on the force of transvaluation itself. The basic point for Locke is simply *that* we are able, and indeed are compelled, to move from something like a feckless sentimentality—even one weighed down by "hectic exhibitionism" and "patronizing curiosity"—toward something else. Whether you call that "artistry" or "genius," and insist that the aesthetic merits of the object will eventually make themselves self-evident to all, is less interesting than the basic grasp of the valuational shift.

To put it simply: art obliges our attention because it needs us to become ready to change the way we value.

What, then, becomes of the spiritual? In fact, the more urgent question in Locke's "The Negro Spirituals" is not so much, how has the genre been reevaluated since 1875? but instead, what should we expect of the prospective value of the spiritual as a genre in 1925 and after? And this naturally implies that the "thing" itself, the genre and its representative repertoire, has a certain instability that goes hand in hand with the instability of its evaluation. So it is not only that Locke wants us to appreciate how folk art, though "despised and rejected at first," nonetheless, "generations after, flowers again and transcends the level of its origin" such that the "slave songs . . . have come to be recognized as artistically precious things." And it is not only that it "still requires vision and courage to proclaim their ultimate value and possibilities," although eventually those alive to such possibilities "will treasure the Spirituals for their true musical and technical values."[68]

Unlike for Cohn, the point is not just to pay attention to what is valuable, to give value its due, to bring it to its clearest apprehension. Rather, the modernist Locke, no "salvage ethnographer," is aware that he is not describing a tradition on the verge of fading away and in need of a retrospective (and hence conservative) appreciation, but one that is ongoing and continues to be woven immediately into his own cultural sphere. Thus he distances himself from the notorious negrophile writer Carl Van Vechten, who wants new arrangements and performances of spirituals to adhere to a simplistic folk ethos, and who therefore, with good intentions but misplaced motives, "insists in the name of the folk spirit upon his preference for the 'evangelical renderings' of Paul Robeson's robust and dramatic style as over against the subdued, ecstatic and spiritually refined versions of Roland Hayes." In doing so, Van Vechten "overlooks the fact that the folk itself has these same two styles of singing, and in most cases discriminates according to the mood, occasion and song type, between them. So long as the peculiar quality of Negro song is maintained," Locke concludes, "there can be no set limitation."[69]

And it emerges implicitly from this comment and more explicitly elsewhere, that as far as Locke is concerned, there really is no essential thing that can be characterized as "*the* spiritual." Instead, he points to the way elements of the spiritual of the past and present are always held in potentiality. Neither the "folk" versions inherited from the mid-nineteenth century, nor the second-generation choral arrangements by groups like the Fisk Jubilee Singers, nor Harry T. Burleigh's harmonically experimental settings with piano for living room and concert stage, nor indeed any of the growing number of recorded renditions from the 1910s and 1920s actually *are* the thing itself. Yet the genre somehow exerts a continuous pressure toward relatively open-ended revision and renewal: It "is at the very least potentially polyphonic," "can therefore undergo, without breaking its own boundaries, intricate and original development," and is in fact approaching the "line of advance in modernistic music."[70]

To be described as "modernist" is high praise coming from Locke. As he would write in a talk on value delivered at Columbia University in December 1945, musical modernism (which for him at this point means things like Stravinsky and Hindemith as well as recent "good jazz," and so forth) has been responsible for bringing about a set of "enlarged criteria of evaluative judgment" and a "widening of the range of appreciation and participation" in public life, all of which amounts to "as good an example as we can find of what democratization can mean in a value field." And while there is, again, the appearance of an easy liberal idealization of mere "tolerance" for diverse expressive modes, I think we should take Locke at his word when he suggests that these virtues are the sort of things that are not just nice to have, in the manner of a self-satisfied aesthetic connoisseurship, but are in fact necessary for survival within a patently and pervasively conflictive midcentury environment: "I merely throw out the suggestion that through modernism and its enforced but not normatively chaotic relativism we have forged a psychological key for the active and simultaneous appreciation of diverse styles within our own culture. . . . In an approaching world interchange of culture it is just such widening of taste to a cosmopolitan range and level which seems most desirable, if indeed not imperative."[71]

The "imperative" to value the post-Reconstruction spiritual is due to a multiplicity of overlapping factors: its inherent aesthetic qualities, the possible ongoing compositional uses one might make of it, its participation in the refinement of modern religious experience, what it does for the interests of racial progress and racial pride, and the way it might increase public awareness and recognition of the cultural contributions of Afrodiasporic folkways and African American composers and performers. All of these are motives, kinds of imperatives in themselves, and of course all for Locke are laudable. But also, more narrowly in the vein of the preceding discussion, we might note that the spiritual stands as a privileged example of an art form whose historical position enables it to exact certain ethical demands. It requires you to acknowledge and modulate among the range of conflicting values that it gathers together, whether or not they are claimed as your own: true religious feelings, deeply felt (even if misplaced) sentimentalisms, quests for authenticity, appreciations of abstract beauty, and so forth—all converge within what is otherwise a seemingly modest expressive form. The spiritual is important for Locke because it establishes the aptest available scene for a transvaluation that is both culturally plural and also, on account of that pluralism, imperative.

Conclusion

What do Cohn and Locke give us to work with? In some ways, it is a long distance from the all but explicit call of obligation to attend in the one, to the less explicit yet no less forceful sense of obligation toward transvaluation in the other. And the motives driving these two imperatives are clearly distinct. On the one hand, there is the phenomenological pressure the world places on us, merely by virtue of *being* world. For a phenomenologist, we are not neutral toward the world, nor it to us. "Attentiveness" in this perspective is only a special case of the otherwise generalized state of inherent interestedness that is living. And what, on the other hand, is attentiveness for Locke? It is more than a merely "appreciative" attunement to diverse aesthetic qualities, despite what his rhetorical self-positioning may at times suggest. The sheer fact of aesthetic appearances showing up for us continues to be as remarkable for Locke as it is for Cohn. Locke's attention is not just the response to the shock of the new, but the response to the more easily overlooked shock of the newly valued. "Behold!" means not just, "Pay attention to this!" but also, "What are we to make of this?" Yet there is a more encompassing and fundamental imperative underlying and supporting these imperatives, which is that, insofar as all their various objects of attention are *made* objects, artifacts of a certain historical experience and of a shared persisting world, they enjoin our valuation because we need these cultural artifacts in order to renew the world.

Notes

1. Yves Citton, *The Ecology of Attention* (2014), trans. Barnaby Norman (Cambridge, UK: Polity Press, 2017), 1, 48.

2. Many of the contributions to *The Oxford Handbook of Music Listening in the 19th and 20th Centuries*, edited by Christian Thorau and Hansjakob Ziemer (New York: Oxford University Press, 2019), hinge on this dynamic, including, in an especially clear case, Thorau, "'What Ought to Be Heard': Touristic Listening and the Guided Ear" (207–28). A broader view of the issue, focusing on the juncture at which the distinction between market value and aesthetic value acquired a salience that has more or less endured to the present, is Nicholas Mathew, "Interesting Haydn: On Attention's Materials," *Journal of the American Musicological Society* 71, no. 3 (Fall 2018): 655–701.

3. Friedrich Nietzsche, *On the Genealogy of Morals* (1887), trans. Michael A. Scarpitti (New York: Oxford University Press, 2014), Third Essay, Section 27, 142; as well as the unfinished work foretold there: Nietzsche, *The Will to Power: Selections from the Notebooks of the 1880s*, trans. R. Kevin Hill and Michael A. Scarpitti, ed. Hill (Penguin Books, 2017), whose subtitle was to have been "An Attempted Transvaluation of All Values."

4. Relevant highlights include Christian von Ehrenfels, *System der Werttheorie*, 2 vols. (Leipzig: O. R. Reisland, 1897–98); William James, *Talks to Teachers on Psychology, and to Students on Some of Life's Ideals* (New York: Henry Holt, 1899); Georg Simmel, *The Philosophy of Money* (1900), trans. Tom Bottomore and David Frisby (London: Routledge, 2004); Max Scheler, *Formalism in Ethics and Non-Formal Ethics of Values: A New Attempt Toward the Foundation of an Ethical Personalism* (1913), trans. Manfred Frings and Roger L. Funk (Evanston: Northwestern University Press, 1973); Alexius Meinong, *On Emotional Presentation* (1917), trans. Marie-Luise Schubert Kalsi (Evanston: Northwestern University Press, 1972); Wilbur Urban, *The Intelligible World: Metaphysics and Value* (London: George Allen & Unwin, 1929); Wolfgang Köhler, *The Place of Value in a World of Facts* (New York: Liveright Publications, 1938); and John Dewey, *Theory of Valuation* (Chicago: University of Chicago Press, 1939).

5. This is to say nothing of the partially overlapping body of specialist literature on economic value theory, which is at once narrower in scope and also more venerable, though it saw periods of reinvigoration with the series of economic and political crises beginning in the 1840s, which of course inspired Karl Marx's watershed adaptation of existing labor theory of value. For an overview, one could do worse than revisit Robert L. Heilbroner's 1953 classic, *The Worldly Philosophers: The Lives, Times, and Ideas of the Great Economic Thinkers*, 7th ed. (New York: Simon and Schuster, 1999).

6. Julian Johnson, *Who Needs Classical Music? Cultural Choice and Musical Value* (New York: Oxford University Press, 2002); Lawrence Kramer, *Why Classical Music Still Matters* (Berkeley: University of California Press, 2007); and Kramer, "Introduction," in Keith Chapin and Lawrence Kramer, eds., *Musical Meaning and Human Values* (New York: Fordham University Press, 2009), 1–9.

7. Simon Frith, *Performing Rites: On the Value of Popular Music* (Cambridge, MA: Harvard University Press, 1996). The International Association for the Study of Popular Music had been founded in 1981 (jointly by Frith and Charles Hamm), and its official journal, *Tracking*, launched in 1988, renamed *Journal of Popular Music Studies* in 1993.

8. Ralf von Appen, *Der Wert der Musik: Zur Ästhetik des Populären* (Bielefeld: Transkript Verlag, 2007); Theodore Gracyk, *Listening to Popular Music: Or, How I Learned to Stop Worrying and Love Led Zeppelin* (Ann Arbor: University of Michigan Press, 2007); David Hesmondhalgh, *Why Music Matters* (Oxford: Wiley Blackwell, 2013); and Alison Stone, *The Value of Popular Music: An Approach from Post-Kantian Aesthetics* (Cham: Palgrave Macmillan, 2016); Timothy D. Taylor, *Making Value: Music, Capital, and the Social* (Durham, NC: Duke University Press, 2024). And finally, with diametrically opposed titles that frame things as unmistakably as can be, see also Christopher Washburne and Maiken Derno, eds., *Bad Music: The Music We Love to Hate* (New York: Routledge, 2004); along with John J. Sheinbaum, *Good Music: What It Is and Who Gets to Decide* (Chicago: University of Chicago Press, 2019).

9. Tom Smucker, *Why the Beach Boys Matter* (Austin: University of Texas Press, 2018); Donna Gaines, *Why the Ramones Matter* (Austin: University of Texas Press, 2018); Karen Tongson, *Why Karen Carpenter Matters* (Austin: University of Texas Press, 2019); Fred Goodman, *Why Lhasa de Sela Matters* (Austin: University of Texas Press, 2019); Adele Bertei, *Why Labelle Matters* (Austin: University of Texas Press, 2021); Stephanie Phillips, *Why Solange Matters* (Austin: University of Texas Press, 2021); Charles L. Hughes, *Why Bushwick Bill Matters* (Austin: University of Texas Press, 2021); Tanya Pearson, *Why Marianne Faithfull Matters* (Austin: University of Texas Press, 2021); Caryn Rose, *Why Patti Smith Matters* (Austin: University of Texas Press, 2022); Steacy Easton, *Why Tammy Wynette Matters* (Austin: University of Texas Press, 2023); Andrew Chan, *Why Mariah Carey Matters* (Austin: University of Texas Press, 2023); Lynnée Denise, *Why Willie Mae Thornton Matters* (Austin: University of Texas Press, 2023); and Allyson McCabe, *Why Sinéad O'Connor Matters* (Austin: University of Texas Press, 2024).

10. The tendency is hardly limited to music studies: witness Rita Felski, *Uses of Literature* (Oxford: Wiley, 2008); Martha Nussbaum, *Not for Profit: Why Democracy Needs the Humanities* (Princeton: Princeton University Press, 2010); Helen Small, *The Value of the Humanities* (New York: Oxford University Press, 2013); and most recently, Wendy Brown, *Nihilistic Times: Thinking with Max Weber* (Cambridge, MA: Harvard University Press, 2023), all of which contrast sharply with an older skepticism toward the notion of "value" as such in literary theory and cultural studies, represented by, for example, Barbara Herrnstein Smith, *Contingencies of Value: Alternative Perspectives for Critical Theory* (Cambridge, MA: Harvard University Press, 1988); and John Frow, *Cultural Studies and Cultural Value* (New York: Oxford University Press, 1995); not to mention the classic social theorization of value developed across essentially the entire oeuvre of Pierre Bourdieu. For a general historical assessment, see Hans Joas, *The Genesis of Values* (1997), trans. Gregory Moore (Chicago: University of Chicago Press, 2000).

11. This includes two short monographs: Arthur Wolfgang Cohn, *Das Tonwerk im Rechtssinne* (Berlin: Julius Springer, 1917); and Cohn, *Kann das Geld abgeschafft werden?* (Jena: Gustav Fischer, 1920).

12. Alain Locke, ed., *The New Negro: An Interpretation* (New York: Albert and Charles Boni, 1925).

13. In fact, Locke's time as a student in Berlin in 1910 and his regular return visits for many years afterward put him at only one or two degrees of separation from Cohn in terms of social and intellectual circles. On Locke's studies with Hugo Münsterberg and Georg Simmel and others at the University of Berlin (then Friedrich-Wilhelms-Universität), see Leonard Harris and Charles Molesworth, *Alain L. Locke: Biography of a Philosopher* (Chicago: University of Chicago Press, 2008), 91–99. Meanwhile, Tamara Levitz establishes Locke's place within the

larger context of transnational music-philosophical history in Levitz, "The Twentieth Century," in Tomás McAuley, Nanette Nielsen, Jerrold Levinson, and Ariana Phillips-Hutton, eds., *The Oxford Handbook of Western Music and Philosophy* (New York: Oxford University Press, 2020), 225–62.

14. An exception that proves the rule, largely by virtue of its ironic frankness concerning the audacity of its central question, is Ted Cohen, "Liking What's Good: Why Should We?," in William Irwin and Jorge J. E. Gracia, eds., *Philosophy and the Interpretation of Pop Culture* (Lanham, MD: Rowman & Littlefield, 2007), 117–30. If an appeal to aesthetic imperatives seems both unfashionable and improbable in contemporary music studies, it is somewhat less so in environmental ethics, which is perhaps less remote from present concerns than one might think. See, for example, Holmes Rolston III, "From Beauty to Duty: Aesthetics of Nature and Environmental Ethics," in Arnold Berleant, ed., *Environment and the Arts: Perspectives on Environmental Aesthetics* (Hampshire, UK: Ashgate, 2002), 127–41; and the subsequent edited volume inspired by Rolston's 2002 essay: Allen Carlson and Sheila Lintott, eds., *Nature, Aesthetics, and Environmentalism: From Beauty to Duty* (New York: Columbia University Press, 2008).

15. See Benjamin Steege, *An Unnatural Attitude: Phenomenology in Weimar Musical Thought* (Chicago: University of Chicago Press, 2021); on Cohn specifically, 61–66.

16. If the pursuit seems somewhat particular to its moment, note that similar thematics continue to appear as well in nonphenomenological literature: for example, Akeel Bilgrami, "The Visibility of Value," *Social Research* 83, no. 4 (Winter 2016): 917–43.

17. But see Rudolf Boehm, "Husserl und Nietzsche," in *Vom Gesichtspunkt der Phänomenologie: Husserl-Studien* (The Hague: Martinus Nijhoff, 1968), 217–36.

18. Edmund Husserl, *Ideas for a Pure Phenomenology and Phenomenological Philosophy* 1 (1913), "General Introduction to Pure Phenomenology," trans. Daniel O. Dahlstrom (Indianapolis: Hackett, 2014), §31, 54; original emphasis here and throughout, unless noted.

19. Husserl, *Ideas* 1, §27, 49.

20. The terrain can get uninvitingly technical around this matter, but see David Woodruff Smith, "Constitution Through Noema and Horizon: Husserl's Theory of Intentionality," in Jeff Yoshimi, Philip Walsh, and Patrick Londen, eds., *Horizons of Phenomenology: Essays on the State of the Field and Its Applications* (Cham: Springer, 2023), 63–80, at 71–72.

21. Husserl, *Ideas* 1, §37, 64–65.

22. Husserl, *Ideas* 1, §37, 65.

23. On attention as a moment of change, see Benjamin Steege, "The Problem of Attention," in *Helmholtz and the Modern Listener* (Cambridge: Cambridge University Press, 2012), 80–122.

24. Biographical information comes from Th. M., "Personalia," *Im deutschen Reich: Zeitschrift des Centralvereins deutscher Staatsbürger jüdischen Glaubens* 26, no. 10 (October 1920): 325–26; and Adolf Weber, "Zum Gedächtnis an Arthur Wolfgang Cohn," in Cohn, *Kann das Geld abgeschafft werden?* (Jena: Gustav Fischer, 1920), 141–42.

25. Cohn, "Die Erkenntnis der Tonkunst. Gedanken über Begründung und Aufbau der Musikwissenschaft," *Zeitschrift für Musikwissenschaft* 1, no. 6 (March 1919): 351–60; Cohn, "Das Erwachen der Ästhetik. Neue Bücher und Schriften," *Zeitschrift für Musikwissenschaft* 1, no. 11 (August 1919): 669–79; and the posthumous Cohn, "Das musikalische Verständnis. Neue Ziele," *Zeitschrift für Musikwissenschaft* 4, no. 3 (December 1921): 129–35.

26. Cohn, "Die Erkenntnis der Tonkunst," 352. "'Erkenntnis' heißt: Befaßtsein eines Erkennenden (des Subjekts) mit einem Gegenstand (dem Objekt). Oder umgekehrt: Zur-Geltungkommen des Objekts durch das Subjekt."

27. Cohn, "Das Erwachen der Ästhetik," 672. "Das 'Erscheinende' (Phänomen) wird als 'selbstgegeben' erfaßt."

28. Max Scheler, "Der Formalismus in der Ethik und die materiale Wertethik," *Jahrbuch für Philosophie und phänomenologische Forschung* 1 (1913): 405–565; and Moritz Geiger, "Beiträge zur Phänomenologie des ästhetischen Genusses," *Jahrbuch für Philosophie und phänomenologische Forschung* 1, no. 2 (1913): 567–684. Scheler's complete treatise, an extensive critique of Kantian ethics, was translated as *Formalism in Ethics and Non-Formal Ethics of Values* (Evanston, IL: Northwestern University Press, 1973).

29. Scheler, *Formalism in Ethics and Non-Formal Values*, 490; translation modified. Scheler writes, similarly, in a posthumous essay that "a value-signal experienced as coming from things, not from us, announces, as though with a trumpet flourish, that 'Something is up!'" Scheler, "*Ordo Amoris*," in *Selected Philosophical Essays*, trans. David R. Lachterman (Evanston, IL: Northwestern University Press, 1973), 98–135, at 101.

30. In some accounts, love is itself a kind of attention; as, for example, in Simone Weil, *Waiting for God* (1950), trans. Emma Craufurd (London: Routledge, 2021), and especially the essay, "Forms of the Implicit Love of God," 99–155. I thank my co-participants in the workshop that led to this volume for pointing out this resonance.

31. Scheler, *Formalism in Ethics and Non-Formal Values*, 197. Cf. Adolf Reinach, *Zur Theorie des negativen Urteils* (Halle: Max Niemeyer, 1911); and Dietrich von Hildebrand, "Die Idee von der sittlichen Handlung," *Jahrbuch für Philosophie und phänomenologische Forschung* 3 (1916): 126–252.

32. Cohn, "Das Erwachen der Ästhetik," 672; ". . . erfährt weder deduktiv noch induktiv, treibt nicht Ästhetik von oben oder unten"; paraphrasing Geiger, "Beiträge zur Phänomenologie des ästhetischen Genusses," 570–73.

33. The table appears in Cohn, "Die Erkenntnis der Tonkunst," 357.

34. Cohn, "Das Erwachen der Ästhetik," 672. He refines the view somewhat in the 1921 essay, "Das musikalische Verständnis" (131), drawing a contrast there between "understanding" as a reflective disposition and more "spontaneous" dispositions, such as loving and hating, in which he thinks it clearer that *value* (or lack of value) does arise within the act.

35. Present throughout the literature, that insight is most fully developed recently in the work of Jean-Luc Marion. See, for example, Marion, *Being Given: Towards a Phenomenology of Givenness*, trans. Jeffrey L. Kossky (Stanford: Stanford University Press, 2002).

36. Cohn, "Die Erkenntnis der Tonkunst," 356. ". . . ebenso Wertschätzung und Wertschöpfung, als Forschung, Wissen, Mitteilung. Die Erkenntnis in diesem Sinne, als kulturelle Verwirklichung der Wahrheitsidee, heißt *Wissenschaft*." In this context, "science" is the best we can do for *Wissenschaft*, which of course connotes a holistic sense of learning and study, rather than any particular program for (natural) scientific research.

37. Although *Die Schöpfung* (as in Franz Josef Haydn's oratorio by that name) is indeed the term for God's "creation" of the world as recounted in Genesis, the German verb *schöpfen* equally means "scooping" or "ladling"—as, for example, of water or soup—or, more generally, "drawing forth" ("pro-ducing"), and hence immanently deflates a pretense to generation of the radically new.

38. Cohn includes Karl Grunsky, August Halm, Hermann Kretzschmar, Ernst Kurth, Hans Pfitzner, Hugo Riemann, and Arnold Schering—in other words, a broad cast of the leading figures in German musicology circa 1920—as models for the incipient direction of scholarship. I am confining my discussion to an account of Cohn's theoretical position without addressing the

extent to which it is in fact born out in the models. I am skeptical, to say the least, that Cohn's estimation of actual scholarly practice was correct.

39. Cohn, "Die Erkenntnis der Tonkunst," 359. ". . . besagt nichts anders als ein musikalisch wertvolles Tun, gleichgültig, ob dies ein Hervorbringen oder Empfangen, ein Schaffen, Wiedergeben oder Aufnehmen ist."

40. Cohn, "Das Erwachen der Ästhetik," 678; quoting Scheler, *Formalism in Ethics*, 204.

41. Cohn, "Das Erwachen der Ästhetik," 678.

42. Jonathan Crary, *Techniques of the Observer: On Vision and Modernity in the Nineteenth Century* (New York: Zone Books, 1991), 5–6.

43. Scheler, *Formalism in Ethics and Non-Formal Values*, 105–110.

44. Scheler exercised an enormous influence on later Catholic philosophers, including most famously Karol Józef Wojtyła, Pope John Paul II, who wrote his 1954 dissertation in theology on Scheler. See Michael Gubser, *The Far Reaches: Phenomenology, Ethics, and Social Renewal in Central Europe* (Stanford: Stanford University Press, 2014), 80–100 (on Scheler), and 188–210 (on Wojtyła); and Edward Baring, *Converts to the Real: Catholicism and the Making of Continental Philosophy* (Cambridge, MA: Harvard University Press, 2019), 116–47.

45. Locke, "The New Negro," in *The New Negro*, 3–16, at 15.

46. Locke, "The New Negro," 3–4.

47. For a related interpretation, see Leonard Harris, "Alain Locke and Community," *The Journal of Ethics* 1, no. 3 (1997): 239–47, at 245, which in general also offers a succinct but substantive discussion of the interrelations among Locke's theoretical approaches to value, race, and aesthetics.

48. Locke, "The New Negro," 6.

49. It has been defended perhaps most influentially by Houston A. Baker, Jr., in Baker, *Modernism and the Harlem Renaissance* (Chicago: University of Chicago Press, 1987), 72–98. For discussion of its music historiography in particular, see Paul Allen Anderson, *Deep River: Music and Memory in Harlem Renaissance Thought* (Durham, NC: Duke University Press, 2001), 113–66; and Grant Olwage, "Listening B(l)ack: Paul Robeson After Roland Hayes," *The Journal of Musicology* 32, no. 4 (2015): 524–57.

50. Locke, "The New Negro," 15.

51. W. E. B. Du Bois, *The Souls of Black Folk: Essays and Sketches* (Chicago: A. C. McClurg 1903).

52. Locke, "The New Negro," 4; emphasis added.

53. Charles S. Johnson, "The New Frontage on American Life," in *The New Negro*, 278–98, at 297.

54. Locke, "Cultural Relativism and Ideological Peace" (1944), in Leonard Harris, ed., *The Philosophy of Alain Locke: Harlem Renaissance and Beyond* (Philadelphia: Temple University Press, 1989), 95–102.

55. Locke, "The Problem of Classification in Theory of Value or, An Outline of a Genetic System of Values" (PhD dissertation, Harvard University, 1917).

56. Locke, "Values and Imperatives" (1935), in *The Philosophy of Alain Locke*, 31–50, at 34.

57. Locke, "Values and Imperatives," 36.

58. For the relation to pragmatism, see Locke, "Pluralism and Intellectual Democracy" (1942), and "Pluralism and Ideological Peace" (1947), in *The Philosophy of Alain Locke*, 51–66, and 95–102; and the assessment in Neil W. Williams, "Absolutism, Relativism, and Anarchy:

Alain Locke and William James on Value Pluralism," *Transactions of the Charles S. Peirce Society* 53, no. 3 (Summer 2017): 400–424.

59. Locke, "Values and Imperatives," 37.

60. Locke, "Values and Imperatives," 34 and 36.

61. Locke, "Values and Imperatives," 40.

62. Locke, "Values and Imperatives," 34.

63. Kobena Mercer, *Alain Locke and the Visual Arts* (New Haven: Yale University Press, 2022), 5. Mercer writes further: "Locke's core premise—that value is never intrinsic to objects or activities deemed to be socially important, but the outcome of acts of valuation that are historically conditioned and thus susceptible to change—led directly to his view that modern America's racial identities were changeable, too" (2). This is fundamental, yet also omits the spotlight Locke places on value's obligatory character, even amid historical change.

64. For broader background to Locke's project, interpreting it as relatively "apolitical," see Henry Louis Gates, Jr., "The Trope of a New Negro and the Reconstruction of the Image of the Black," *Representations* 24 (Fall 1988): 129–55, at 147. For a recent defense of the continuing political promise of Locke's aesthetic theory, see Michelle K. L. Rose, "'His Is a Reverent Vandalism': Alain Locke's Aesthetics and Fugitive Democracy," *Political Theory* 51, no. 4 (2023): 703–35.

65. Cecilia Sjöholm, "Arendt, Poetry and Music: On Sounding," in Jonathan De Souza, Jessica Wiskus, and Benjamin Steege, eds., *The Oxford Handbook of the Phenomenology of Music* (New York: Oxford University Press, 2023), https://doi.org/10.1093/oxfordhb/9780197577844.001.0001.

66. Locke, "The Negro Spirituals," in *The New Negro*, 199–213, at 202–203.

67. The canonical counterpoint is Du Bois's insistence that art perform the work of propaganda on some level. See W. E. B. Du Bois, "Criteria of Negro Art" (1926), in Henry Louis Gates, Jr., and Gene Andrew Jarrett, eds., *The New Negro: Readings on Race, Representation, and African American Culture, 1892–1938* (Princeton: Princeton University Press, 2007), 257–60; followed in the same volume by Locke's response, "Art or Propaganda?" (1928), 260–61, which includes this representative formulation: "Negro things may reasonably be a fad for others; for us they must be a religion. Beauty, however, is its best priest and psalms will be more effective than sermons" (261).

68. Locke, "The Negro Spirituals," 199–200.

69. Locke, "The Negro Spirituals," 208. The point is reiterated in Locke, *The Negro and His Music* (Washington, DC: The Associates in Negro Folk Education, 1936), 124. Locke quotes from Carl Van Vechten, "The Folksongs of the American Negro," *Vanity Fair* (July 1, 1925): 52, 92, at 92. Van Vechten's complaint essentially hinges on the thinly veiled suggestion that Hayes's singing is "too white."

70. Locke, "The Negro Spirituals," 209.

71. Locke, "A Functional View of Value Ultimates" (1945), in *The Philosophy of Alain Locke*, 79–93, at 91–92.

PART II

Collectivity

CHAPTER 4

Omniaudience and the History of the Good Listener: William Crotch's *Specimens of Various Styles of Music* (1807)

Nicholas Mathew

The liberal ethic of listening is the other side of the liberal ethic of voice. Here's what I mean by this: If *having a voice* means having agency, autonomy, and personhood, then *listening* is what the best subjects or institutions must do as soon as a voice pipes up.[1] Listening helps to secure a functioning politics of recognition—the receptiveness and close attention that, as Charles Taylor says in his now classic essay on the subject, is "not just a courtesy we owe people," but "a vital human need."[2]

Enter the phrase "we are listening to your voices" into a search engine and you'll turn up countless hits: announcements from local and national governments, press releases from lobby groups and corporations, solidarity statements from academic bodies and university departments, credos from unions and activist organizations. Taken together, these institutions clearly express one of the foundational contradictions that structures popular philosophies of listening: Attentive listening maximizes productivity, even as it models a humanizing resistance to such instrumental ways of thinking. This contradiction was apparent in the business literature on listening as early as 1957, when Ralph Nichols and Leonard Stevens complained in the pages of the *Harvard Business Review* that "people in general do not know how to listen." Their conclusion was nakedly instrumentalizing: "The improvement of listening can be of great value in today's business."[3] Nearly sixty years later, the UK-based think tank INTRAC and two English universities inaugurated a multiyear research collaboration entitled *The Listening Zones of NGOs* with much the same premise as Nichols and Stevens: The power of listening (so the project claimed) has been untapped—although, in this case, the goal was a mutually enriching form of intercultural cooperation. Listening "is key to the effectiveness of partnerships, programmes and ultimately to challenging the very power dynamics that are at the root of poverty and marginalisation," asserted Vicky Brehm, one of the project's research associates.[4] She thus implied that receptive listening not only promotes more functional institutional relationships, but an institution's heightened attention might even remedy the most intractable problems of wealth distribution and economic exploitation.

Claims like these—which are everywhere nowadays—infuse political rhetoric with the language of popular therapeutic practice, where *being heard* has for some time been the other side of talking cures.[5] M. Scott Peck, the high priest of modern couples counseling and author of the 1978 bestseller *The Road Less Traveled*, built his career on the concept of "true listening"—a technique of profound other-directedness that promised nothing less than the "unification of speaker and listener." "Listening is by far the most important form of attention" in any loving relationship, he insisted.[6] The idea that listening is an unending moral labor—what Peck called "the work of attention"—and the related view that people have become indolent and inattentive have since pervaded the genre of self help, from Madelyn Burley-Allen's *Listening: The Forgotten Skill* to Michael Nichols's *The Lost Art of Listening*.[7] People, it seems, have always just forgotten how to listen. The newest TED-talk-ish interventions on political dialogue, such as Celeste Headlee's 2017 *We Need to Talk* and Nihal Arthanayake's 2022 *Let's Talk*, explicitly mix political and therapeutic conceptions of auditory attention, especially in response to the age of culture wars and polarization, not to mention the regular public shamings, social media pile-ons, and twitterstorms of Web 2.0.[8]

These competing yet related conceptions of attentive listening may appear to stand at a metaphorical remove from the kinds of listening that musicians, music teachers, and music scholars have historically cultivated and theorized. Yet, as James Davies, Martha Feldman, and others have shown with respect to the concept of voice, the figurative voices of the liberal political imaginary have never been wholly separable from the voices that musicians and music lovers embody and enjoy.[9] In the 1980s and 1990s, opera critics were quick to hear the voice of the prima donna as a source of liberation, even amid the kind of ritual narrative "undoing" denounced by feminists such as Catherine Clément.[10]

Women in opera "sing with an authority equal to that of their male oppressors," wrote Paul Robinson in 1989; "while the operatic text tells the story of woman's undoing, the music tells just the opposite story: of her empowerment."[11] Around the same time, many scholars of race were theorizing the radical grain of the voice in twentieth-century popular music. The Black voice, argued Lindon Barrett, complicated the European division of "singing" and "signing," and so helped to subvert the exploitative value forms of advanced capitalism.[12]

Similar continuities between the figurative and the fleshly suffuse scholarly theories of listening. One prominent strand within these theories originates in the postwar Euro-American avant garde, in particular John Cage's belief in a pervasively sounding world, to which the artist should listen with an unprejudiced ear—a heightened attention to the sonic "inner life" of all things.[13] This belief, which Douglas Kahn has called "panaurality," gathered ethical urgency in its subsequent experimentalist iterations, not least in Pauline Oliveros's philosophy of "deep listening"—"a way of listening in every possible way to everything possible," as she described it.[14] According to its present-day advocates at the Center for Deep Listening in Troy, New York, deep listening "promotes experimentation, improvisation, collaboration, playfulness, and other creative skills vital to personal and community growth."[15] Oliveros's vertical axis of aural inclusivity in turn inspired the horizontal one of Annea Lockwood and Ruth Anderson's experimentalist handbook *Hearing Studies*, developed in one of Oliveros's seminars at the University of California, San Diego. The manifesto consists of exercises to encourage "extended listening"—the other side of "extended technique" in musical performance, perhaps. Like "deep listening" advocates, Lockwood and Anderson promise to nurture a scattered range of personal and political virtues: "Extended listening," they argue, is "an extension of one's aural vocabulary [that] brings with it an awareness of the different purposes of different musics, and is valuable personally within the context of our current culture."[16] You might have noticed by now that the ethical position-taking of most listening theories is typically signaled by pairing the word *listening* with an urgent adjective. True listening (Peck), deep listening (Oliveros), extended listening (Lockwood and Anderson): Listening is always less than it could be and could always include more.

Musicologists have tended to be less brazen than self-publicizing sound artists or self-help gurus in advertising the ethical merits of their listening habits. Still, the idea that attentive listening foments spiritual elevation is deeply rooted in European aesthetics, so it should be no surprise that scholars have continued to claim that better listening is both viable and politically improving.[17] As the core technique of recognition, attentive listening has also been essential to musicological comparativism and the study of musical contact, encounter, and exchange. In his 1993 study *Music in Renaissance Magic*, pointedly subtitled *Toward a Historiography of Others*, Gary Tomlinson professed his desire to "turn away from the easily heard voices from [the] past and to listen hard to other voices that have seemed almost too distant to hear"—a mission steeped in the anthropology of Clifford Geertz and the literary New Historicism that Geertz helped to inspire.[18] Stephen Greenblatt's *Shakespearean Negotiations* of 1988, a key text of New Historicist theory, had represented the business of historical literary inquiry as a "straining to listen" to voices of the dead.[19] In recent years, though, the idea that listening in a historically or cross-culturally sensitive way involves strain—a kind of aural heavy lifting—has yielded to ideals of listening either as a kind of dazzling technical wizardry or a more cautious, almost bureaucratic collaboration. In the introduction to their brilliant collection of essays on historical sonic encounter, *Acoustemologies in Contact*, Emily Wilbourne and Suzanne Cusick lionize listening that proceeds "consultatively" as well as being "virtuosic"; they borrow the second of these words from the anthropologist Martin Daughtry, who uses it to describe the skilled auditors of wartime landscapes.[20] Wilbourne and Cusick's listening vocabulary is thus at once managerial and theatrical. It seems to me that this language responds to two sides of a pincer movement, which is currently squeezing arts and humanities disciplines within the neoliberal academy: on the one hand, the powerful incentive to demonstrate activist political credentials and, on the other hand, the imperative to deliver the measurable metrics of so-called impact. In many circumstances, these pressures, apparently issuing from opposing ends of the political spectrum, are barely distinguishable. We shouldn't be surprised, then, when musicians and music scholars are tempted to claim that being "good listeners," in the widest ethical sense, is their disciplinary strong suit.

The view that I advance in this essay is that the claim to good listening has historically almost always depended on a politically problematic fantasy of attentiveness, which ultimately serves to strengthen and sequester communities of "good listeners." By analogy with the megalomaniac delusion of omniscience, and with a nod to the Cagean ideal of panaurality, I call this fantasy "omniaudience." Omniaudience refers to the belief that people—certain classes of people, it almost always turns out—can and should listen to everything. As we will see, omniaudience has a long history, extending at least as far back as the taxonomic projects of liberal musician-scholars in late eighteenth-century England and their earliest ventures in musical comparativism. This essay is devoted to exploring one small corner of this history—a corner that I take to be representative—so that I can propose an answer to two questions. First: When did Euro-American music scholarship start to conceive of its mission as cultivating good listeners, and to equate good listening with a heightened attention to and recognition of sonic differences? And second: Has the disciplinary objective of attending to differences always been yoked to the liberal project of recognizing and valuing difference?[21]

* * *

There are many places that a music scholar could begin the history of good listening, but there are several reasons to start with William Crotch's lectures on music at the Royal Institution in London, which he began in 1804.[22] They were something of a landmark in the history of modern musical disciplinarity. Crotch was in demand as a speaker in the capital. He was Heather Professor of Music at Oxford University, a supremely talented keyboard player and respected composer, an aesthetic theorist and music historian, and a leading proponent of so-called ancient music. The Royal Institution was a newish body, founded in 1799 by the American Sir Benjamin Thompson. It was one of a generation of learned societies in London and the provinces that aimed to disseminate knowledge to a broadly conceived public, to teach through "regular courses of philosophical lectures and experiments, the application of the new discoveries in science," as a description from 1809 put it.[23] Notably, Crotch inaugurated music as a subject worthy of study at the Royal Institution, following lectures that had, until then, been devoted largely to the natural sciences: His predecessors included the chemist Thomas Garnett and the professor of natural philosophy Thomas Young, who covered subjects such as hydrodynamics, electricity, and optics.[24] This context is crucial in the story of listening as a didactic framework, since Crotch's lectures proceeded from the idea that he was charged with teaching a learned constituency of nonmusicians how to listen better. By the early nineteenth century, musical novices or aspirant chapel masters could buy or borrow plenty of instructional books on reading, performing, and composing music. But barely any of these books primarily addressed or sought to improve *listeners*—that is, a group of people whose goal was to listen to music attentively and knowledgeably, rather than to help make it. The listener—and in particular the *good listener*—as a viable social role was partly a function of changing institutional configurations, represented precisely by such bodies as the Royal Institution and Crotch's university employer.

Though distinctively shaped by British colonialism, as we will see, these institutional changes were by no means exclusively British. Parallel trends are discernable in early nineteenth-century Berlin, where, as Desmond Sheehan has shown, a succession of Napoleonic-era disruptions of the city's old learned societies displaced music-making institutions such as Carl Friedrich Zelter's *Singakademie* into the modern research university. This process of displacement more or less created a career for Adolf Bernhard Marx, whose many publications, issuing from an evolving Prussian arts curriculum, helped to determine what *good listening* was—at least where a subcategory of elite European music is concerned—for the next century or more.[25]

In Britain, the scholarly publication to which Crotch responded most directly was Charles Burney's *General History of Music*, completed in 1789—one of the earliest narrative histories of European music. Burney provided detail about composers, performers, institutions, genres, and musical techniques, and was forthrightly opinionated about them. But by and large Burney did not prescribe, approve, or even theorize ways of listening. To inculcate listening techniques supposedly adequate to the many kinds of music he described was not Burney's main concern. Like most eighteenth-century writers, he proceeded on the assumption that music was on an upward trajectory and had been constantly improving until his own day: Volume One of his history presumed the "contented ignorance of barbarous times" and the "slow progress of human knowledge."[26] This premise caused a minor spat between Burney and Crotch following the first course of Royal Institution lectures. The venerable historian drafted a letter to his young contemporary in 1805, admonishing him for daring to criticize Joseph Haydn, a composer still widely lauded in London since the epochally successful Hanover Square concerts of the mid 1790s.[27] Burney knew Haydn personally and regarded his latest music—which by 1805 included the much celebrated oratorios *The Creation* and *The Seasons*—as the apogee of music history.

Crotch had indeed expressed doubts, shared by a number of contemporaries in England and beyond, about word painting in *The Creation*.[28] But his quibbles, and the boldness with which he asserted them, were evidence of a deeper disagreement with Burney. Crotch's values were those of the Concert of Antient Music. To him, music history was less teleology than cycle, and every age was characterized by episodes of growth and decline—episodes of the kind that Edward Gibbon had described in his *Decline and Fall of the*

Roman Empire, almost precisely contemporary with Burney's history.[29] Musical decadence was portended, in every age, by the preponderance of music that Crotch called "ornamental": diminutive, charming, and playful.[30] In a tripartite hierarchy adapted loosely from the popular art theories of Uvedale Price,[31] the musically ornamental, Crotch explained, always existed alongside music that was sublime and beautiful. But this universalizing musical typology was, in its way, historically relativizing: In Crotch's view, any music from any era or place could achieve the sublime (his highest-status category) and even the most recent music could be counted pleasing yet trivial. Hence his notional displacement of Haydn—whom he also liberally praised, by the way—from the pinnacle of music history.

In 1807, Crotch published, along with a short preface, Volume One of all the musical illustrations he had played during his Royal Institution and Oxford lectures—an immense range of music, transcribed for keyboard and notated on the grand stave. The book was arranged not chronologically but according to the aesthetic terms of Crotch's discussions. This meant that a movement from a recent symphony by Haydn appeared cheek by jowl with operatic excerpts by Gluck, "ancient" liturgical music by Orlando Gibbons, Hungarian dance tunes, a cobbler's hornpipe, and a good deal of "national music" (Crotch's term, and the term of choice among his English contemporaries) from places ranging from South Africa to Denmark. As Howard Irving points out, the title of Crotch's publication is telling, given the institutional context of his lectures: *Specimens of Various Styles of Music*.[32] As in the theatrical experiments and taxonomic displays of the natural sciences—the main focus of the Royal Institution lectures—Crotch presented a collection of specimens, each one representing a genus or species. Each specimen was supposed to be audibly distinct. Crotch's "various styles" are without doubt an emergent version of the biocultural "audiotypes" that Rachel Mundy has theorized in her antihumanist critique of Guido Adler's late nineteenth-century musical style concept.[33] Crotch's compendium of music drawn from a newly imagined globe aimed above all to teach listeners to detect and understand differences. The preface of the *Specimens* numbered among the collection's objectives the education of the musician's taste, so as to "prevent his being bigoted to particular sorts of music."[34]

The early nineteenth-century texts of Crotch's Royal Institution lectures survive only in scattered manuscript form, though they were adapted for publication in the 1830s as the *Substance of Several Courses of Lectures on Music*. By this time, Crotch confidently dismissed the now old-fashioned outlook of Burney's Whig musical history—the view, as Crotch skeptically put it, that "music is continually improving."[35] Crotch contrasted this with his own vision of "just criticism," which he sought to impart to the "musical student": the art of "discriminating and comparing the several styles, and of appreciating their relative value." Good listening now equaled discrimination.[36]

I want to pause over Crotch's concept of "discrimination" here, partly to uncover the buried relationship between the word's more recent, negative meaning and its older (but still current) positive one. To *discriminate*, as Johnson's 1755 dictionary had put it, is to "mark with notes of difference"—to appreciate differences, which is usually a sign of knowledge or discernment.[37] The negative sense of discrimination as an unethical and sometimes unlawful practice within a liberal polity is no less concerned with the perception of difference: To discriminate reprehensibly is also to notice differences—maybe even the very ones that others notice with a purer intent. But, rather than celebrate the fact of difference, the bad discriminator invests these differences with unwarranted values and hierarchies.

No reader of this essay will be surprised when I say that the first (positive) kind of discrimination encouraged by Crotch, despite his professed desire to eliminate bigotry, is also of the second (negative) kind. As the argument of this essay unfolds, it will become ever clearer that every liberal attempt to appreciate difference with texture and understanding, no matter how well meaning, is vulnerable to a critique (no less liberal) that shows how this difference could have been appreciated with still greater texture and understanding. Acute musical attention, in other words, reveals the need for more acute or discerning audition. Crotch is, in this respect, an easy target. More important to my argument, though, is to observe that any conception of aural discrimination of the sort that Crotch tried to inculcate already assumed his and his audience's right to listen to, and potential to listen competently to, more or less any kind of music in existence. The contents pages of the *Specimens of Various Styles of Music* provide a startling summary of the ambitious scope of Crotch's roving ear (see Figure 4.1). Here, then, is the foundational principle of omniaudience: The discriminating ear can always be turned receptively toward a world of distinct musical voices, "marking the notes of difference." This is surely the acquisitive aural orientation that Dylan Robinson, elaborating his ethic of Indigenous sound studies, has called hungry listening.[38] For if Crotch's contents page expresses a primitive version of the universalist fantasy of deep listening—"listening in every possible way to everything possible"—it also displays with unusual clarity the ethical hazards that have always been part of ethnomusicological self-examination, but

Figure 4.1. Crotch indexes the world.

have lately plagued the discipline with particular force: how to distinguish desirable pluralism from mere food court multiculturalism; how to avoid converting music into a delectable fruit salad of mingling culture-portions; how to escape a research environment in which scholars have extracted sonic specimens to fuel an economy of Euro-American disciplinary prestige; and, crucially, whether or not to endorse the relentless making-legible of music to the omniaudient Euro-American ear.

Kahn, in his interrogation of sound across the twentieth-century arts, points out that Cagean panaurality became thinkable partly because of technologies such as the microphone, which created the impression that listening could be extended prosthetically into formerly inaudible corners of the universe. But once this radical conceptual extension had taken place, the technological mediators that made it possible disappeared from view.[39] Something similar happened in the case of Oliveros's deep listening. Oliveros traced her early revelations in listening to a tape recorder bought for her by her mother in 1953. "I noticed that the microphone was picking up sounds that I had not heard while the recording was in progress," she recalled.[40] Her subsequent mantra, "listen to everything all the time," was thus modeled on the supposedly expanded receptivity of the microphone.

In Crotch's case, the media technologies that, like the experimentalists' microphone, rendered the world's music legible are even less obvious, perhaps, because they are these days so unexceptional. There is the book itself, of course, which blended the ethos of the contemporary miscellany and anthology with the lexicographical and completist ambitions of the encyclopedia. And inside is the two-stave notational format to which all the specimens were adapted—a technology that allowed the widest access within an extensive European and colonial market for musical prints directed at amateurs. Crotch explicitly addressed this constituency in his preface, explaining that his specimens furnished "good subjects for practice."[41] The stave was inseparable from the standard topography of keyboard instruments, the interface through which every one of the specimens became comparable, and all differences appreciable. The preface to the *Specimens* also gestured toward several ancillary tools of legibility, which, to the present-day scholar, can only seem portents of the standardizing processes of nineteenth-century industrial modernity. For instance, Crotch's opening table, entitled "Time of the Specimens," assimilated each of his examples to the logic of his simple pendulum metronome (see Figure 4.2).[42] Institutions may have created the viable category of *listener*, then, but it was this bundle of media technologies that

TIME OF THE SPECIMENS.

Number of the Tune.	Length of the Note.	Number of Inches.	No.	Note.	Inch.	No.	Note.	Inch.	No.	Note.	Inch.	No.	Note.	Inch.	No.	Note.	Inch.
1	quaver	12	63	crot.	18	126	crot.	9	180	minim	30	241	crot.	14	302	d. crot.	16
2	quaver	10	64	d. crot.	18	127	quaver	10	181	crot.	32	242	d. crot.	18	303	crot.	22
3	crotchet	10½	65	crot.	24	128	crot.	32	182	minim	18	243	crot.	6	304	crot.	18
4	crot.	30	66	d. crot.	36	129	quaver	10	183	quaver	14	244	crot.	16	305	crot.	12
5	crot.	24	67	crot.	32	130	quaver	9	184	quaver	9	245	d. crot.	30	306	crot.	21
6	quaver	18	68	crot.	15	131	crot.	12	185	crot.	7 or 12	246	crot.	6	307	crot.	8
7	quaver	21	69	d. crot.	12	132	d. crot.	36	186	crot.	10	247	crot.	24	308	crot.	16
2nd Mov.	crot.	26	70	d. crot.	18	133	crot.	30	187	crot.	10	248	crot.	10	309	minim	30
3rd Mov.	crot.	18	71	crot.	14	134	crot.	14	188	crot.	10	249	crot.	8	2nd M.	minim	24
8	crot.	33	72	crot.	24	135	crot.	30	189	crot.	11	250	d. crot.	27	3rd M.	minim	14
9	quaver	6	73	crot.	28	136	crot.	36	190	crot.	6	251	quaver	10	4th M.	minim	10
10	crot.	24	74	crot.	24	137	minim	17	191	crot.	14	252	quaver	10	5th M.	minim	22
11	crot.	18	75	crot.	8	138	crot.	24	192	crot.	16	253	quaver	10	2nd Ed.	minim	30
12	crot.	9	76	quaver	12	139	crot.	6	193	minim	14	254	quaver	10	2nd M.	minim	24
13	crot.	12	77	crot.	28	140	crot.	32	194	minim	20	255	minim	24	3rd Ed.	minim	18
14	minim	24	78	crot.	12	141	crot.	8	195	minim	20	256	d. crot.	36	310	minim	28
15	crot.	15	79	d. crot.	12	142	d. min.	30	196	minim	17	257	d. crot.	36	311	crot.	10
16	quaver	18	80	crot.	24	143	crot.	14	197	minim	36	258	minim	16	312	crot.	28
17	crot.	18	81	crot.	8	144	crot.	36	198	minim	20	259	crot.	21	2nd M.	d. crot.	36
18	crot.	18	82	crot.	32	145	quaver	12	199	minim	20	260	crot.	13	313	crot.	13
19	crot.	30	83	crot.	36	146	quaver	12	200	crot.	11	261	crot.	13	2nd M.	crot.	9
20	crot.	9	84	quaver	10	147	crot.	36	201	minim	36	262	crot.	18	314	crot.	32
21	crot.	8	85	d. crot.	33	148	crot.	36	202	d. min.	30	263	crot.	18	315	crot.	13
22	quaver	12	86	crot.	12	149	quaver	16	203	d. min.	36	264	crot.	18	316	crot.	12
23	crot.	32	87	crot.	10	150	crot.	15	204	d. min.	36	265	crot.	18	317	crot.	24
24	crot.	32	88	quaver	10	151	minim	13	205	d. min.	24	266	minim	36	318	minim	36
25	quaver	6	89	crot.	15	2nd Ed.	crot.	12	206	minim	12	267	minim	36	319	crot.	18
26	dott. crot.	20	90	crot.	24	3rd Ed.	minim	18	207	minim	18	268	minim	36	320	crot.	24
27	quaver	9	91	quaver	13	152	crot.	7	208	minim	16	269	minim	36	321	crot.	12
28	crot.	24	92	quaver	30	2nd Ed.	quaver	7	209	minim	16	270	crot.	14	321*	crot.	14
29	crot.	24	93	crot.	14	3rd Ed.	crot.	7	210	minim	16	271	d. crot.	22	322	d. crot.	36
30	crot.	24	94	d. crot.	18	153	quaver	6	211	minim	12	2nd Ed.	crot.	22	323	crot.	24
31	crot.	6	95	quaver	12	2nd Ed.	crot.	6	212	d. crot.	18	272	d. min.	30	324	d. crot.	36
32	crot.	7	2nd Ed.	crot.	30	154	crot.	27	213	minim	11	273	crot.	18	325	quaver	12
33	crot.	5½	96	d. crot.	30	155	quaver	12	214	d. min.	18	274	d. min.	36	326	crot.	17
34	quaver	8	97	d. crot.	20	156	crot.	16	215	minim	18	275	crot.	21	327	crot.	24
35	quaver	15	98	d. crot.	18	157	crot.	33	216	quaver	9	276	crot.	16	328	crot.	16
36	crot.	18	99	crot.	7	158	crot.	21	217	crot.	14	277	crot.	12	329	crot.	28
37	quaver	5	100	crot.	8	159	crot.	21	218	crot.	17	278	crot.	16	330	crot.	26
38	quaver	18	101	quaver	12	2nd M.	d. crot.	21	219	crot.	13	279	crot.	7	331	d. crot.	18
39	quaver	14	102	d. crot.	36	160	crot.	21	220	crot.	12	280	quaver	6	332	crot.	24
40 1st Ed.	crot.	22	103	quaver	12	161	d. crot.	18	221	d. crot.	30	281	crot.	18	333	d. crot.	36
2nd Ed.	crot.	30	104	quaver	12	162	d. crot.	30	222	minim	24	282	crot.	28	334	crot.	16
41	d. crot.	18	105	crot.	14	163	crot.	24	223	crot.	12	283	crot.	12	335	quaver	12
42	d. crot.	18	106	crot.	19	164	crot.	24	224	d. crot.	10	2nd M.	crot.	18	336	crot.	22
43	d. crot.	18	107	crot.	36	165	d. crot.	20	225	crot.	12	3rd M.	quaver	7	337	crot.	26
44	crot.	12	108	crot.	36	166	crot.	21	226	minim	18	284	crot.	15	338	quaver	14
45	crot.	24	109	crot.	6	2nd M.	d. crot.	21	227	minim	36	285	crot.	15	2nd M.	quaver	17
46	crot.	18	110	minim	30	167	quaver	8	228	minim	22	286	d. min.	36	339	quaver	14
47	quaver	18	111	crot.	18	168	d. crot.	12	229	crot.	12	287	d. crot.	11	340	d. crot.	20
48	crot.	36	112	quaver	12	2nd Ed.	crot.	12	230	crot.	12	288	crot.	7	341	quaver	15
49	crot.	32	113	minim	22	169	crot.	12	231	crot.	7	289	crot.	7	342	crot.	18
50	crot.	20	114	crot.	36	170	crot.	12	232	minim	16	290	crot.	14	343	d. crot.	27
51	crot.	6	115	crot.	18	171	crot.	36	233	crot.	8	291	crot.	8	344	crot.	24
52	d. crot.	18	116	crot.	14	172	crot.	36	234	minim	14	292	crot.	7	345	crot.	10
53	quaver	5	117	quaver	12	2nd M.	quaver	21	235	minim	20	293	crot.	13	346	quaver	19
54	d. crot.	26	118	quaver	10	173	minim	24	236	d. crot.	15	294	crot.	14	347	crot.	15
55	crot.	24	119	quaver	16	174	crot.	24	237	crot.	8	295	minim	24	348	quaver	18
56	crot.	8	120	quaver	12	175	crot.	30	238	quaver	12	296	d. crot.	30	349	crot.	20
57	crot.	11	121	crot.	9	176	crot.	16	2nd M.	d. crot.	8	297	minim	24	350	crot.	23
58	quaver	32	122	quaver	9	177	crot.	20	2nd Ed.	crot.	18	298	crot.	9	351	crot.	14
59	crot.	16	123	crot.	12	178	crot.	30	2nd M.	minim	18	299	crot.	22	352	crot.	12
60	crot.	24	124	crot.	30	2nd M.	d. crot.	22	239	crot.	6	300	crot.	17	353	crot.	21
61	crot.	32	125	crot.	8	179	crot.	24	240	crot.	12	301	crot.	16	354	crot.	8
62	crot.	21															

Figure 4.2. Crotch regularizes the world.

produced the terrain of equivalence that modeled and enabled discriminating listening—and so arguably created Crotch's audibly differentiated "various styles" themselves. Appreciating difference meant enforcing an inaugural equivalence.

Crotch's specimens had dramatically different journeys toward this terrain of equivalence, after all, which involved divergent patterns of labor, extraction, and exploitation. The symphonic music by Haydn was native to Crotch's media landscape: The Allegretto from the Symphony No. 82 would already have been known to most music lovers via printed keyboard reductions, which had been produced and sold concurrently with performances of Haydn's symphonies in England since the 1780s. In this case, Crotch's volume, like the reprints and excerpts that made up most eighteenth- and early nineteenth-century printed material, largely reaggregated existing media of the same kind. Items such as the Sanctus by Orlando Gibbons, by contrast, had been disseminated and gathered together in more diverse handwritten and printed forms over centuries, and gradually entered the eighteenth-century printscape via piecemeal transcriptions or adaptations, mostly by ancient music enthusiasts such as Crotch himself.

How "national music" ended up among Crotch's specimens, especially examples from beyond Europe, was more obviously fraught. Much of the Indian music, such as the sample of *tarana* that Crotch singled out for special praise (see Figures 4.3a and 4.3b), was derived from a single source: William Hamilton Bird's *Oriental Miscellany*, first published in Calcutta in 1789. Bird had gathered his materials from various people and places across British-dominated India, but, as research by Ian Woodfield showed some time ago, he also drew substantially on transcriptions prepared by prominent Anglo-Indian women, including Sophia Plowden, who, in 1787, made a trip from Calcutta up the Ganges to the court at Lucknow with her harpsichord and her husband, the East India Company officer Richard Plowden.[43] Besides reproducing songs from Bird's *Miscellany*, Crotch also consulted Plowden's private collection of poetic and musical transcriptions.

It is hard to avoid the conclusion that the European musical-technological frameworks into which Plowden and Bird squeezed the songs they heard, and the wider British craze for "Hindostannie Airs" in the 1780s, were straightforward misrecognition, or even a species of colonial epistemic violence—violence ultimately sublimated into one of the tidily presented specimens of Crotch's omniaudient compendium.[44] But I want to be clear about what this conclusion also implies about listening: To deplore such musical

Figure 4.3a. Crotch's specimen of Indian music.

Figure 4.3b. The source of Crotch's Indian tune: William Hamilton Bird's *Oriental Miscellany* (1789), 28.

misrecognition is to say that none of the actors involved was listening well enough—that is, that they were listening without the requisite virtuosity or consultative procedures. This is not a new criticism. Bird's published transcriptions were condemned by several scholars at the time. The dedicatee of the *Oriental Miscellany,* Warren Hastings—who, by the time of Crotch's

Specimens, had, following a long and sensational trial, been acquitted of crimes committed during his time as governor-general of Bengal—had long insisted that British traders and rulers should immerse themselves in Indian texts and religious practices, and explicitly protested against what he called "the Interpolation of European Taste into the Recital of the Music of Hindostan." A supposedly uncontaminated intimacy with Indian cultures was integral to the colonial knowledge project.[45]

A related but more nuanced position available to the present-day scholar—one adopted by Katherine Butler Schofield in a multifaceted creative project on Bird's *Miscellany* alongside the harpsichordist Jane Chapman—is cautiously to celebrate products of the sonic contact zones of the late Mughal and early colonial period, such as the so-called *nautch*: fashionable parties at which troupes of high-status North Indian courtesans danced, recited, and sang.[46] In these zones, Bird, Plowden, and others listened to and gathered their material. This position moves historical listening of the sort recorded by Bird's *Miscellany* farther from mere misrecognition and closer to the "experimentation, improvisation, collaboration, [and] playfulness" envisioned by the Center for Deep Listening. Butler Schofield and Chapman invite us to hear, via Crotch's *Specimens*, not only Bird and Plowden but also the compelling voices of non-Europeans, such as the legendary Kashmiri courtesan Khanum Jan, from whom Plowden learned many songs and poems. From the former silence, more voices resound. Lurking within even a colonial document such as Bird's *Miscellany* is an untapped, subversive auditory richness. That's the idea, anyway. But the point is this: The politically responsible scholar always aims at yet more attentive, yet more discriminating listening—an extended listening to "everything possible," now both with and through the unvariegated musical mediascape of Crotch's *Specimens*.

* * *

My foregoing discussion of Crotch's *Specimens* implies that there is no form of listening that does not take its place on a liberal spectrum extending from good to bad, open minded to bigoted, extended to narrow, deep to shallow, true to false. Crotch's inadequacies as a listener may be manifest, but to address them, so my discussion has suggested, is not to change his listening orientation as much as it is to refine and improve it.

There are alternatives to this view, though. A broadly decolonizing politics has led a generation of scholars, especially within Indigenous studies, to reject the liberal politics of recognition altogether—and the forms of attention it encourages—as the very opposite of political liberation. Recognition by the powerful is precisely how "*colonial relations are produced and maintained*," asserts Glen Coulthard in his Frantz Fanon-inspired polemic against the liberal discourse of Indigenous rights.[47] Lester Hu borrows this insight to theorize the sonic governance projects of eighteenth-century France and Qing-era China, arguing that conferring voices that can be listened to—"making the mute(d) submit themselves to empire's phono-extractions"—was one of the main technologies of imperial domination.[48] Good listening, as Crotch and his liberal successors envisioned it, is, in this context, no longer an unimpeachable good in itself. This is implicit when David Garneau queries "the belief that everything should be accessible to those with the means and the will to access" it, that "everything should be comprehensible"—the very idea, in other words, that Euro-American disciplines can or should, as Oliveros insisted, "listen to everything all the time."[49]

From this perspective, people truly value difference not by improving the techniques or expanding the horizons of attention but by dwelling on what is, or ought to remain, unintelligible or inaccessible—by celebrating the insurmountable opacity that Édouard Glissant saw as essential to any self-determination: "Opacities must be preserved; an appetite for opportune obscurity in translation must be created."[50] In this spirit, Emily Apter's anticosmopolitan intervention, *Against World Literature*, focuses not on moments of receptive self-examination, mutual transformation, or dialogic exchange—that is, the very stuff of comparativism—but the blockages of the untranslatable, even the dark glamor of mutual incomprehension.[51] In music studies, Olivia Bloechl has similarly urged scholars to pay attention to the uncanny, out-of-place objects of colonial history, and so to those who are "prevented from reaching recognizable articulation in music," whose presence is measurable only as awkward silence or spectral trace.[52] Legibility is not desirable; it's the problem.

And yet it seems to me that when scholars emphasize or even celebrate opacity they still risk reinforcing an old collection of romantic-modernist values, which elevate—in this instance, as epistemologically responsible and intellectually exciting—things that are formally intractable, semiotically unfathomable, and ontologically supersensible. Nothing is more familiar to scholars in the humanities, and nothing makes them feel more at home, than being presented with a runic fragment of culture that is hard or even impossible to understand—something so open-ended and unreachable that it perpetually goads the critic into discursive action.[53] Even the prescient and transformative intellectual contributions of Tomlinson, starting out from his earliest mission to listen hard to what is "almost too distant to hear" in the European Renaissance, have since described an almost avant-gardist trajectory whereby the scholarly ear is compelled to bend to ever more dramatically inaccessible voices—from the half-century-old Aztec song traditions from which "no performable music writing survives" to the proto-music-making capacities that congealed "800,000 years before the advent of sapient humans." No matter what the Euro-American scholar listens to, the Keatsian unheard melodies will always be sweeter.[54] Something of this ethic and its suppressed aesthetic informs Glenda Goodman's otherwise persuasive and energizing theory of "Music Transcription as Imperial Technology" in the eighteenth century, which posits a realm of absent listening that nonetheless remains potentially redemptive—a realm that stymies "full capture" by European notation and dramatizes "the tension between what persists and what is lost."[55] My view is that settler-colonial scholars should be careful when they imply that there is or was an Indigenous aural plenitude, forever out of colonial earshot. Rather than avoid imposing a Euro-American aesthetic on their object of study or appreciation, with its all-too-legible misrecognitions, this might just as well involve substituting one familiar aesthetic category for another—circumscribable beauty becomes ungraspable sublime. Opacity is the consequence of self-determination, of calculated indifference to an acquisitive, attentively grasping audience; for the same reason it is the primary formal feature of the romantic-modernist art work.

Besides, any conceptual scheme that opposes the presence of notation to an absent aurality does not challenge the assumption that notation is always a tool of sonic capture and preservation. This is an idea that became plausible only in the late eighteenth century, and may even have depended to some degree on processes of colonial extraction themselves. Just as the anthropologist Brian Hochman maintains that European ethnography naturalized twentieth-century sound reproduction technologies as a way of capturing the real, investing particular media forms with European ideologies of vanishing Indigeneity, so you could argue that eighteenth-century proto-ethnographic projects such as Plowden's and Bird's helped to secure the association—by no means self-evident—between notational technologies and sonic preservation.[56] It is no accident, I would speculate, that only a few years after the first edition of Bird's *Miscellany*, Breitkopf and Härtel began their complete works editions of Mozart and Haydn, the earliest preservationist projects of their kind.

The question of how historians might listen with, through, and against colonial technologies of inscription to recover traces of the aural domain of Indigenous voice is fundamental to a newer wave of sound-focused scholarship inspired by anthropologists of the ontological turn. Proponents of the ontological turn, rather than proceeding on the disciplinary premise that the world hosts a multiplicity of local "cultures," each of which places its construction on a single nature, propose that human societies do not merely construe but make worlds, and so repeatedly create "nature" anew.[57] Things that appear untranslatable, according to this view, are not the result of especially messy culture clashes as much as irreconcilable worlds. For Ana María Ochoa Gautier, writing mostly about nineteenth-century Colombia, the impasse of the sonically untranslatable can be surmounted, if at all, not through an unprecedentedly discriminating technique of intercultural understanding but through the scholar's almost shamanic conjuring of other acoustic worlds.[58] It is not clear to me yet whether Ochoa's position truly conveys music studies beyond the liberal ethic of "good listening"—which is to say, beyond a radicalized version of Crotch's premise—or whether it is a theoretically turbocharged version of the same thing, propelling the ideal of mutual proximity past some conceptual event horizon into the ecstatic merging of selves. Dissolving the subject-object binary has long been the goal of more conventional ethical programs of attention, after all (recall Peck's promised "unification of speaker and listener"). Even Greenblatt's New Historicist listening practice culminated in the startling revelation that he had realized while writing that "I could hear only my own voice, but my own voice was the voice of the dead"—a conception of historiography as a kind of spiritualism.[59]

Experimentalist panaurality and the older liberal ethic of good listening both idealize moments in which listening subject and sounding object merge. This partly explains why the language of vibration and resonance has become something of a disciplinary vernacular in recent musical ethnography and contemporary sound art—an echo of the vibrational ontologies in philosophy, which treat music, sound, and listening as contingent epiphenomena of a more fundamental, intermaterial motion.[60] So Wilbourne and Cusick predicate their ethic of acoustemological encounter on the "conviction that sound—as vibrational force—necessitates bodies in sonic contact." And in his trenchant essay "Against Listening," the artist Brian House invokes the paradox of an "experiential knowledge not delineated by the conventional bounds of the subject" in his desire to "reach towards a political consciousness that includes the nonhuman." Vibration allows listening to be dislodged from subjectivity itself. True omniaudience transcends listening altogether.[61]

Vibrational ontologies nonetheless tend to diverge from equally dispersed conceptions of distributed cognition or the extended mind, where "attention" is less a faculty of individual subjects than a shared cultural technique, mediated through a variety of materials and technologies.[62] The main reason is this: The monist universalism of vibration swallows up and obliterates not only voices and listeners but technologies and bodies too.[63] Everything is reducible to vibration. It strikes me that, as with the Schopenhauerian dissolution of the self in the Wagnerian music drama or Wordsworthian intimations of the symphonic connectedness of all nature, such vibrational fantasies—popularized by sound artists and scholars, so experienced mostly in the context of white cube galleries and written histories and ethnographies—are better understood as aesthetic effects, and signally romantic-modernist ones at that.[64] "Like harmony in music; there is a dark / Inscrutable workmanship that reconciles / Discordant elements, makes them cling together / In one society," rhapsodized Wordsworth in *The Prelude*.[65] The posthuman music aesthetic goes back a long way.

No wonder, then, that the scenario of distinctively musical listening keeps coming back, despite having been theorized away into vibrational everythingness. Even in the most recent wave of posthuman political ecologies, listening to music, as in Wordsworth's self-dispersing nature worship, remains a strikingly frequent metaphor for an ethically discriminating orientation toward the world. Anna Tsing's plangent study of life amid capitalism's broken global assemblages, *The Mushroom at the End of the World*, begins by invoking her own "revelation in listening" when following the "polyphonic" textures of "madrigal and fugue" in order to elaborate what she calls the "arts of noticing": the capacity to discriminate marginal or underrepresented things such as fungi.[66] Tsing's landscapes are posthuman, postcolonial, even postapocalyptic. In them, subjectivity is always scattered across vast material networks. And yet we are still enjoined to deploy our listening skills to discriminate it all—skills acquired, in Tsing's case, in the college seminars where she first encountered what Crotch would have called ancient music. Traces of a venerable liberal listening tradition are not hard to detect.

The aestheticized "world that speaks," and the virtuous theorist who cultivates attention with special sensitivity, is an old trope of eco-criticism, prominent in works such as David Abram's *The Spell of the Sensuous*. "That which we term 'language' is as much a property of the animate landscape as of the humans who dwell and speak within that terrain," he writes.[67] Jacob Smith's *Eco-Sonic Media* is only the most recent instance of the eco-theorist as benevolent conferer-of-voices, turning his keen ear, in this case, to the "noise" of the shellac record, through which he hears the voices of the "kusum trees and reddish insects" that labored to produce its raw materials.[68] To a student of eighteenth-century cultural history, the prototype of this ecological omniaudience is plain. It is a concept that Crotch would have recognized and endorsed: sympathetic resonance—that motif or perhaps master trope of the eighteenth century that, more even than the ideal of good listening, was foundationally bound up with music and sound. Like Peck's true listening, sympathy promised to blend subject and object. Haydn's English Canzonetta "Sympathy" expressed this modern ethical philosophy pithily, with a melody full of subtle repetitions and mirrorings: "each affection of thy heart / By sympathy is mine."

But thinkers who have criticized the sympathetic ideal, including its postwar therapeutic recasting as empathy—not least historians of the transatlantic slave trade, from George Boulukos to Saidiya Hartman—have observed that its specular structure produces not only identification but also distancing.[69] Sympathy collapses yet reinforces the distinction between self-consciously receptive "good listeners" and the gratefully listened to. As Lynn Festa argues in her profound analysis of the colonial dynamics of sympathy, the objects of sympathetic listening can never fully be absorbed into the larger collective, partly because, as the instigators of feeling attention, they serve to "consolidate our affiliation with others who feel the same way about a shared object." The fully vibrational inclusivity of liberal fantasy will remain permanently out of reach.[70]

According to the liberal political paradigm I have been examining in this essay, bad listening is responsible for all the political problems of misrecognition; all misrecognition happens in the very moment that we listen. (How could it be otherwise?) For that reason, listening must always be extended, deepened, made truer, more consultative, more virtuosic. This was also the case with eighteenth-century sympathy. Brycchan Carey describes how the publicity campaigns of British abolitionists in the 1780s and after, whose stated goal was to arouse sympathy among the well-heeled British public for enslaved African people, obsessively sought out and denounced instances of false sentiment—a crime that could be repaired only by a still more acute and authentic form of sympathetic fellow-feeling.[71] Likewise, any good listening depends on constant vigilance in the face of worse listening. It would seem that listening, in doses of ever greater purity, is the only possible cure for its own political shortcomings.

* * *

Walter Benn Michaels and Adolph Reed Jr. distinguish the liberal politics of *discrimination* from the Left politics of *exploitation*. The former is concerned with expunging a system's biases, the latter with remaking exploitative systems themselves.[72] The solution to discrimination, as we have seen, primarily involves changes in attitude: Rather than perceive difference and act unjustly, people should perceive difference and celebrate it. This was Crotch's concern, and it remains a central concern of those in music studies today: We teach and learn, as rigorously as we can, how to discriminate sonic differences—partly, these days, in order that we do not end up as one of Crotch's bigots, discriminating against a sound because of its difference. This is emblematically liberal insofar as it plays out its politics on the site of the self and within the domain of signs and culture. I don't want to dismiss this kind of politics. Cultural practices are the main concern of music studies, after all (or at least they were, until the dawning incredulity with the nature-culture opposition, whose disciplinary consequences are yet fully to unfold). I want only to suggest that taking care to contemplate the frequently murky relationship between discrimination and exploitation—especially the ways in which concerns about the former can sometimes displace or conceal the reality of the latter—might inculcate greater humility about the limits of our disciplinary politics, especially "fantasies about the exchange rate between the attention economy and the material economy," as Olúfémi Táíwò bracingly puts it.[73] Contemplating the relation of discrimination to exploitation might also remind even the most scrupulously attentive of scholars in music studies of what Stuart Hall once called the "necessary modesty of theory."[74] And it could help to dispel our discipline's recurring colonial dream of omniaudience: the entitled idea that we are able—or even have the right—to hear "everything possible," not to mention the narcissistic assumption that our listening has transformative political consequences.

Seeking to resist the subtle misrecognitions and extractive logic of the settler-colonial ear, Robinson, via Garneau, insists on the "irreconcilable spaces" of Indigenous knowledge.[75] It is no coincidence, I think, that this strategically bounded conception of Indigenous sonic culture ends up looking formally similar to the polemical indifference and withdrawal of the

Euro-American romantic-modern art work—perhaps even like an indigenized version of Milton Babbitt's notorious insistence that he didn't care who listened. Once again, the artwork, even in its uncommunicative refusal, models the best kind of human self-determination. And such refusal, therefore, does not prevent the extractive ear from aestheticizing as usual; it may even encourage it. Like Keats before his Grecian urn, the settler-colonial would-be listener, when confronted by a newly resistant and sublimely sovereign version of Indigenous sonic culture, is compelled to deploy the rhetorical strategies of apophasis or aposiopesis (that is: saying by theatrically not-saying or theatrically falling silent)—strategies that recreate the traditionally gendered dynamic that Barbara Johnson once called "muteness envy." The "power of reclaimed silence," writes Johnson, is "not unrelated to the idealization of muteness found in the aesthetic tradition"—a tradition that she traces to (where else?) Keats's Grecian urn and its unheard melodies.[76]

At the same time, to exclude the extractive Euro-American ear from Indigenous spaces, no matter how discriminating and sympathetic it claims to be, is also, I would argue, to diminish the importance of listening altogether—by showing that the careful discriminations of the Euro-American attention economy may have fewer consequences for the exploitative relations of the material economy than many good listeners might wish. And at this point I become uncertain. For one thing, if good listening is less politically impactful than good listeners might fantasize, then it is surely also less urgent that today's music scholars condemn previous generations of musicians, music-lovers, and settler-colonial scholars who—out of fascination, love, or the best of political intentions—once sought what used to be called cross-cultural understanding. The political and epistemological failings of giant comparativist projects such as Alan Lomax's unfortunately named Global Jukebox—perhaps the closest modern descendant of Crotch's *Specimens*—are much easier to address than the exploitative relationships that made them possible.[77] And this makes me worry that the liberal politics of listening—good listening, bad listening, extended listening, hungry listening, true listening, bigoted listening, deep listening, consultative listening, or even the politically gracious refusal of listening altogether—can only be practiced within an infrastructure produced by exploitation, an infrastructure that may remain largely untouched, even as scholars, still in the tradition of Crotch, reorganize, expand, or deaccession their sonic specimens and recalibrate their feelings about them. Crotch, you might say, was there at the start, building the room in which good discrimination could be developed and bad discrimination expunged. But, as Táíwò writes, "Bad roommates aren't the problem, for the same reason that being a good roommate isn't the solution: the problem is that we are still trapped in the room."[78]

Consider, once again, Sophia Plowden, sailing up the Ganges to Lucknow—a city, that, as all the scholars who have described this episode feel bound to point out, was to witness, in 1857, the beginning of the Indian Uprising, in the course of which Lucknow and the surrounding villages were brutally razed by British soldiers. It ought to go without saying that behind all of the Indian music that ended up among Crotch's specimens was not just the "cultural logic" but the violent material exploitation of imperialism—a form of exploitation that was the very precondition for the better or worse listening practices of Plowden, Bird, Crotch, and indeed all subsequent generations of Euro-American music scholars. The point at which we are able to begin listening well—and so to start remediating the problems of discrimination—is so often the point at which exploitation has already happened. With this in mind, making distinctions between better or worse listeners—even between the most ravenous of hungry listeners and those who perform the painful Orphic turn away from the object of their love lest they destroy it—looks less like a matter of political urgency and more like the narcissism of small differences. Apprehending the exploitative work of often inaudible systems[79] is rarely about pleasure or sentiment or self-improvement, and changing these systems or building new ones always involves more risk and material upheaval than changing how you listen or adjusting how you feel. The kind of discriminating listening that music scholars seek to cultivate, and the previously unheard voices and "various styles" that they accordingly promote, are politically vital. I would go so far as to say that this is the sort of cultural politics that music scholarship is best equipped to pursue. But exploitation demands reparation, and I am not sure that being listened to will ever be enough.

Notes

1. The inseparability of the politics of voice and the politics of listening is the premise of Daniel Gross, *Being-Moved: Rhetoric as the Art of Listening* (Oakland: University of California Press, 2020).

2. Charles Taylor, "The Politics of Recognition," in *Multiculturalism: Examining the Politics of Recognition*, ed. Amy Gutmann (Princeton: Princeton University Press, 1994), 25–73, here 26.

3. Ralph Nichols and Leonard Stevens, "Listening to People," *Harvard Business Review* 5 (September 1957), 65. The most influential elaboration of this premise in the disciplines of

organization management and political theory was Albert O. Hirschman's *Exit, Voice, and Loyalty* (Cambridge, MA: Harvard University Press, 1970), which attempted in part to show how businesses and states could attend more effectively to the (disgruntled) voices of stakeholders and forestall mere abandonment and disillusion ("exit," in Hirschman's terms). Thanks to Brian Kane for this citation.

4. Vicky Mansuco Brehm, "Towards a 'Listening Organisation'? Listening, Power, and Language in International Development NGOs," INTRAC blogpost, April 11, 2017. https://www.intrac.org/towards-listening-organisation-research-update/.

5. See, for instance, Gross's conception of psychoanalytic listening in *Being-Moved*, passim.

6. M. Scott Peck, *The Road Less Traveled* (New York: Touchstone, 2003), 127, 129. On "the work of attention," 120–131.

7. Madelyn Burley-Allen, *Listening: The Forgotten Skill* (New York: Wiley, 1982); Michael Nichols, *The Lost Art of Listening: How Learning to Listen Can Improve Relationships* (New York: Guilford Press, 1995).

8. Nihal Arthanayake, *Let's Talk: How to Have Better Conversations* (London: Orion, 2022); Celeste Headlee, *We Need to Talk: How to Have Conversations That Matter* (New York: Harper Wave, 2017).

9. See the essays in the colloquy convened by Martha Feldman, "Why Voice Now?" *American Musicological Society* 68, no. 3 (2015), 653–85; *The Voice as Something More: Essays Toward Materiality*, ed. Martha Feldman and Judith Zeitlin (Chicago: University of Chicago Press, 2019); James Davies, *Romantic Anatomies of Performance* (Berkeley: University of California Press, 2014); and esp. James Davies, "'I Am an Essentialist': Against the Voice Itself," in *The Voice as Something More*, 142–68. Also Gross, *Being-Moved*, 12–15.

10. Catherine Clément, *Opera or the Undoing of Women*, trans. Betsy Wing (Minneapolis: University of Minnesota Press, 1988).

11. Paul Robinson, review of *Opera or the Undoing of Women* by Catherine Clément, *New York Times*, January 1, 1989. Already in 1996, Roger Parker expressed a weary skepticism at the "empowerment of the umpteenth prima donna," mostly via the concept of voice. See Parker, *Leonora's Last Act: Essays in Verdian Discourse* (Princeton: Princeton University Press, 1997), 6. Marcus Pyle provides the most recent example in opera scholarship, arguing that the operatic femme fatale (Carmen and Salomé are his examples) "uses her voice to exploit the symbolic boundaries of authorial agency." See Pyle, "The Rhetoric of Seduction, or Materiality Under Erasure," *Nineteenth-Century Music* 41, no. 3 (2020), 194–208, here 195.

12. Lindon Barrett, *Blackness and Value: Seeing Double* (Cambridge: Cambridge University Press, 1999).

13. John Cage cited in Frances Dyson, "The Ear That Would Hear Sounds in Themselves: John Cage, 1935 –1965," in *Wireless Imagination: Sound, Radio, and the Avant-Garde*, ed. Douglas Kahn and Gregory Whitehead (Cambridge, MA: MIT Press, 1992), 386. Edmund Mendelssohn observes that the expansive Cagean conception of listening ultimately turns every sound into "a living voice"—a voice that must, therefore, in Cage's liberal ethic of sonic inclusion, be heard. See Mendelssohn, *White Music Mythologies: Sonic Presence in Modernism* (Stanford: Stanford University Press, 2023), 181.

14. Douglas Kahn, *Noise, Water, Meat: A History of Sound in the Arts* (Cambridge, MA: MIT Press, 1999); Pauline Oliveros cited at the Center for Deep Listening: https://www.deeplistening.rpi.edu/deep-listening/pauline-oliveros/.

15. https://www.deeplistening.rpi.edu/deep-listening/.

16. Recently published as Ruth Anderson and Annea Lockwood, *Hearing Studies* (Red Hook, NY: Open Space, 2021). Quotation on p. 1 of the manuscript.

17. See, for instance, Matthew Riley, *Musical Listening in the German Enlightenment: Attention, Wonder and Astonishment* (Aldershot: Ashgate, 2004).

18. Gary Tomlinson, *Music in Renaissance Magic: Toward a Historiography of Others* (Chicago: University of Chicago Press, 1993), 43. See also Clifford Geertz, *The Interpretation of Cultures* (New York: Basic Books, 1973).

19. Stephen Greenblatt, *Shakespearean Negotiations: The Circulation of Social Energy in Renaissance England* (Berkeley: University of California Press, 1988), 1.

20. Suzanne Cusick and Emily Wilbourne, Introduction to *Acoustemologies in Contact: Sounding Subjects and Modes of Listening in Early Modernity* (Cambridge: OpenBook, 2021), 9, 4. See also J. Martin Daughtry, *Listening to War: Sound, Music, Trauma, and Survival in Wartime Iraq* (New York: Oxford University Press, 2015).

21. On the changing politics of difference in music scholarship in particular, see the compendious and insightful introduction by Olivia Bloechl, with Melanie Lowe, to their edited volume *Rethinking Difference in Music Scholarship* (Cambridge: Cambridge University Press, 2014), 1–52.

22. On Crotch's status in elite European musical culture, see Howard Irving, *Ancients and Moderns: William Crotch and the Development of Classical Music* (Aldershot: Ashgate, 1999).

23. Cited in Jamie C. Kassler, "The Royal Institution Music Lectures, 1800–1831: A Preliminary Study," *Royal Musical Association Research Chronicle* 19, no. 1 (1985), 1–30, here 4.

24. Kassler, "Royal Institution Music Lectures," 14–21.

25. Desmond Sheehan, *Sacred Harmonies: Music and Religion in Berlin, 1760–1840* (PhD diss., University of California, Berkeley, 2021), Chapter 4. To be as clear as I can: In sketching these British and Prussian institutional histories, I am proposing that the ideal of "good listening" that music historians might associate with the formalist theories of A. B. Marx or the early classicism of Crotch is not easily separated from the kinds of cross-cultural "good listening" that scholars more usually situate in the history of musical comparativism. We will see in the work of Crotch how these modes of "good listening" overlapped. In the Prussian case, which I don't consider here, Marxian "good listening" could be placed in a tradition that includes the major figures of musical anthropology such as Carl Stumpf.

26. Charles Burney, *A General History of Music*, vol. 1 (London, 1776), 406.

27. See Irving, *Ancients and Moderns*, 38–40.

28. Irving, *Ancients and Moderns*, 193.

29. Like Burney's history, volume one of Gibbon's *Decline and Fall* was published in 1776.

30. William Crotch, Preface, *Specimens of Various Styles of Music*, vol. 1 (London, 1807), 2.

31. Uvedale Price, *An Essay on the Picturesque as Compared with the Sublime and the Beautiful* (London, 1794).

32. Howard Irving, "Empiricism, Ideology and William Crotch's *Specimens*," *Nineteenth-Century Music Review* 9, no. 2 (2012), 237–53.

33. Rachel Mundy, "Evolutionary Categories and Musical Style from Adler to America," *Journal of the American Musicological Society* 67, no. 3 (2014), 735–68.

34. Crotch, Preface, 1.

35. William Crotch, *Substance of Several Courses of Lectures on Music* (London: Longman et al., 1831), 15.

36. Crotch, *Substance*, 14–15.

37. Samuel Johnson, *A Dictionary of the English Language*, 2 vols. (London: Knapton et al., 1755) s.v. "discriminate."

38. Dylan Robinson, *Hungry Listening: Resonant Theory for Indigenous Sound Studies* (Minneapolis: University of Minnesota Press, 2020).

39. Kahn, *Noise, Water, Meat*, 192.

40. Pauline Oliveros, *Quantum Listening* (Ignota Books, 2022), 33.

41. Crotch, Preface, 1.

42. Crotch, Preface, 15.

43. Ian Woodfield, "Collecting Indian Songs in Late 18th-Century Lucknow: Problems of Transcription," *British Journal of Ethnomusicology* 3, no. 1 (1994), 73–88.

44. For a discussion of these dynamics in the case of Hindostannie airs, see Nicholas Cook, "Encountering the Other, Redefining the Self: Hindostannie Airs, Haydn's Folk Song Settings and the 'Common Practice' Style," in *Music and Orientalism in the British Empire, 1780s–1940s: Portrayal of the East*, ed. Martin Clayton and Bennett Zon (Aldershot: Ashgate, 2007), 13–37; and Ian Woodfield, "The 'Hindostannie Air': English Attempts to Understand Indian Music in the Late Eighteenth Century," *Journal of the Royal Musical Association* 119, no. 2 (1994), 189–211.

45. Cited in Woodfield, "The 'Hindostannie Air'," 192.

46. https://scroll.in/article/884483/when-an-englishwoman-visited-mughal-courts-and-wrote-their-music-in-european-notation.

47. Glen Coulthard, *Red Skin, White Masks: Rejecting the Colonial Politics of Recognition* (Minneapolis: University of Minnesota Press, 2014), 17 (italics in original).

48. Zhuqing Lester Hu, "Chinese Ears, Delicate or Dull? Toward a Decolonial Comparativism," *Journal of the American Musicological Society* 74, no. 3 (2021), 501–569, here 559.

49. David Garneau, "Imaginary Spaces of Conciliation and Reconciliation," *West Coast Line* (Summer 2013), 28–38, here 29.

50. Édouard Glissant, *Poetics of Relation*, trans. Betsy Wing (Ann Arbor: University of Michigan Press, 1997), 120.

51. Emily Apter, *Against World Literature: On the Politics of Untranslatability* (New York: Verso, 2013).

52. Olivia Bloechl, *Native American Song at the Frontiers of Early Modern Music* (Cambridge: Cambridge University Press, 2008), 10–11.

53. Even Glissant, consistent with the linguistic concerns of contemporary Francophone philosophy, occasionally seems to treat the political challenge of Creole cultures as continuous with the interpretative challenge of the modernist text; see, for instance, Glissant, *Poetics of Relation*, 113–116.

54. Gary Tomlinson, *The Singing of the New World: Indigenous Voice in the Era of European Contact* (Cambridge: Cambridge University Press, 2009), 5; Tomlinson, *A Million Years of Music: The Emergence of Human Modernity* (New York: Zone Books, 2015), 26.

55. Glenda Goodman, "Sound Heard, Meaning Deferred: Music Transcription as Imperial Technology in the Eighteenth Century," *Eighteenth-Century Studies* 51, no. 1 (2018), 39–45, here 44.

56. Brian Hochman, *Savage Preservation: The Ethnographic Origins of Modern Media Technology* (Minneapolis: University of Minnesota Press, 2014).

57. The anthropologist most closely associated with the ontological turn is Eduardo Viveiros de Castro; see, for instance, the essays collected as Castro, *The Relative Native* (Chicago: University of Chicago Press, 2015).

58. Ana María Ochoa Gautier, *Aurality: Listening and Knowledge in Nineteenth-Century Colombia* (Durham, NC: Duke University Press, 2014).

59. Greenblatt, *Shakespearean Negotiations*, 1.

60. In music studies, the major text remains Nina Eidsheim, *Sensing Sound: Singing and Listening as Vibrational Practice* (Durham, NC: Duke University Press, 2015).

61. Cusick and Wilbourne, "Introduction," 1–2; Brian House, "Against Listening," *Contemporary Music Review* 36/3 (2017), 157–170, here 160.

62. On distributed cognition, see Edwin Hutchins, *Cognition in the Wild* (Cambridge, MA: MIT Press, 1996); on the extended mind, see Andy Clark and David Chalmers, "The Extended Mind," *Analysis* 58, no. 1 (1998), 7–19.

63. For a robust critique of vibrational ontologies in music studies as "vibrational monism," see Brian Kane, "Do Vibrations Make Decisions?" *Kalfou* 9, no. 2 (2022), 318–327.

64. Another musical practice that has given aesthetic shape to vibrational ontologies, making them appear more plausible, is EDM (and the associated technology of speakers and subwoofers). Prominent contributions to vibrational sound studies by DJs include Steve Goodman, *Sonic Warfare: Sound, Affect, and the Ecology of Fear* (Cambridge, MA: The MIT Press, 2010) and Luis Manuel Garcia-Mispireta, *Together, Somehow: Music, Affect, and Intimacy on the Dancefloor* (Durham, NC: Duke University Press, 2023).

65. William Wordsworth, *The Prelude: 1799, 1805, 1850*, ed. Jonathan Wordsworth, M. H. Abrams, and Stephen Gill (New York: W. W. Norton, 1979), 47.

66. Anna Tsing, *The Mushroom at the End of the World: On the Possibility of Life in Capitalist Ruins* (Princeton: Princeton University Press, 2015), 24, see 17–25 for the "arts of noticing."

67. David Abram, *The Spell of the Sensuous* (New York: Vintage, 1996), 139.

68. Jacob Smith, *Eco-Sonic Media* (Oakland: University of California Press, 2015), 38.

69. George Boulukos, *The Grateful Slave: The Emergence of Race in Eighteenth-Century British and American Culture* (Cambridge: Cambridge University Press, 2008); Saidyia Hartman, *Scenes of Subjection: Terror, Slavery, and Self-Making in Nineteenth-Century America* (New York: Oxford University Press, 1997).

70. Lynn Festa, *Sentimental Figures of Empire in Eighteenth-Century Britain and France* (Baltimore: Johns Hopkins University Press, 2006), 54.

71. See Brycchan Carey, *British Abolitionism and the Rhetoric of Sensibility: Writing, Sentiment, and Slavery, 1760–1807* (New York: Palgrave Macmillan, 2005), 37–9, 72–4, et passim.

72. See the essays in Walter Benn Michaels and Adolph Reed Jr., *No Politics But Class Politics* (London: Eris Press, 2022), esp. 101–114; also Benn Michaels, "The Beauty of a Social Problem (e.g. Unemployment)," *Twentieth-Century Literature* 57, nos. 3–4 (2011), 309–27, here 316–17, et passim. Their critique is to an extent indebted to the analysis by Barbara and Karen Fields in *Racecraft: The Soul of Inequality in American Life* (London: Verso, 2012), esp. chapters 4 and 5.

73. Olúfẹ́mi Táíwò, *Elite Capture: How the Powerful Took Over Identity Politics (And Everything Else)* (Chicago: Haymarket Books, 2022), 74.

74. Stuart Hall, "Cultural Studies and Its Theoretical Legacies [1992]," in *Stuart Hall: Essential Essays*, vol. 1 (Durham, NC: Duke University Press, 2019), 84.

75. Robinson, *Hungry Listening*, 24, et passim.

76. Barbara Johnson, "Muteness Envy [1995]," reprinted in *The Barbara Johnson Reader: The Surprise of Otherness*, ed. Melissa Feuerstein, Bill Johnson González, Lili Porten, and Keja Valens (Durham, NC: Duke University Press, 2014), 200–216, here 213.

77. See https://www.culturalequity.org/resources/gjb.

78. Táíwò, *Elite Capture*, 75.

79. Relatedly, Gavin Williams points out in his review of the volume *Audible Empire: Music, Global Politics, Critique*, ed. Ronald Radano and Tejumola Olaniyan (Durham, NC: Duke University Press, 2016) that oppressive structures such as empire need not always be audible at all, and that it is by and large scholars of music and sound who would like them to be. See Williams, Review of *Audible Empire*, *Twentieth-Century Music* 15, no. 1 (2008), 137–143, here 142.

CHAPTER 5

Listening, Technique, and Attention

Brian Kane

For Cat Slowik[1]

Here is an innocuous sentence:

> Similar to how in nineteenth-century medicine the stethoscope introduced new techniques of listening and precisely thereby produced new ways of understanding the human body, Stockhausen employed his microphones to enable new modes of listening and unlock otherwise unheard vibrations for compositional processes. (Lutz Koepnick, 2020)[2]

As a scholar of sound studies, I have a few shelves in my office full of books on sound, sound art, listening, sound media, audio technology, and auditory culture. I would wager that a large percentage of those books include an innocuous sentence of this variety. Perhaps the authors do not mention Stockhausen—maybe they mention some other composer, or sound artist, or even none at all—but they likely assert something about the intertwined historical development of sound technologies and listening. And when it comes to listening, they probably do exactly what this author does, which is to use the terms "techniques of listening" and "modes of listening" interchangeably.

I cannot reproach anyone for doing so since there has been little critical reflection in sound studies concerning what differentiates a mode of listening from a technique of listening. Yet I have a nagging feeling that there is something at stake in the distinction; in particular, I feel that there is something crucial to a technique that would distinguish it from a mode, orientation, or disposition of listening.

To develop this intuition, I will focus much of this chapter on Jonathan Sterne's account of "modern audile technique," as presented in *The Audible Past* (2003) and in other contemporaneous writings. I do so because Sterne has written the most extensive and detailed account of what it means for listening to be a technique. While I will be rehearsing aspects of Sterne's argument, I do so to articulate some reasons why his account of technique—as good as it is and as much as I admire it—derails on issues of embodiment, practical knowledge, and action. As I argue, Sterne's commitment to technique as embodied, practical knowledge leaves him inattentive to the role of attention in technique, and thus causes him to understate the complex ways that techniques are amalgams of both pure and practical knowledge. An account of technique that does not draw *attention to attention* cannot, I argue, be sufficient.

* * *

In *The Audible Past*, Sterne describes "modern audile technique" as a set of orientations toward listening that cohered in the beginning of the nineteenth century around the work of René Laennec (1781–1826), French physician and inventor of the stethoscope. The technique then developed and diffused, helping to establish "modern audile technique," which Sterne describes as a set of orientations toward listening to the communications technologies of the telegraph, phonograph, and other audio media that emerged in the latter half of the nineteenth and early twentieth centuries.[3]

Before discussing the specific auditory orientations that comprise "modern audile technique," I want to situate Stern's thinking about the nature of technique within a broader intellectual-historical framework. In a short essay published in 2006, entitled "Communication as *Technē*," Sterne notes that the ancient Greek word *technē* is the root for "two terms that designate some of the most important aspects of communication: *technique* and *technology*." The knowledge that stems from *technē* is "embodied knowledge, not formal or logical knowledge," that is, it is the kind of knowledge found in skills and abilities, the capacities that humans employ, cultivate, and exploit in their everyday actions. *Technē* designates "a practical art and practical knowledge."[4] For Sterne, *technē* is a matter of know-how. By making this claim, Sterne is also invoking the longer tradition of contrasting *technē* with *epistēmē*, which denotes an explicit form of knowledge or "knowing-that."[5]

By defining technique and technology within the conceptual frame of the distinction between *technē* and *epistēmē*, Sterne focuses on the difference between practical and theoretical knowledge. To illustrate, he offers the example of the musician. Unlike knowing the correct number of flats in the key of G-flat—an example of theoretical knowledge, or *epistēmē*—the musician's fluency with an instrument is a form of embodied, practical knowledge. As such, it requires training and habituation. In *The Audible Past*, Sterne builds upon the work of French sociologist and anthropologist Marcel Mauss, especially his celebrated essay on "Techniques of the Body," to establish the groundwork for technique as a form of embodied, practical knowledge. Mauss considers practices like swimming, diving, sitting, and walking in order to observe how participants within different cultures, or even different generations within the same culture, train and condition their bodies.[6] Sterne is interested in Mauss's ideas as a way of accounting for the propagation of techniques, and how skillful users of their own bodies become habituated to sequences of actions and gestures such that that they recede to the level of tacit practical knowledge. These techniques are imprinted on the body through the manifold forces of social pressure, norms, and custom.

In addition, Sterne also relies on the French sociologist Pierre Bourdieu, who develops Mauss's ideas on body techniques and their role in the *habitus*. In a passage cited by Sterne, Bourdieu writes, "One might say that arms and legs are full of numb imperatives," full of "practical logics" that don't follow the rules of formal reasoning but have an embodied logic.[7] These "numb imperatives" are also *dumb imperatives* in the sense that *technē*, as know-how, is inarticulate; it is a form of knowledge inscribed so completely into the impressible flesh that it becomes a second skin or second nature, a tacit and habitual protocol of gestures and actions.

By focusing on *technē* as embodied knowledge, Sterne aims to "to develop an account of communication as a social phenomenon . . . [to] examine what people actually do when they are communicating—not what they *say* they are doing or what they *think* they are doing, but what they do."[8] Everyday communicative exchange is thus a kind of art, skill, or practical knowledge in which we are all performers. This holds not just for techniques of communication but for communication technologies as well. Technologies such as the telegraph or the phonograph are understood as concretizations of practical knowledge. Thus, "Machines and technological systems are an extension of the logic of . . . practical knowledge, and realized action hidden in *technē*, because they are essentially crystallized sets of repeatable activities and relationships." It follows that techniques and technologies, although sharing an etymological root, are not equivalent. Techniques *precede* technologies because "technologies are crystallized bits of practical art and practical reason—they are techniques externalized and delegated to machines."[9]

Sterne makes this argument throughout *The Audible Past* and provides an extended version of it in his book *MP3: The Meaning of a Format* (2012). Because technologies are crystallized techniques, Sterne's account of the origins of an audio technology is rarely a story about genius inventors, but rather an account of the practices (or techniques) that concretize into a technological object. For instance, his chapter on the phonograph presents Alexander Graham Bell and Francis Blake's grisly ear-phonautograph—that sound writing machine that incorporates the ear of a human cadaver into its machinery—in order to investigate the various social forces *that lead to* the incorporation of a human ear into such a machine. Sterne brings together a set of independent cultural practices, conceptualizations, scientific discoveries, and social aspirations. The factors at work are many: the interest in visualizing or writing sound for the purposes of deaf education, linguistics, and poetics; the rise of otology as a medical discipline; the increasing knowledge of the physiology of the ear, which is tied to practices of autopsy and scientific instrumentation; the production of a scientific view about the discreteness of the senses and the mechanical nature of the ear's operation, which permitted a conceptualization of the ear (and particularly the eardrum) as a specific type of transducer. These various factors, each in a sense independent and detachable from the rest, are articulated together and legibly inscribed in the phonograph. Sterne's analysis of the phonograph and its process of crystallization enables him to develop a more general theory of technology, where technologies are best understood as the result of techniques that precede them.

The same argument holds for constitution of techniques, but on a different register from technologies. Just as a set of techniques crystallize into a novel technology, Sterne sees techniques as themselves crystallizations of practices (or combinations of practices) that precede them. The most extended treatment of technique appears in the two chapters of *The Audile Past* dedicated to an analysis of "modern audile technique." Sterne writes,

> Listening involves will, both conscious and unconscious—perhaps a better word than *will* would be *disposition* or *feel*.[10] Orientations toward and styles of listening are part of what sociologists and anthropologists have come to call *the habitus*. Following Pierre

> Bourdieu, *habitus* denotes a set of dispositions, what he calls *a feel for the game*. The *habitus* is socially conditioned subjectivity . . . a mix of custom, bodily technique, social outlook, style, and orientation. . . . Modern audile technique, combines a relatively stable set of practical orientations toward sound and listening.[11]

In particular, Sterne identifies six orientations, common to "medicine, telegraphy, and . . . sound-reproduction technologies," as follows:[12]

> 1) Listening becomes a scientific skill, associated with rational and scientific practices of using the senses.
> 2) Listening is separated from other senses, and constructed as a discrete activity, a distinct faculty of hearing.
> 3) Listening, now separated, reconstructs acoustic space.
> 4) Listening no longer focuses on the privileged cases of voice or music; rather, it attends to these as instances of a more general category of sound.
> 5) Listening now occurs in mediated contexts, through stethoscopes, listening tubes, and so forth, whereby distant sounds become proximal.
> 6) Finally, listening gains symbolic currency; it becomes a form of specialization, which can signal to others that the listener possesses knowledge, skill, or social status.

As I understand them, these orientations are not to be taken as sufficient or necessary conditions of modern audile technique; rather, modern audile technique can be understood as the "coming together" of these various orientations in varying degrees. They are intended to be broad enough to appear repeatedly in many distinct contexts, yet also be specific enough to identify what is particularly modern about "modern audile technique," as opposed to some other modality of audile technique or even audile technique in general.

* * *

Across "Communication as Technē" and *The Audible Past*, Sterne remains consistent in his views: In the former, he explicitly theorizes technique within the *technē/epistēmē* distinction, while in the latter he assumes this framework to explicate audile technique. My aim is not to challenge the various orientations that Sterne articulates as central to this technique, but rather to consider how it, as a whole, relates to the issues of pure and practical knowledge introduced in "Communication as Technē." What is the status of modern audile technique vis-à-vis pure and practical knowledge?

On first glance, the answer to the question would seem obvious: Modern audile technique is a form of practical knowledge, a kind of know-how possessed by modern listeners, or a kind of embodied knowledge of the sort that Sterne sees in everyday actions. As a combination of "a relatively stable set of **practical** orientations toward sound and listening," modern audile technique is a matter of training and cultivation, a way of culturally forming habitual dispositions that shape the "feel for the game" of listening.[13] This is why Sterne invokes Bourdieu in the passage cited above. Bourdieu often uses the metaphor of a "feel for the game" to describe the mastery of the immanent rules of a game by those who play it. This is a "mastery acquired by experience of the game" and, importantly, "one which works outside conscious control and discourse (in the way that, for instance, techniques of the body do)."[14] Modern audile technique is not a self-conscious construct that historical listeners employed knowingly, explicitly, and intentionally. Rather, it is a descriptive framework employed by the historian to capture what modern listeners actually did—or so Sterne argues.[15]

The individual listening techniques that cohere into "modern audile techniques" are also theorized as akin to body technique or an interlaced set of body techniques. Sterne writes, "My use of the word *technique* in relation to listening is derived from Marcel Mauss's notion of 'techniques of the body.'" Mauss's own explication of techniques of the body focused on bodily actions and ways of manipulating them, not—as Sterne himself notes—on sensory or perceptual actions: "Although [Mauss] does not include sensory activities—looking, listening, tasting, smelling, touching—these are certainly implied and even occasionally mentioned in the context of other techniques. So my argument makes a very short leap from Mauss's list of techniques to a history of techniques of listening in modernity."[16]

Let us consider what is involved in the leap from body techniques—like walking, running, sitting, and diving—to "sensory activities," or what I will call "perceptual techniques." From the perspective of training, the leap may indeed be small in that both body techniques and perceptual techniques must be produced through training and pedagogy (whether formal or informal, institutional or cultural). Repetition and practice are key components of training. Both body techniques and perceptual techniques are predicated on the acquisition of some particular skill. In that respect, we might reasonably imagine body techniques and perceptual techniques as located on a single continuum, differing only by degrees.

However, there are differences of kind that we cannot ignore. One key difference concerns the evidence used to judge whether a technique has been successfully acquired. While the incorporation of a body technique like Mauss's example of diving is legible from viewing the body of the diver in action—and thus on display for all to see—the same cannot be said for perceptual techniques. For instance, how does a teacher or evaluator know that a medical student learning mediate auscultation has in fact correctly assimilated the technique? There is no immediate evidence of the student's perception, as there is for the diver's action. Rather, evidence is available only through the transformation (or mediation) of perceptions into actions—in other words, through the production of traces. In the case of modern-day mediate auscultation, the medical student is evaluated through testing and clinical apprenticeship; the student's public diagnostic judgments are used as signs by which to judge whether the perceptual technique has or has not been correctly incorporated.

My qualm is not with the idea that people learn perceptual techniques—because clearly they do. Rather, my qualm is with too hastily equating body techniques with perceptual techniques. I think that perceptual techniques are interesting and important *precisely because their mode of evaluation is different from body techniques*. Perceptual techniques require the production of traces subsequent to the act of perception in order to be evaluated. These traces exteriorize—through action, language, symbols, or signs—the presence of perceptual techniques in mediated forms. There will always remain a potential indeterminacy about whether *my* perceptual technique and *your* perceptual technique are in fact *the same technique*, even if they produce identical or nearly identical traces. This raises a challenge for any cultural theorist who wants to rely on perceptual techniques as the glue that binds together a culture of perceivers. If the evaluation of such techniques necessarily contains a moment of indeterminacy, how can one maintain as a consistent object of study something like modern audile technique (in the singular), which would rely on the stability and wide diffusion of the perceptual techniques it comprises? For once we scratch at the surface, how much variation and difference will be found?

To push this argument further, we might also consider whether body techniques are not also subject to a similar issue of mediation. There is an unfortunate tendency in Anglophone scholarship to detach Mauss's essay on the techniques of the body from the rest of his work on techniques and technology, and to read it through the influence of Pierre Bourdieu's reception of Mauss.[17] Body techniques begin to look like the site where the coercive power of culture impresses itself into the flesh, forming disciplined subjects whose bodies are the pages on which are written the annals of their domination. The connection between body techniques and techniques in general is lost, as is Mauss's understanding of techniques as complex indices of the condition of a society or of a civilization.

Moreover, when body techniques are reduced to "numb imperatives" that forcefully and tacitly constrain the body's actions, they miss the full valence of Mauss's specific interest in them. As Carrie Noland argues, Mauss includes the possibility of moments of resistance to such discipline in his own theorization of body techniques. Noland, in her thorough reading of Mauss's "Techniques of the Body" essay, makes this point by homing in on a moment when Mauss discusses a particular body technique of walking.[18] Borrowing from Elsdon Best's ethnography of the Maori, Mauss focuses on a body technique called "doing the *onioni.*" Mauss, citing Best, says that Maori women "adopted a gait that was acquired in youth, a loose-jointed swinging of the hips that looks ungainly to us, but was admired by the Maori. Mothers drilled their daughters in this accomplishment, termed *onioni*, and I have heard a mother say to her girl: 'Ha! Kaore koe e onioni' [you are not doing the *onioni*] when the young one was neglecting to practice the gait."[19]

While one can read this as the moment in which the daughter is being disciplined into the *habitus*, it can also be interpreted as a moment when the daughter is resisting such discipline. For Noland, moments of resistance can be felt at the kinesthetic level of apperception—where the daughter both knows how the *onioni* is supposed to feel and yet also enjoys the feeling of resisting it, of stylizing it, of altering it. And those feelings of resistance might transform how we understand body techniques in general, opening up knots of difference within the *habitus* and its totalizing social formation of subjects.

The story of the Maori daughter is helpful for considering the ways that the use of the body within the *habitus* need not conform wholly to the categorical distinction between pure and practical knowledge. If the daughter both knows how to conform her body to the norms as well as to resist those norms, then we have a situation where action in accord with the *habitus* does not necessarily lead to tacit, silent habituation. The daughter both has a "feel for the game" of doing the *onioni* at the same time that she can distance herself from it in order to project her body's movements along divergent paths. If the daughter both knows how the *onioni* is supposed to feel and yet also enjoys the feeling of doing it differently, then she also has the potential to make the difference between those two explicit.

Mauss defines a technique as "an ensemble of movements or actions, in general and for the most part manual, which are organized and traditional, and which work together towards the achievement of a goal known to be physical or chemical or organic."[20] This definition weaves together a few distinct threads. The sequence of actions that defines a technique is not random or haphazard, but organized according to its effectiveness. Techniques are acquired through teaching, passed on from to an apprentice from a master, or learned through imitation. Techniques are also telic in that they involve a moment of prolepsis as the technique is geared toward its goal. That goal is "known" in that the actor is explicitly aware of the end to which a technique is aimed. A technique is an effective action, learned from others, that has an explicit purpose.

The possibility of the articulation of technique disappears when technique is understood only in terms of habitual, "practical" knowledge. *Technē*, as characterized by Sterne, is tacit and skillful but unable to articulate itself, to know itself, to theorize itself; it requires the silence of drastic, effective action and risks failure when forced into the rational, reasonable, gnostic light of explication. In contrast, Noland's reading of Mauss—which builds on Mauss's own understanding of technique—places emphasis on the moment of kinesthetic resistance in order to show how the girl's conscious toggling between techniques reveals the *explicitness* of technical knowledge. For the Maori girl knows both how to do the *onioni* and how not to do the *onioni*. Thus she has a knowledge that is not simply tacit, but both habitual *and* volitional, felt *and* understood, silent *and* articulate. This is a kind of knowledge that is neither wholly pure nor wholly practical, but through resistance and toggling, a kind of practically pure knowledge or impurely practical knowledge.

* * *

I would argue that knowledge of this sort—neither wholly pure nor wholly practical—is characteristic of techniques in general. In the Greek context where the word *technē* originates, the person who possesses a technique (the *technitēs*) is not the same as a person who can simply perform an action.[21] The *technitēs* not only knows how to perform some action, but also knows the reasons why they are performing that action. Technique involves the ability to project into the future the use of a skill, the performance of an action, or the production of an object, in line with a reason or purpose. In other words, a technique is applied for the sake of some reason, aim, or goal. We can assert that technique always involves a moment of prolepsis, a projection of a project into the future, which then fixes the goal toward which an action is aimed. This is one of the reasons why *technē* is a mode of knowledge and more than just an act of habit.[22]

Although we have Mauss to thank for the restoration of the word *habitus*, he uses it in a sense that is quite different from the habitual, unconscious, socially shaped actions that constitute Bourdieu's *habitus*. When introducing the Latin word *habitus* to his Francophone audience, Mauss writes,

> The word [*habitus*] translates infinitely better than "*habitude*" (habit or custom), the "exis," the "acquired ability" and "faculty" of Aristotle (who was a psychologist). . . . These 'habits' do not just vary with individuals and their imitations, they vary especially between societies, educations, proprieties and fashions, prestige. In them we should see the techniques and work of collective and individual practical reason rather than, in the ordinary way, merely the soul and its repetitive faculties.[23]

If techniques involve ensembles of organized, traditional actions, such actions constitute a Maussian "habitus" that is irreducible to a series of habituated, repetitive gestures. This is why Mauss is at pains to distinguish the *habitus* from "habit" or merely repetitive, habitual actions. Where the latter are unconscious acts, the former are goal-oriented instances of effective knowledge. While techniques might be learned through processes of imitation, training, or apprenticeship—insofar as "the individual borrows the series of movements which constitute it [i.e., a technique] from the action executed in front of him or with him by others"—imitation should not be understood as slavish, nor should training be understood as disciplining.[24] In learning a technique, the student is also engaging an act of "borrowing," a central term in Mauss's theory of technique.

Borrowing is one of the ways that cultures intermix and develop; techniques are "eminently liable to borrowing," in that they can be learned and then transported from one milieu to another. Moreover, borrowing can happen on levels from the individual to the social. Recall that the practices and techniques found in the Maussian *habitus* "do not just vary with individuals and their imitations . . . , they vary especially between societies, educations, proprieties and fashions, prestige."[25] Techniques exhibit multiple features of a culture or a society all at once, "to the point of characterizing it, indeed of standing for it, so to say, like a sign. Nothing manifests the difference between two social traditions than the difference . . . between the implements and the crafts of two societies."[26] But as cultures interact, they borrow techniques from one another, hybridizing and transforming them even as they are reproduced.[27] "Techniques," Mauss argues, "are the means . . . which a society possesses to act upon its milieu."[28] Through technique, then,

> the human actor identifies himself with the mechanical, physical and chemical order of things. He creates and at the same time he creates himself; he creates at once his means of living, things purely human, and his thought inscribed in these things. Here is true practical reason being elaborated.[29]

One would be mistaken to read this invocation of practical reason in the way that Bourdieu does. Practical reason, for Mauss, is not the same as Bourdieu's unconscious "practical logics," nor is Mauss's habitus the same as Bourdieu's habitus. It is a kind of reason that is applied, effective, and explicit. What ultimately distinguishes Mauss from Bourdieu on the question of techniques and the *habitus* can be boiled down to the status of the word "practical." Mauss's thinking about technique and its relation to modes of knowledge was influenced not only by his anthropological and sociological research, but also by his understanding of the ancient Greek discourses on *technē* and their influence on his contemporaries. Addressing the role of techniques in art and in science, Mauss writes, "When the arts and the sciences and their historical relations are studied concretely, the division between pure reason and practical reason seems scholastic, scarcely realistic, barely psychological, and even less sociological. . . . Indeed, it might be better . . . not to distinguish *techné* and *epistémé*."[30]

An unfortunate consequence of the Anglophone reading of Mauss is that body techniques have become identified with tacit, silent, habituated action. What this leads to is a strange situation where Mauss's readers eschew the difference between the incorporation of a technique and virtuosity. If body techniques are simply the ways that a culture disciplines the uses of bodies in conformity with its norms, then it would appear that everyone who uses their bodies in conformity with these norms would also be something of a somnambulant virtuoso. Tautologically, we are all virtuosi of the ways that we have been trained to use our bodies. However, the difference between being a virtuoso and a *technitēs* is that the *technitēs* can articulate the reasons for and ways that they employ a technique. They can teach it, they tell others why they are doing it, and they can anticipate its outcome. The virtuoso only needs experience but does not need self-awareness. By relying too heavily on the Anglophone interpretation of body techniques, one remains inattentive to the difference between technique and virtuosity. Techniques become silent habits, rather than an articulate, explicit relation of pure and practical knowledge.

By casting techniques as a form of "practical knowledge" in the narrow sense—as know-how distinct from knowing-that—Sterne reifies the categorial distinction between pure and practical knowledge and downplays the role of attention in technique. *There is no attention to attention*. The idea that actors in the flow of action function unconsciously, tacitly, and habitually (i.e., virtuosically) shunts attention to the side. It is as if to attend to phenomena within the midst of action is to break one's immersive and effective relation with the world and to force a kind of impractical syncopation in one's flow.

My argument is not that a person in possession of technique never enters into flow states or into sequences of habituated action below the level of tacit awareness and attention. There is no conflict between being a virtuoso and possessing technique, but there is a difference. The *technitēs* has the ability to attend to their actions and make moments of attention explicit through acts of mediation and exteriorization. *They both know what they are doing and they know how to do it.* The *technitēs* may not think of all the things they do when they are in the flow of action, but if there is a technique involved, then it can be made explicit, articulate, and exteriorized. One marker of the succession from virtuoso to *technitēs* is that the latter can *attend to attention*. To put it concisely, audile techniques are not reducible to the orientations and dispositions comprised by the *habitus* of listening; rather, they are conscious, intentional, explicit engagements with (and within) the *habitus*.[31]

If technique is an alloy of know-how and knowing-that—a braiding of *technē* and *epistēmē* rather than their strict separation—then it would support a surprising conclusion: that "modern audile technique" as described by Sterne fails to be a technique at all. It would not cross the line that distinguishes technique from habituated action, no matter how virtuosic such habituated action might be. For it to actually attain the status of technique, audile technique would have to avoid reifying the categorical distinction between pure and practical knowledge, or knowing-that and knowing-how. My suggestion is that the relation of *technē* and *epistēmē* should be thought of as *practically pure knowledge* or *impurely practical knowledge, where attention is given to attention.*[32]

* * *

To develop this suggestion, I want to consider a case study: the transition from wired to wireless telegraphy that occurred at the turn of the twentieth century. The reasons for choosing this example are twofold. First, Sterne discusses wired and wireless telegraphy under the heading of "sound telegraphy" and treats it as a central moment in his account of modern audile technique. Second, the transition from wired to wireless telegraphy offers a well-documented, archivally extensive source for reconstructing audile technique along the lines that I am suggesting—as a form of *practically pure knowledge* or *impurely practical knowledge, where attention is given to attention*. As a result, it meets the criteria for being a technique and not just an act of habituated virtuosity.

Wired telegraphy, which came to prominence in the middle of the nineteenth century, was a system of direct, point to point communication, where electric pulses were sent along long wires from station to station. Operators on a line could hear all the messages traveling up and down it, since the line formed a closed circuit. When operators needed to send a message they would break the circuit. By tapping on a "key"—a simple momentary switch—they could send pulses to all the others on the line. The pulses represented letters, numbers, and other symbols by means of a coded system of long and short durations, what we now refer to generally as Morse code. Originally the pulses sent down the line were inscribed with ink on a long thin strip of paper—by a machine called an inker—but many of the operators learned to "read" the code directly by attending to the clicking sound of the inker's relays as pulses were received. Reading messages "by ear" was more efficient and became the normal mode for most telegraph operators. Wired telegraphy became a form of communication that, for the most part, relied on sound and on the attentive capacities of its operators. To aid them, special sounders were developed to make the clicks of the code clearer, more resonant, and directional. Often these sounders were built into resonator boxes that made the clicks more distinct and directional (see Figure 5.1).

Figure 5.1. Telegraph sounder in a resonator box. (Source: from author's collection)

In contrast, wireless telegraphy, which used radio waves to send Morse code, only emerged at the end of the nineteenth century, when various scientific experiments and devices were combined into a commercially workable system by Guglielmo Marconi. Early wireless telegraphy emerged as a challenge to the monopoly of the telegraph system and first took hold in maritime communication, where the freedom from wires meant that a network of vessels could stay in communication as they traveled their routes. Powerful industrial spark gap transmitters sent wideband, damped blasts of electromagnetic energy into the "ether," which could be detected at large distances without the use of wires. Given a proper radio frequency detector and rectifier, these coded blasts could be heard using high-impedance headphones or telephone earpieces. Today, spark gaps are no longer in use. The modulated carrier wave, which is the basis of voice transmission over the air, came into use in the 1910s and 1920s.

In the transition from wired to wireless telegraphy, we have two practices that are extraordinarily close; we might even be tempted to think of them as siblings. At the pedagogical level, the manuals used for training wired telegraphers into fluency with Morse code were also used by wireless operators. Figure 5.2 is a photograph of a Marconi wireless training school in New York City, taken in 1916. It is also the image reproduced on the cover of *The Audible Past* and discussed in the body of the text. The setup and arrangement of the classroom are modeled on vocational schools for wired telegraphy from the nineteenth century. Morse code is written on the chalkboard in the back of the room. Students are listening to code being sent over their headphones and transcribing the messages on a pad. (There is even an old telegraph sounder on the second desk on the left.) Figure 5.3, from a Marconi training manual produced during World War I, displays a "key" for training operators to tap out their messages. Here, recruits on the right side of the tables tap out

Figure 5.2. Marconi training school, NYC, 1916. (Source: National Museum of American History, Washington, D.C., USA)

Figure 5.3. U.S. Army Signal Corps in telegraph training at the Pratt Institute, Brooklyn, New York during World War I. (Source: Library of Congress Prints and Photographs Division)

messages written on a printed sheet to operators listening on the left side, transcribing the messages on a pad.

Sterne reads these kinds of images as depicting the characteristic comportment and posture of "modern audile technique." In *The Audible Past*, Figure 5.2 is preceded just a few pages earlier by another illustration (Figure 5.4), published in 1890, of a wired telegraph operator working on a train. In both cases, the comportment is similar: Heads are "turned down in concentration" and the bodily dispositions represent "a separation from the environment."[33] The operator is in the pose of relaxed, but attentive listening; they are using their specialized skill to interpret the sounds they hear; they are each being "individuated" by the use of audio technological mediation to inhabit a private sphere of sounds, a new kind of acoustic space. The main difference, argues Sterne, is that the older illustration depicts solitary headphone listening (as the train operator does) compared with collective headphone listening, "alone together" (as in the Marconi school).[34] In the latter, the students attend to the same signal, but they remain atomized, individuated, and separated, possessing their own personal mediated auditory space.

Figure 5.4. Train telegraph operator, ca. 1890. (Source: *Inventors, Philip Gengembre Hubert*, New York: Scribners and Sons, 1896)

For Sterne, headphones play an important role in this story. Not only are they interpreted as a technology where the dispositions and orientations of modern audile technique are crystallized, they also serve as an icon, allowing Sterne to detect the presence of audile technique from an archive of images, not sounds. Where there are headphones, Sterne infers the presence of a new configuration of acoustic space, of this new acoustic space's availability as private property of the bourgeois listener, of the need to make new sonic phenomena into signs, and so forth.[35] It is the technological common denominator that connects mediate auscultation and "sound telegraphy"—and, consequently, it is one of the planks in Sterne's argument about how modern audile technique made its way from a specialized listening practice to the broad public.

In fact, wired telegraphers almost never used headphones except in very specific situations. Figure 5.4 depicts the use of a "secret sounder." This was a device that was invented to cope with the problem of the publicity of telegraphic communication, the fact that the clattering of the sounders could be heard (or overheard) by others, and that its messages could be stolen. Most telegraphy happened in offices and stations. In busy locales, a telegrapher's office might have many operators working the lines, and the room might be divided into compartments or partitions, each containing a sounder. In those cases, "secret sounders" offered a solution. Consisting of a very small electromagnetic clicker, a secret sounder could be wired into the telegraph circuit and then held to the ear. In some cases, "secret sounders" were manufactured in small, hand-held metal enclosures that could be placed against the ear (see Figure 5.5). In other cases they could be wired directly into the earphone cups of a headset. Replacing the flexible diaphragms that were normally installed in the earphone cups, a secret sounder limited the volume and spatial extension of the relay's click.

Figure 5.5. Secret sounder with lid on and off. (Source: from author's collection)

Where there is volume, there is publicity. Since business transactions relied on the telegraph for speed, transactions in banking and the stock market were often telegraphed. This meant that rival interests could potentially intercept messages or eavesdrop on them. There are many accounts in the archives about spies hanging around outside the window of a telegrapher's office or sitting in the front office in disguise (as a member of the public working on a telegram) to capture messages. Others might try to tap the telegraph lines with the use of inductor coils.

Messages sent on a telegraph were available to all others listening on that line, thus there was little privacy in interoffice communication between operators. The telegrapher's office was a place of persistent overhearing, not a place where the ostensible privacy of the headphone listener was on display. Although Sterne characterizes modern audile technique as promoting the privatization of sound—where a group of headphone-clad listeners reside "alone together," each in their own isolated sonic world—wired telegraphy never fit neatly into this scheme.[36] It was a modern sound technology that put listeners *together at a distance.*

Wireless telegraphy, in contrast, depended entirely on the headphone. Before vacuum tubes could amplify small audio signals, high impedance headphones were crucial for listening to the tiny currents of radio waves. Early wireless telegraphy had no loudspeakers, no mode for acoustic public address, and thus no way to be publicly overheard. Many amateur radio builders, especially children, would steal the earpieces off public telephones for their home crystal sets.

The necessity of the earpiece was just one crucial difference across the technical divide of wired and wireless telegraphy. Another difference was that wireless technology brought a wholesale change to the sound of Morse code and created problems for its intelligibility. In wired telegraphy, messages are percussed on the sounder as a series of onsets and offsets. When the telegraph key is depressed and current flows through the circuit, the electromagnet in the sounder is activated, a sounding bar is pulled sharply downward, and a click results. But the same happens when the key is released, when current stops flowing, and the sounding bar moves back up to its original position. The duration of the code is marked by the onset and offset of each letter.

In early wireless telegraphy, however, messages are sent through the ether by means of a spark gap. A "spark gap" is a high voltage transformer with two closely spaced electrodes on it. When current is sent into the transformer, it is converted into a high voltage potential at one of the electrodes. When the voltage is high enough, a spark jumps the gap from one electrode to another, producing both an audible snap and a burst of electromagnetic energy in the radio frequency spectrum. By adding a little vibrating switch to the circuit, a spark gap can be used to create a continuous, buzzy tone when a telegraph key is depressed and current flows in the circuit. The "pitch" of the buzz is dependent on the tension placed on a little spring metal switch that vibrates back and forth on the edge of the spark gap. This little switch "makes" and "breaks" the circuit continuously, and—in essence—chops up a direct current into a rapid series of pulses. The sonic result is that the code sent by spark gap is marked not by the onset and offset of each letter, but by the duration that the telegraph key is depressed.

The difference between the way that wired and wireless telegraphy mark the duration of each letter is a consequence of the distinct operational technologies upon which they were based. The difference between the sounder (with its marking of duration by onsets and offsets) and the spark gap (with its marking of duration by a continuous buzz) created a gap in the intelligibility of the codes, as electrical engineer and historian James Calvert notes:

> In radiotelegraph operation [i.e., wireless telegraphy], the code is received with an earphone as . . . a tone, not as clicks. This is true of the earlier spark-gap transmitters as well as later beat-frequency reception, which produces a clean tone. A radiotelegraph operator will not understand sounder clicks [i.e., wired telegraphy], and an operator used to the sounder will not be able to read tone code. The sounds are too far different for this, and each must be separately learned.[37]

The difference between hearing onsets and offsets versus hearing the steady state portion is not, say, perceptually analogous to seeing a visual outline versus a shaded-in portion. Although conceptually akin, they are perceptually distinct. Operators had to retrain themselves to attend to this new code. So even with the same code and the same kind of training, the same habitual acts of pressing the key, there was a gap of intelligibility between these sibling practices of wired and wireless telegraphy. This important fact is missed if we lump wired and wireless telegraphy together under the heading of "sound telegraphy."

Other "sonic" and perceptual differences abound. Instead of the distinct, percussive clunk of the sounder, the soundscape of wireless telegraphy was layered with noise: there was the noise and buzz of the spark gap transmitter itself, which sent bursts of damped waves across the frequency spectrum; but also the electromagnetic noise caused by environmental conditions such as weather, lightning, and the sun. Wired telegraphy had its own share of environmental disturbances like the increase in stray currents caused by the aurora borealis or the occasional direct lightning strike that might cause sparking or even fuse metal components together. Yet none of these were as common—or particularly sonic—as the disturbances heard over the wireless. Weather conditions would cause loud pops, hisses, or crackles, or produce broadband static that could blot out wireless reception entirely.

This phenomenon of atmospheric energy inducing currents in electrical wires has been described by Douglas Kahn under the heading of the aelectrosonic or "natural radio." One of his earwitnesses is Thomas Watson, Bell's assistant in the development of the telephone. In his autobiography, Watson describes how he used to

> spend hours at night in the laboratory listening to the many strange noises in the telephone and speculating as to their cause. One of the most common sounds was a snap, followed by a grating sound that lasted two or three seconds before it faded into silence, and another was like the chirping of a bird. My theory at the time was that the currents causing these sounds came from explosions on the sun or that they were signals from another planet. They were mystic enough to suggest the latter explanation but I never detected any regularity in them that might indicate they were intelligible signals.[38]

Reflecting on the cause of these mysterious sounds, Watson observes, "These currents were probably from the same source as the static that afflicts the modern radio. . . . I, perhaps, may claim to be the first person who ever listened to static."[39]

While the pedagogy of sending and receiving Morse code was preserved from wired to wireless telegraphy, only with wireless did operators have to develop listening techniques for segregating out noisy signals from within a field of static. To make matters even more complicated, early wireless telegraphy did not use finely tuned circuits or carrier frequencies, so the sonic field was often cluttered with interference as multiple operators tapped out code simultaneously. Not only did wireless operators need to adjust their receivers to distinguish between unintentional atmospheric noise and intentional signals, they also had to track messages against a background din of other messages. New techniques were developed for distinguishing these simultaneous audio streams.

First of all, different kinds of transmitters—such as spark gap, quenched spark gap, rotary spark gap, or Poulsen arc—produced variations in the sounds transmitted. Thus, differences in timbre or register aided in segregating one auditory stream from another. Moreover, as in wired telegraphy, different operators could be recognized by their "touch" or "hand." As Francis Arnold Collins, author of *The Wireless Man*, put it, "It is as easy to recognize an operator from his touch as you pick out a familiar voice in a crowd. . . . They sound much alike to you, but you will soon get to know a man's speed, and touch of the key, whether light, strong, or hesitating."[40]

In addition, operators soon developed attentive techniques for tuning in on particular messages and cutting out others. Collins offers a good description to his reader of the attentive work of the wireless operator in selecting messages through the manipulation of tuning capacitors and other controls. From out of the field of humming or clicking, an operator homed in on a particular message intended for him. "He bent quickly over the complicated series of dials and levers, turning the arrows this way and that. Instantly the wireless chatter became blurred and gradually died away, as one message seemed to leap up and drown out all the rest."[41] At the same time, a high wattage station in close proximity might simply mask all other signals entirely, effectively erasing the field's layered depth and differentiation for the duration of its transmission. Yet, these sonic phenomena of blurring, dying away, leaping up, and drowning out are unique to wireless telegraphy.

Finally, in addition to scanning the field for distress signals or routine messages, wireless operators also employed listening techniques akin to the diagnostic and monitory modes described in Karin Bijsterveld's work on sonic skills.[42] As operators listened to messages over the ether, they simultaneously attended to the sound of received transmissions in order to monitor the state of their radios. They had to know when their own devices were not working optimally and how to diagnose and correct the situation. In order to properly adjust their transmitters and receivers they might twist variable capacitors, finely adjust the placement of inductor coils to change their amount of coupling, or manipulate the size, placement, and orientation of their antennas.

Here we see why an account of audile technique that rests on a reified distinction between knowing-how and knowing-that is inadequate for narrating a history of listening that is something other than an articulation of regimes of listening, or a historical sociology of listening.

Fortunately, there is a huge archive of firsthand accounts, amateur wireless publications, Marconi training materials, radio-themed fiction, and other sources from which we can reconstruct the cultural and individual audibilities of wired and wireless telegraphy. In the actions of telegraphic operators we see the difference between a silent, habituated practice and what it means to possess a technique: These operators in their acts of listening, manipulating, and adjusting the apparatus are also demonstrating the relation of pure and practical knowledge that is the hallmark of technique. *They not only know what to do, they know why they are doing it.* I do not think it is an exaggeration to say that their *practically pure knowledge* or *impurely practical knowledge* of the apparatus made them into contemporary *technitēs* whose knowledge of the device's operation is evinced in the very act of listening and in the physical manipulations of the technology designed to optimize their attention. To unlock these moments we need to consider the soundscape of wireless telegraphy, to consider the sonic aspects that solicited listeners' attention, and to trace (as best we can) the inception of perceptual techniques and the acquisition of skills for developing them. In other words, we must attend to attention if we want to understand audile technique *as a technique* and not just a habit, orientation, or disposition.

Notes

1. I dedicate this essay to Cat Slowik, whose thinking about the history and meaning of technique has profoundly shaped my views. Some of the ideas presented in this paper were developed in tandem in our course "Audile Techniques," taught at Yale University in fall 2019.

2. Lutz Koepnick, *Resonant Matter: Sound, Art, and the Promise of Hospitality* (New York City: Bloomsbury Academic, 2020), 51–52.

3. Jonathan Sterne, *The Audible Past: Cultural Origins of Sound Reproduction* (Durham: Duke University Press, 2003), 93*ff.*

4. Jonathan Sterne, "Communication as Techné," in *Communication as . . . : Perspectives on Theory*, ed. Gregory J. Shepherd, Jeffrey St. John, and Ted Striphas (London: Sage Publications, 2006), 91–98 (at 91–92).

5. While it is not unusual to define *technē* in contrast to *epistēmē*, many influential accounts of technique and technology to emerge since the turn of the twentieth century—like those of Martin Heidegger, Lewis Mumford, Siegfried Gideon, and Marshall McLuhan—have relied on contrasting *technē* with a different term, *physis*. The distinction between *technē* and *physis* captures the ancient Greek distinction between artifice and nature. If *physis* describes the immanent

and internal properties of each thing—a plant's or a creature's "nature" in the sense of what it simply does and possesses by virtue of being a member of its species—*technē* is external and secondary, something that a creature employs to augment its powers but that it does not possess indigenously.

6. Marcel Mauss, "Techniques of the Body," in *Techniques, Technology and Civilisation*, ed. Nathan Schlanger (New York: Durkheim Press/Berghahn Books, 2006).

7. Pierre Bourdieu, *The Logic of Practice* (Stanford: Stanford University Press, 1990), 69, quoted in Sterne, "Communication," 93.

8. Sterne, "Communication," 93.

9. Ibid., 95.

10. Note the draining off of intentionality, will, explicit volition, and conscious decision for more and more tacit and affective states—for "disposition" and "feel." I will have more to say about this below.

11. Sterne, *The Audible Past*, 92.

12. Ibid.

13. Ibid., 91. Emphasis added.

14. Bourdieu, *Logic*, 61.

15. For an even-handed overview of Bourdieu's arguments against rational actor theory and defense of tacit, unconscious, socially shaped action, see Chapter 4 of Richard Jenkins's *Pierre Bourdieu* (Hove, East Sussex, United Kingdom, 2002).

16. Sterne, *The Audible Past*, 91.

17. The history of Anglophone reception of "Techniques of the Body" is relevant. I believe that Mauss's essay first appeared in English in the journal *Economy and Society*, 2, no. 1 (1973). The article was then reprinted in Jonathan Crary and Sanford Kwinter, eds., *Incorporations* (New York: Zone, 1992) as part of an upsurge of interest in "the body" and biopolitics that was influenced by the American reception of Foucault. Hubert Dreyfus and Paul Rabinow's book, *Michel Foucault. Beyond Structuralism and Hermeneutics* (Chicago: University of Chicago Press, 1983), was important in making Foucauldian discipline look like power impressed into habits, self-regulation, etc. This reading was shaped by Dreyfus's longstanding interest in the phenomenology of embodiment and in thinkers like Merleau-Ponty and Heidegger who rejected explicit Husserlian intentionality in favor of more embodied and less self-reflective modes of action. Jonathan Crary, the editor of *Incorporations* and co-founder of Zone Books, employed Maussian "body techniques" in his influential book *Techniques of the Observer* (Cambridge, MA: MIT Press, 1990), which links the idea of body techniques and new technologies to shifts in observational regimes. This is an amalgam of Mauss, Foucault, and Bourdieu—but it is also promoted by Bourdieu's own reception in Anglophone scholarship and his citation of "Techniques of the Body" as examples of practical logics that are hidden to consciousness. *The Audible Past* was itself modeled in many respects on Crary's *Techniques of the Observer*—it focuses on techniques of the listener and their crystallization into new technologies, to a shift in auditory regimes from pre-tympanic to tympanic, etc. One cannot hold Sterne responsible for aligning Mauss, Bourdieu, and Foucault, since (I believe) it was the norm at the time when *The Audible Past* was written. Anglophone scholarship follows Bourdieu's winnowing of Mauss. In contrast, I would point the reader to Nathan Schlanger's introduction to *Techniques, Technology and Civilization*, which puts Mauss's thinking about technology into context.

18. Carrie Noland, *Agency and Embodiment: Performing Gestures/Producing Culture* (Boston: Harvard University Press, 2009), chap. 1 (at p. 28).

19. Quoted in Mauss, "Techniques of the Body," 81.

20. Mauss, "Techniques and Technology," in *Techniques, Technology, and Civilization*, 149.

21. The artisan (*cheiortechnês*) acts without knowledge (Aristotle, in *The Metaphysics*, 981a30–b5).

22. Aristotle, in *The Metaphysics*, notes that a general sign of knowledge is the ability to teach (Aristotle 981b10). The possession of technique allows the *technitēs* to be a teacher to others, because they can articulate the reasons why a particular technique is being employed and to what end.

23. Mauss, "Techniques of the Body," 80.

24. Ibid., 81.

25. Ibid., 80.

26. Ibid., "The Divisions of Sociology," in *Techniques, Technology, and Civilization*, 52.

27. Mauss's emphasis on "borrowing" undergirded his critiques of purist racial theory—recall, he fought in World War I and died in 1950—and his optimism concerning technical development.

28. Mauss, "The Divisions of Sociology," in *Techniques, Technology, and Civilization*, 52.

29. Ibid., 53.

30. Ibid., 54.

31. From this claim, it would follow that *technē* is unequally distributed. Unlike the *habitus*, which everyone participates in, not everyone is a *technitēs*. The skills and knowledge that people possess will differ quite dramatically, even as no one will have a monopoly on all the *technai*.

32. Perhaps we might want to think about this in terms of a dialectic of concretization, in the Simondonian sense, where theoretical knowledge is put to work in practical contexts and where challenges emerging from practical contexts solicit further theoretical knowledge. But theoretical knowledge can never fully anticipate what will happen in its materialization, and materialization solicits specific questions that can be answered only under the rubrics of novel theories.

33. Sterne, *The Audible Past*, 158.

34. Ibid., 165.

35. For example, "The iconography of listening to early sound-reproduction technologies, especially around ear tubes and headphones, suggests a direct line of descent from the stethoscope and the telegraph to the telephone, phono-graph, and radio" (Ibid., 98).

36. Ibid., 163.

37. James Calvert, "Receiving by Ear." James Calvert, "The Electromagnetic Telegraph," accessed November 26, 2025, https://web.archive.org/web/20120106212148fw_/http://mysite.du.edu/~jcalvert/tel/morse/morse.htm.

38. Thomas Watson, *Exploring Life: The Autobiography of Thomas Watson, with Birth and Babyhood of the Telephone* (New York: D. Appleton, 1926), 81.

39. Ibid., 82. Aside from the astonishing claim that he may have been the first person to hear static, Watson's text is also an example of the way that something as simple as a long wire can function in two distinct ways electrically and produce radically distinct sonic phenomena. The telephone wire that Bell and Watson used to send messages during the day functions similarly to the wires used in wired telegraphy. Electrical currents, whether they be the long and short pulses of telegraphy or the varying amplitude waves of transduced voices, are introduced into a closed circuit where a long wire is being used as a transmission line. At the same time, the long wire acts—unintentionally—as an antenna. It induces stray electrical currents from

the atmosphere. By day, the wire functions like a wired telegraph, but at night it functions like a wireless telegraph. The wired telegraph is a closed system where signals are intentionally introduced from the inside. The wireless telegraphy couples to the atmosphere, picking up energy from the surroundings and making it available for detection.

40. F. A. Collins, *The Wireless Man: His Work and Adventures on Land and Sea* (New York: Grosset & Dunlap, 1912), 13–14.

41. Francis Collins, "An Evening at the Wireless Station," *St. Nicholas*, v. 39 part 2 (1912), 1111–12.

42. Karin Bijsterveld, *Sonic Skills: Listening for Knowledge in Science, Medicine and Engineering (1920s–Present)* (London: Palgrave Macmillan, 2019).

CHAPTER 6

Listening in the *Sĕrat Cĕnthini*

Tony Day and Sarah Weiss

The *Sĕrat Cĕnthini*, a poem composed in Surakarta, Central Java by Ki Ng. Ronggasutrasna, R. Ng. Yasadipura II, and Ki Ng. Sastradipura under the direction of the Crown Prince of Surakarta (later Pakubuwana V, r. 1820–1823), was written in 1815 during the British occupation of Java (1811–1816). Its Canto 363 features an unusually detailed account of sound and listening as a Javanese village awakens, preparations for a wedding get underway, and musicians begin to play. Although these scenes are preserved in a written text, the multilayered sonic experiences of life in nineteenth-century village Java are clangorously and melodiously present throughout the text, both in its description and in the words themselves. Written in verse to be sung and heard, the *Sĕrat Cĕnthini*'s abundant use of alliteration, assonance, and onomatopoeia invites the listener to pay attention to all of God's creation, described in sonic, haptic, and kinesthetic ways.

Our analysis distinguishes three stages of increasingly attentive listening in the *Sĕrat Cĕnthini*. We further argue that the poem invites us to consider adding religious "intention" (Jav. *niyat*, Arabic *niyya*) to the attention that we as readers, like the villagers it describes, are paying to the *sound* of the poem—both the text itself and the performances described therein. This process ultimately transforms listening into a religious exercise of "audition" that the Sufis call *samā*ʿ (Arabic).

The Sonic World of the *Sĕrat Cĕnthini*

The 722 cantos and 247,766 lines of poetry in the *Sĕrat Cĕnthini* are arguably the greatest expression of Javanese literary art; indeed, the eminent philologist R. M. Ng. Dr. Poerbatjaraka deemed the poem "*de onvolprezen Tjĕntini*" (the unsurpassed *Cĕnthini*).[1] But what kind of inimitable literary work is it? The Eurasian translator C. F. Winter Sr.'s mid-nineteenth-century fictional informant, Parakawi, called the *Sĕrat Cĕnthini* a "*perlambang ngelmi*" ("oracle of knowledges"[2]), an enigmatic phrase that points to both the work's encyclopedic character and its interest in esoteric thought and techniques of all kinds.[3] More recently, Theodore Pigeaud classified the *Sĕrat Cĕnthini* as a "romance of vagrant students containing encyclopedic passages," grouping it with three other works.[4] Nancy Florida coined the phrase "*santri lĕlana* story" to expand the family of texts to which the *Cĕnthini* belongs, and Tim Behrend adopted and refined Florida's definition of the *santri lĕlana* (wandering student of Islam) genre to clarify its common characteristics. According to Behrend, all *santri lĕlana* stories have (1) at least one *santri* (pious Muslim, student of religion) as protagonist, (2) a plotline in which this figure, in search of a lost relative as well as religious knowledge, wanders (*lĕlana*) through forests and populated countryside far from courts and cities in a landscape that may or may not be identifiable as contemporary Java, (3) a story of adventure related by means of recurring episodes that are combined or omitted in various configurations, (4) an end in which the protagonist's mystical religious ideas get him into trouble, and he either escapes punishment or is killed.[5]

The *Sĕrat Cĕnthini* certainly fits but also exceeds Behrend's criteria in an astonishingly complex way. The earliest version of the *Sĕrat Cĕnthini* comes from Cirebon, a town on the northwest coast of Java, at the beginning of the seventeenth century. During the seventeenth and eighteenth centuries the poem was copied and read in Muslim religious schools (*pĕsantren*) in West and East Java. When the poem reached Surakarta in Central Java in the late eighteenth/early nineteenth centuries, Sufi court poets rewrote it, greatly expanding the number of episodes and characters, as well as the kinds of information and literary styles it contained. The version we will discuss, the so-called *Major Cĕnthini* or *Cĕnthini Kadipaten*, can be dated to the household of the Crown Prince of Surakarta on January 7, 1815. It is *no less than forty times longer* than the original Cirebon version.[6]

Musical performance is one of the many aspects of Javanese culture described at length in the *Sĕrat Cĕnthini*, where it is represented as a favorite pursuit of some of its itinerant characters. Performances in the poem involve playing on gamĕlan instruments or frame-drums, both as part of Muslim ritual events and simply for pleasure, or as accompaniment to dancing by male, female, or transvestite performers. Throughout, musical performance is fundamental to the very nature of the poem itself, which is written in poetic

Figure 6.1. Listening as a young soldier recites from the *Sĕrat Damar Wulan*. Illustration from British Library MSS89, 96r.

meters intended to be sung, either silently by solitary readers as they read the text to themselves, or in public performance to attentive listeners, who may even be lulled to sleep, as in Figure 6.1, by its poetic melodiousness.[7] "Sounding" is fundamental to many experiences represented within the text, as well as to its reception by those who read and hence listen to it.

Several scholars of Javanese music have translated and discussed descriptions of musical performance from the *Sĕrat Cĕnthini*.[8] Generally, however, these authors have not focused on "sounding" as such, or on what the poem can teach us about the experience of sound in Java.[9] Our goal in this essay is not to construct a unifying metaphor for understanding Javanese culture through musical structure (or vice versa), but rather to allow the poets of the *Sĕrat Cĕnthini* to comment on the nature of sound and listening in their own words.

What can we learn about listening by simply reading an extended passage from a poem? The passage we have in mind comes in the middle of the *Sĕrat Cĕnthini* in a long episode devoted to festivities surrounding the marriage of Amongraga, a wandering Sufi sheikh (master, senior religious scholar), to the beautiful, pious, and talented daughter of the village head of Wanamarta, Tambangraras. Canto 363, in the *Pocung* meter,[10] opens in the morning after a night of religious instruction, performances, and lovemaking, in which the sacred theme of higher, spiritual "union" of man with God is parodied in a detailed and energetic description of anal sex between Jayengraga,[11] Tambangraras's younger brother who is a skillful musician, and two youths, Surat and Senu, who have aroused him by their dancing during the evening's festivities.[12]

On hearing the first verse of the canto, the listener is likely to chuckle because the text presents Surat and Senu, the saucy (but reluctant) sexual partners of a few moments and verses ago, as the singers of a melodious early morning prayer in Arabic, whose words are "Prayer is better than sleeping." These devout sounds in Arabic are soon amplified by Jayengraga intoning a sequence of obligatory Islamic prayers. We join him at verse 7 where, as praying draws to a close, all the voices of the village world begin to sound forth.[13]

7. *Swara barung rěmpěg nganyut-anyut arum* *sabawaning desa* *kang barungan kang sarěnti* *ingkang lagya dhikir myang wus sělawatan*[14]	7. His even tones gently transported you away, with every village voice,[15] in unison or on its own, sounding God's name or singing in chorus with drums.
8. *Wěneh kang wus ngaji turutan gumrumung* *anulya gya adan* *sabawane desarja-di* *wor krěngketing pipikul sami pěpasar*	8. Others recite lessons in a collective buzz. Then comes the call to prayer. All the sounds [*bawa*] of the prosperous town mix with the squeak-squawk of loads shouldered to market.
9. *Wěktu Subuh wong sadesa tanginipun* *samya nambut-karya* *malah bangun keh kang tangi* *olah ratěngan wěneh mangkat pěpasar*	9. At the hour of *Subuh*[16] the villagers arise, all to work at their jobs. What's more, many, once out of bed, roast food while others set out for the marketplace.
10. *Kang (ng)galěpung ting gěrandhung myang kang nglěmbur* *pandhe gěndhing sayang* *tan sěpěn raina wěngi* *aměwahi sěněnging desa raharja*	10. Those grinding meal "crunch-crunch"[17] or working through the night, gamělan-makers and smiths, are not quiet at any hour, increasing pleasure in the prosperous village.

11. *Těka sěmu yen Subuh kweh swaranipun* *bangun bang-bang wetan* *barung ingon-ingon muni* *jaran kěbo sapi bebek miwah ayam*	11. Thus it would seem that at *Subuh* many voices awaken at sunrise, just when animals make noises, horses, water buffaloes, cows, ducks, and chickens.
12. *Mrak manguwuh nyangungung prěkutut manggung* *myang putěr běnggala* *sikatan munya cing-go-ling* *aneng kaywan kolik munya aklikikan*	12. Peacocks let out a shrill call and turtle doves coo, with red-eyed Bengal doves. Black wagtails sing "cingcinggoling." In the dark trees a cuckoo sounds "aklikikan."[18]
13. *Pěksi tuhu neng gurda angganter asru* *cocak kěkroyokan* *kuthilang imbal kudhasih* *bethet jalak sauran menco kěpodhang*	13.Her mate sings loudly back from the *waringin* tree. Blue *cocak* swarm en masse.[19] *Kuthilang* and *kedhasih* bicker. *Bethet, jalak* converse with *menco, kěpodhang*.[20]
14. *Swara umyung unine pěksi ambarung* *swaraning pujiyan* *kang pěksi lir angrakěti* *kang amuji myang kadya tutuku jajan*	14. Voices resounding, the bird sounds accompany[21] the prayerful voices. It's as if the birds make new friends of those who pray and are shopping, looking for snacks.

The *Sěrat Cěnthini* consistently alludes to the tenets of Islam along with the political realities of early nineteenth-century Java and social norms more generally, and it playfully contests the stricter constructions of law-abiding Muslims, articulating a Sufi critique of legalistic, conservative Islamic practice.[22] Raunchy sex is juxtaposed with pious behavior; gorgeous singing by talented and trained performers with that of country bumpkins chanting out of religious conviction or a false pride in their voices; order with disorder. The point is not invidious comparison so much as a demonstration of an inclusive tolerance for all forms of Islamic and Javanese cultural ways of being. This intervention occurred at a time when identities were becoming increasingly polarized just before the outbreak of the Muslim Prince Dipanagara's rebellion against the Dutch in 1825.[23] This historical background forms one important aspect of the sonic humor in the *Sěrat Cěnthini*.

The playfulness of Canto 363 in particular is announced by its meter, *Pocung*, described by various Javanese handbooks surveyed by Ben Arps in an appendix to his study of *macapat* (1992) as "frivolous," "coaxing by making a joke," or "without tension."[24] The light-hearted nature of the *Pocung* meter signals soundings of all sorts, both unstructured fun as well as serious, anti-structural, Sufi play with religious ideas. In the verses translated above, the frenzy of the preparations can only be dimly sensed, but the sounds of the bustling, early-morning world of villagers and birds anticipate it. The routines of village life every morning are described in detail: *Subuh* prayers are recited, sellers head to market, smithies resume work at their forges, peddlers amble through neighborhoods offering sweet cakes, and birds and domestic animals attend to their morning ablutions while chattering to their mates. Strikingly, every image in these verses communicates the sonic world of a Javanese village dawn by connecting sound, touch, and movement. Praying voices are melodious. The buzzing of Arabic lessons being recited in unison mingles with the squeaking of the poles of merchants carrying goods to market. Repetitive work sounds, such as the grinding of meal or the smelting of metal, accompany the conversations of animals and the twittering of birds, onomatopoetically experienced through the poem, which would always have been sung out loud in the nineteenth century. The articulation of sounds entices the reader onward, beckoning her, or pulling him, through the sung narrative in the same way that the interactions of sounds in the early morning urge the inhabitants of Javanese villages to become ever more awake and engaged with their normal morning routines ahead of the special activities associated with the wedding. Everyone and everything—animal, villager, listener, porter's pole—is drawn toward a goal that is even more significant than participation in the festive day at hand, one that is unnamed yet implicit within and sounded by each word and note of the canto in *Pocung*.

Unsurprisingly, from the very start of the canto, we find an assonantal-alliterative sound play (Jav. *purwakanthi*) that is characteristic of Javanese *macapat* poetry. But the first twenty-seven verses of the canto are also devoted to vocalizations that combine religious with secular sounds. After the "well-tuned" (Jav. *araras*) voices of Surat and Senu singing their morning prayer in what is called a Banten melody in the text,[25] we hear Jayengraga's Arabic invocation transformed into *Pocung* prosody. On and on, through several verses, the sequence of prayers in Arabic, set to poetry in *Pocung* meter, issues forth, until the entire village is "carried away" (Jav. *nganyut-anyut*) by the "gentle"

(Jav. *arum*)[26] voices of various praying individuals, which mingle indiscriminately with other sounds of the awakening village.

A striking feature of the description of sounds in the lines quoted above is their disparate, synesthetic omni-directionality. Some of the sounds are human, others are animal, still others are made by things. The canto begins by introducing one vocal sound and sound-maker after another in sequential fashion, then moves into a representation of village-wide vocalization, at which point, in the last line of verse 8, the sound-making spreads to the squeak-squawk of the bamboo carrying poles on their way to market. In verses 9 and 10 we hear the alliterative and assonant sounding of more early-morning work activity, especially in the first line of verse 10: *kang glĕpung ting gĕrandug myang kang nglĕbur*, "some ground meal with a crunch-crunch, and others smelted." Then the production of sound passes to the morning birds in verses 11–14, whose songs are said to "answer" (Jav. *imbal*) one another.

This is the first moment in the depictions of sound in which we discern that it is heard akin to a *gĕndhing*, a composition of gamĕlan music. The term *imbal* is also used to describe the style of interlocking playing by the *saron* and *bonang* in gamĕlan performance, which has already been alluded to in verse 10, at a moment when the gamĕlan makers are up and at work. Moreover, in everyday usage, an *imbalan* is a message or debt that you receive, then pass on to another. Thus, the root meaning of the word conveys the sense of an ongoing ripple effect that these early sound-saturated verses of the canto render so vividly. Sounds accumulate in one medium, then spill over and become another, until the entire village resounds. But what was once disparate and disorganized is now the coordinated soundscape of the prosperous and productive village itself, which is about to celebrate the union of two of its most pious, eligible, and esteemed young people.

It is worth noting that in music-making events called *klĕnengan*, where musicians gather simply for the pleasure of making music together, the beginnings of pieces are marked by a singer who communicates with the *rĕbab* player to choose the next piece. The *gĕnder* player strikes a few notes to orient the singer to the scale and mode of the *bawa* or introduction to the next selected piece. The singer's melody rises up over the din of musicians as they relax, talk, eat, and smoke. The musicians are gradually drawn to the sound and begin listening with increasing attentiveness as the solo voice reaches the end of the *bawa* and intones the regular rhythms that lead to the stroke of the gong and the beginning of the piece. The description of the village waking up, heard through the soundings of its disparate activities, which gradually join in *imbal* exchanges, mimics the experience of the beginning of a musical piece at an informal musical gathering, something that is described three more times in the canto.

Gamĕlan Sounded and Heard

Throughout verses 28 to 68, Jayengraga ("Master of the Body/Bodies")[27]—the sexy (and certainly oversexed), musical son of the headman and younger brother of the bride—organizes preparations for the wedding ceremony and the performance of comic skits and masked dancing to be held in the late evening. These verses display the social hierarchy of the village: Jayengraga summons (*nimbali*) servants who "appear before his presence" (*prapteng ngarsi*), then "respectfully obey" (*sĕndika turneki*) him before hurrying off, like a flock of little birds, to do the master's will. A temporary pavilion (*tarub*) is erected by a group of villagers and a female servant is given money to buy vegetables at the market. The village world continues to materialize before our eyes and ears, revealing a structure of authority and a hierarchy of masters and servants, men and women, with set roles and places in society. It is in this context, defined now sociologically as well as physically and sonically, that gamĕlan playing begins.

68. *Ingkang mupuh ulam sawĕneh amarut* *ngonceki bĕrambang* *bumbu sate age arih* *Jayengraga pinarak sarwi gambangan*	68. Some are pounding meat, while others are grating things, or peeling some onions. The *sate* sauce is nice and thick. Jayengraga sits there playing on the *gambang*,
69. *Nadyan nuduh ing karya sakaliripun* *sinambi anggambang* *gĕtĕr galĕbĕg kur jawil* *Kulawirya ngĕndhangi cangkĕm dhĕng-dhungan*	69. while giving people instructions in everything and playing the *gambang*, a loud trilling "cluck" with each touch.[28] Kulawirya vocalizes the *kĕndhang* part.[29]
70. *Ngandika rum Jengraga sarwi gumuyu* *Crĕmasana sigra* *mundhuta rĕbab lan suling* *gĕnder kĕndhang payo gagadhon lirihan*	70. Jengraga speaks affectionately, with a laugh: Crĕmasana, be quick, go fetch the *rĕbab* and the *suling*, *gĕnder*, *kĕndhang*. Let's play *gadhon* style, slow and soft.[30]

71. *Kang tinuduh tan adangu prapta ngayun* *Ganda Widiguna* *Widilĕksana tan kari* *Tunjangkara sampun samĕkteng sadaya*	71. Not long it was for those summoned to appear, Ganda, Widiguna, Widilĕksana is close behind, and Tunjangkara. All are now ready to play.
72. *Lingira rum kĕndhangmu aturna gupuh* *mring man Kulawirya* *nulya ingaturkĕn aglis* *angling malih lah Crĕmasana ngrĕbaba*	72. [Jayengraga] says pleasantly, be quick now, offer your *kendhang* to *man*[31] Kulawirya. It was presented straight away. [He] continues, Crĕmasana, you play the *rĕbab*.
73. *Gĕnder iku Gandasana kang anabuh* *si Tunjang nyulinga* *Widiguna kang nyuluki* *wis sĕnggrengĕn rĕbabmu pathĕt manyura*	73. *Gĕnder*, it will be Gandasana who plays it. Tunjang, you play *suling*. Widiguna sings the *suluk*. Good now, bow *pathĕt manyura* on your *rĕbab*.[32]
74. *Yen sĕndhon wus tutug banjur Kuwung-kuwung* *Crĕmasana sigra* *rĕbab tinĕnthĕng*[33] *nĕmneki* *gĕnder suling gambang tinabuh nĕmira*	74. When the prelude is finished, play *Kuwung-Kuwung*.[34] Crĕmasana is quick, the *rĕbab* is stroke-tuned to "six," the "six" on the *gĕnder, suling, gambang* is struck.
75. *Wus pikantuk nyamlĕng sĕnĕng grengek arum* *gya mamet manyura* *tutupan gĕtĕr amiwil* *ngulĕr keket panuduhe ngĕmbat kawat*	75. Once he gets it just right and likes the resonance,[35] he goes to *manyura*. Stopped, the wire vibrates, sound unfurled.[36] Caterpillar grasping,[37] his finger trills the wire,
76. *Cumĕnthong puk angatonggeng masang antup* *jariji pat ganggang* *lir sikile lalĕr kĕsit* *driji manis jĕnthik ting cĕngkĕnuk rikat*	76. curving over, *puk*, on its side about to sting, the four fingers brushing[38] like the legs of a skittish fly, the ring and little fingers making tiny curls,
77. *Anglir sunguting kupu angingsĕp madu* *kosokane lamba* *muput segĕgan lĕstari* *walikane rangkĕp ngĕcĕk rikat rata*	77. like antennae of butterflies sipping nectar, bowing in single strokes, sustained till the end of each phrase, doubled on return with quick, even staccato.

78. *Pacakipun amanjing antěng tumungkul* *dene Gandasana* *anggěnder grěněngan ririh* *angathimil kumrining srěnge nyalupak*	78. The standard[39] of quietude had been fully achieved. As for Gandasana, his *gěnder* mutters and mumbles, stuck together, gesturing angrily, in haste,[40]
79. *Ngěnuk-ěnuk běběngkakane ulěm-puk* *cengkoke clěmpungan* *pancěr sandhungan mincati* *angipinjal jějěl nyělani irama*	79. hands like rice clumps, held wide apart, (make) deep clear sounds. Zither-like melodies find the best ways to move forward, like fleas cramming into spaces between each beat.[41]
80. *Pacakipun lir Uněngan sěmunipun* *Tunjangkara dennya* *anyuling běning cumlěring* *anyaririt tulupane atilěpan*	80. The playing is as elegant as Oněngan. Tunjangkara meanwhile, his *suling* whistles high and clear, the tube spitting out sound like a *tilěpan* bird,[42]
81. *Ngělong tutup papelon aliyu-liyu* *Jengraga anggambang* *něrithil asta kumitir* *tan sulaya kiwa těngěne arikat*	81. less and less, stopping, a stutter, totally done.[43] Jengraga plays *gambang*, trillingly hands keep moving, left and right together, faster and still faster,
82. *Kumalunthung narěthěg gětěr glěběg thur* *pacake wiraga* *aladak ang-Gunungsari* *Widiguna suluk ngěliking manyura*	82. rattling, chattering, thundering, thokkingly[44] sound pours forth, stylishly elegant, arrogant like Gunungsari. *Widiguna's Suluk Ngělik* in *Manyura*[45]
83. *Nganyut-anyut loyop pantěs suluk bangun* *sakeh kang miyarsa* *sadaya karěnan dening* *Jayengraga dennya sasěndhon araras*	83. sweeps away drowsiness, as *suluk* should at dawn. All those who are listening, they all feel pleasure because of Jayengraga's lovely, harmonious final song.[46]
84. *Sawusipun tutug liyu-liyunipun* *pathět Sarayuda* *saběman gong nulya wiwit* *gěndhing Kuwung-Kuwung munya rum araras*	84. After he finishes, with a flourish of the flute, the end of the *pathět* is played, one drum stroke and then gong.[47] *Gěndhing* Kuwung-Kuwung sounds softly and lovely.

85. *Kĕndhangipun ajĕg jĕjĕg lamba kukuh* *ulĕt wilĕt samya* *rĕbut yatmakaning gĕndhing* *apan dangu dennya ngĕs-ngĕs cengkokira*	85. The *kĕndhang* is strictly regular, simple, firm. The interwoven style seized the very soul of the *gĕndhing.*[48] They take their time indulging in variations.
86. *Gĕndhingipun niba kĕndhange angguguk* *kipat akosekan* *bêm sĕbah kĕmpyang nĕlampit* *ambiyantu tan sulaya wilĕtira*	86. The piece "falls,"[49] the *kĕndhang* makes a muffled guffaw, firmly struck with brushed strokes, a sated *bĕm, kĕmpyang* doubling,[50] enhancing, not conflicting with, the weaves of sound [*wilĕt*].
87. *Wusnya dangu nĕsĕg tan antara suwuk* *sigra Jayengraga* *ngasta rĕbab pathĕt ngĕlik* *ngik-ingiking tutup kosoke rangkĕpan*	87. After a while the playing speeds up, then it ends. At once Jayengraga takes the *rĕbab, Pathĕt Ngĕlik,*[51] "*ngik-ngik*" the stroking as he bows double tempo,
88. Tutup bĕsus kosok rangkĕp rikat alus eram kang amulat sĕmbada lan kang nyuluki pan katongton Jengraga kamuktenira	88. polished, the bowing in double time, quick, courtly. The audience is amazed. It fits just right with the singer. It is clear how much Jengraga is loving this![52]

In verse 68, Jayengraga's *gambang* playing emerges, assonantly and alliteratively, out of the word *brambang*, the onions, which are being peeled and chopped in the kitchen. The assonance and alliteration between the two words "*gambang* "and "*brambang*" is not simply a metrical requirement of the *Pocung* verse form, in which the second and fourth lines must end with vowel sound "a." In the first instance the assonance references the actual sound of chopping of onions as rather similar to the resonant, struck-wood sounds of the *gambang*. The modulation from peeling onions to playing *gambang* further exploits the sound structure of the verse form to represent the continuous, interconnected, village-wide sound-making that we observed earlier in the canto. The urgency and energy of this process is picked up in verses 69 to 73 as Jayengraga calls his friends to join him in making music.

Once the music-making commences, Jayengraga contributes to these aforementioned kinesthetic sounds as he improvises on the *gambang* in verse 68 while giving orders to those preparing for the wedding feast, with his uncle Kulawirya vocalizing a drum accompaniment. But after Jayengraga calls out for other musicians to bring instruments and join him in playing a gamĕlan piece, and when the musical sequence begins in verse 74, the poem asks us to attend—both by eye and ear—to the gestures producing the sounds on some of the instruments, which are very precisely described in imagery that connects them to tiny creatures in the natural world. Together, this turns into qualities of the musical sounds of the instruments as communicated by the sung poetic words of the passage.

The ensemble of musicians performs the piece *Kuwung-Kuwung*, moving through the sections of the piece from *sendhon* to *merong* and *inggah*, ending with *Pathĕtan Manyura Ngĕlik*. The focus for those nearby who are attending to the impromptu concert are the individual instrumental performances, described in sharply differentiated terms according to the specific sonic quality of each instrument and its particular mode of performance by one of the named musicians. For the musicians themselves, the focus is on the interplay of *wilĕtan*, "intertwinings" in the performance of *cengkok* variations, which facilitates attainment of the most religiously sought-after goal of ensemble music-making, the musical "soul" (*yatmaka*) of the *gĕndhing*, for those who are capable of hearing it.

Listening throughout the performance, by audience and musician alike, is represented as being continuous and constantly attuned to nuances. These details are captured by the comparisons to different insects and their movement, as well as by suggestive onomatopoeias, and to the ways these nuances intertwine until they reveal the "soul" of the *gĕndhing*. The moment of greatest musical intensity coincides, not with the sounding of the *gong*, but with the arrival of the greatest musical density in verse 85, during which each instrument producing variation after variation is woven together, until the piece "falls" in its approach to the final cadence. Special attention is paid in the passage, in other words, not to structure, but to the "nuances of *textual densification*," as Stephen Feld might put it,[53] or, as Sumarsam has described in his essay "Inner Melody in Javanese Gamelan," to the *wilĕtan*, the ways "each musician relates the part he plays to the melodies played on the other instruments,"[54] which are represented in kinesthetic as well as sonic terms and minutely, metaphorically described, linking musical sounds to God's world at large.

"Intention" (Jav. *niyat*) and Javanese "Listening" in the *Sěrat Cěnthini* as Sufi *samā'*

> Know that hearts and consciences are treasuries of secrets and mines of jewels. Wrapped within them lie their jewels just as fire is enveloped in iron and stone, and hidden like water is concealed under dust and loam. There is no way of extracting such hidden things save by the flint and steel of audition to poetry and music (*samā'*), and there is no entrance to the heart save by the ante chamber of the ears. So musical tones, measured and pleasing, bring forth what is in it and make evident its beauties and defects.[55]

The special power of gamĕlan music can be released and experienced, for those who are playing it and for those who hear it, when making sound and listening to it is motivated by religious "intention." At least, that is how it should happen according to Sufi theory about *samā'*, "listening which is spiritual." Traces of this theory inform the passages we have just been reading in the *Sěrat Cěnthini*.[56] Sufi ideas about *samā'* have long influenced Persian and Turkish music, and from very early on the Sufi love of music and the belief that by playing and listening to music one could achieve union with God were strongly condemned by legalistic Muslims as being akin to "usury, fornication, and intoxication."[57] Ghazālī's Book XVIII from his *Iḥyā' 'ulūm al-dīn*, "Responses Proper to Listening to Music and the Experience of Ecstasy," offers a famous discussion of *samā'*. The writings of Ghazālī, an eminent Sufi scholar from Persia, were well known in Java from the seventeenth century on. Here we make use of the interpretation of *samā'* offered by Ghazālī to elucidate the connections between intentional listening to/playing of music as it appears in the *Sěrat Cěnthini*.[58]

The quotation at the head of this section comes close to summarizing Ghazālī's core idea about "listening that is spiritual." Music is the medium that best conveys divine secrets contained in various kinds of writing, both secular and religious. It enters the "ear of the heart" when a human hears it. The "heart" is where the mystic contemplates—and may even achieve oneness with—God. The mystic capable of intentionally listening to music as a medium of divine revelation (Arabic *samā'*) will be able to hear the mystery that music conceals.

The holy Quran itself has always been sung. Navid Kermani has argued that the singing/recitation of the Quran is essential both to its aesthetics and to its divine message. Kermani quotes Ghazālī from his *Ihyāʾ*:

> I recited the Quran without experiencing any joy in it, until I recited it as if I was hearing the Messenger of God reciting it before his companions. Then I was raised to a higher station, and I recited it as if I was hearing Gabriel conveying it to the Messenger of God. Then God brought on another station (*manzila*), so that now I hear it from Him who speaks it. And I experience an enjoyment and contentment (*naʿīm*) that I can no longer do without.[59]

In Islamic tradition, there are many stories about listeners who become so enraptured by hearing a beautiful recitation of the Quran that they faint, or even pass away. Ghazālī reports four instances of men who died from the pleasure of hearing the Quran recited beautifully, a tiny fraction of the number of such Quranic deaths reported over the centuries.[60] Kermani quotes several sayings attributed to the Prophet, such as: "'Everything has an ornament; the ornament of the Quran is the beautiful voice.'"[61] "If one follows the rules for an ideal recitation as outlined by Ghazālī . . . ," states Kermani, "the musical rendition is not only 'ornament' in the ordinary sense of the term . . . on the contrary, it is basic to the actualization of the message."[62]

And it is not just the Quran that has caused the deaths of those who have become unbearably enraptured from hearing a beautiful recitation. "The number of Sufis who are supposed to have died listening to music or poetry is probably only insignificantly fewer than the number slain by the Quran," Kermani writes.[63] In his remarks on *samāʿ*, Ghazālī and other Sufi authors discuss listening to music, poetry, and the Quran as belonging to the same discourse on *samāʿ*. The Quran is not poetry, but poetry as well as music also belong to the religious sphere. As Kermani puts it, "to the mystics, all three are an organ of religious insight . . . listening is not an end in itself but is directed towards achieving an insight, a 'finding' (*wajd*), as the word for ecstasy aimed at in listening indicates."[64] In other words, listening as *samāʿ* is paying attention to music and poetry with the specific "intention" to discover God's mysteries hidden within sound.

The distinction Ghazālī and other Sufi writers draw between listening merely for pleasure and listening with "intention," as *samāʿ*, is articulated in a subsequent episode in the *Sěrat Cěnthini*. Jayengraga, his older brother Jayengrěsmi, Kulawirya, and their young *santri* companions have set out from Wanamarta in search of Amongraga, who has left his new wife for a series of adventures that will end in his death. The travelers come to a village where they play gamělan and where their learned host, Kidang Wiracapa, instructs them about the difference between playing music the way most musicians do, focusing on the beauty of the music they are making, and playing it as an act of worship to unite with God (Cantos 595–597).[65]

In v. 25 of Canto 595, Jayengraga picks up the *rěbab* and plays the *buka* to the famous piece *Gambir Sawit*. As they play the *gěndhing* the musicians are described (v. 32) as being in a state of (physical and emotional?) "constipation" (*mělět*). But as they listen to the "voice(s) of the *gěndhing*" (*gendheng gěndhing*), they are filled with pleasure and "knead their *wilět*s together" (*atangkěb mulět wilětnya*). Then, threading and weaving their sounds together in the most elegant fashion (*sangsaya angikal běsus*), "they perform the soul of the *gěndhing*" (*olah yatmakaning gěndhing*) (v. 33). In the very next line of the same verse, people who have been watching the performance, women and children, even very old people working in the kitchen, are described "bowling each other over" (*asalang tunjang*) and burning what they are frying as they are "overpowered" (*katungkul*) by the desire to look at those who are playing the *gěndhing*, particularly the handsome Jayengraga (v. 34).

Wiracapa's comment after the playing has ceased is that such skilled musicians could not possibly come from pious Islamic backgrounds. Kulawirya in particular, with his refined playing, could only ever be resistant (*cěgah*) to religion (v. 42): "*Sabab wong gěndhing puniku* | *atine mangeran gěndhing* | *tan ana liyan kacipta* | *namung rasaning ing gěndhing* | *sinungku sajroning nala* | *nalika lakuning gěndhing* ||" (Because a real *gěndhing* person, / his heart only calls *gěndhing* 'Lord,' / that is all he can imagine, / just the feeling /taste of a *gěndhing*, / that is what lives inside his heart/after the *gěndhing* is over //).[66]

But it must not be this way, as Wiracapa elucidates in the following Canto 596, composed in the *Pocung* meter. He explains to his listeners that the "taste, the meaning" (*rasanya*) of the "divine science" (*ngilmu*) requires them to learn to hear and play music differently.[67] The path to "tasting" the music correctly starts with the individual instruments, brought together as a tuned ensemble on which a *gěndhing* is played at different tempi (*irama*). The "soul" (*yatmaka*) of the music emerges when all the instruments are played together as a gracefully interwoven ensemble.[68] Gamĕlan originated with the founder of Javanese culture whose name, "Widayaka," means "a person who sings poetry" (*wong anĕmbang*).[69] Voices sing, hands play instruments; singing (*gendheng*) is joined together (*dhaup*) with musical compositions (*gĕndhing*), whose loveliness originated with voice. Once all these elements combine (*dhaup*) with "divine science" (*ngilmi*), then the tones of the gamĕlan, along with sounds coming from the lips, together with the *gĕndhing*, express the "intention" (*niyat*) of the performance. "Musicians," Wiracapa asserts, "are just a community (*bangsa*) of drunks, but even so, if they get drunk on the teachings of *ngilmi*, then they would commit themselves to abstinence and ascend to Mystery."[70] "Do not give in to the *gĕndhing* and its spell (*surupan*)," he adds; "the sounds of its beauty (*surasaning raras*) are but premonitions of your longing (*kinarya pangresing kapti*), are what guide (*pangatĕre*) the heart as it meditates (*tyasing gagas*)."[71]

Turning back to Sufi theories of *samā*ʿ we can now assess what the description of the performance along with Wiracapa's commentary teach us about how Javanese listen—or should listen—to gamĕlan music. The influence of Sufi ideas on the Javanese poets, probably as formulated by Ghazālī, are evident. To begin with, as is shown in every episode in which travelers play gamĕlan in front of village audiences, there is a difference between the two groups in terms of their "listening" and responding to music, the same distinction Ghazālī and other Sufi theorists of *samā*ʿ make between advanced travelers on the Sufi path and ordinary people. Even though Wiracapa, voicing the opinion that pious Muslims still hold about musicians in Java, calls them drunks, the traveling musicians are sophisticated Sufis capable of seizing the "soul" (*yatmaka*) of a *gĕndhing* when they make their *wilĕt* perfectly intertwined in performance. Kulawirya and his companions do not need Wiracapa to explain this to them. The villagers, on the other hand, are mindlessly, uncomprehendingly thrown into confused awe by what they hear. Ghazālī would not have been surprised. According to him, and he was more insistent on this point than other Sufi theorists of *samā*ʿ, only those who have already attained a high degree of mystical insight can "heart-savour" (*dhawq*) music, "listening to music (*samā*ʿ) [as] a divine mystery (*sirr Allāh*) found within the harmonious relationship of measured tones to the (human) spirits and in the spirits becoming overcome by these melodies and stirred by them."[72] Ghazālī's comment reminds us of Sumarsam's thinking about the "inner melody" of a gamĕlan piece in performance:

> It seems that no single instrument or vocal part in the ensemble is solely responsible for guiding the melodic motion of a gendhing. I believe the real 'authority over melody' is a melodic idea sung in the minds of the individual musicians. This directs the melodic motion. This melodic conception could be called 'inner melody.' Inner melody is the melody that is sung by musicians in their hearts. Inner melody is the essence of melody in Javanese gamelan. My gamelan teacher often advised me '*golekana lan rasakna wosing utawa surasaning lagu*,' that is, 'search for and feel the essence, or the soul, of a melody.'[73]

The moment when Javanese musicians feel the "soul" of the melody seems very much like "the essence of the ecstasy that the mystics pursue in *samā*ʿ," as Kermani puts it, which is an absorption in contemplation, a relinquishment of one's subjectivity.[74] Sumarsam's musicians allow their subjective "inner melodies" to merge with the one essential "inner melody" of a *gĕndhing*, which is its *yatmaka*, its soul, once they have collectively grasped it.

Attending to the *Sĕrat Cĕnthini*

For Sufis, including Sufi poets in early nineteenth-century Central Java, the experience of God's beauty was primarily through sound. For some Sufis in the Middle East, the sounds of the Quran were the most incomparably beautiful.[75] For others, like Ghazālī, it was listening, in the mode of *samā*ʿ, to secular music and poetry rather than to the well-known, frequently memorized verses of the Quran that offered access to the Absolute.[76] The poets of the *Sĕrat Cĕnthini* seem to have agreed with Ghazālī; indeed, Florida reminds us that, near the end of the poem, after a night of recitations and musical performances, "Amongraga pronounces his assessment of their night of devotion" in just these terms:

> Sheikh Amongraga spoke softly,
> 'Brother Ki Malangkarsa,
> as for the faithful of Java,
> to gain understanding in their hearts
> the best of possible ways is
> through the experience (*rasa*) of *suluk* alone
> is reached the end to which we aim.

Different, the effect upon the heart
that comes from the Holy Qur'an–
demanding strenuous struggle,
long it takes to enter the heart,
never reaching the *rasa* of the Word;
for that requires the symbols of song
sung in Javanese macapat verse.'[77]

Up to this point, the situation described in the *Sĕrat Cĕnthini* seems directly connected to the Sufism of Ghazālī. And yet, if we read descriptions in the *Sĕrat Cĕnthini* of villagers listening and reacting to gamĕlan performances, something else seems to be going on other than their noncomprehension of the Mystery contained within its sounds. The sounding of gamĕlan instruments in the early nineteenth century also expressed an autonomous power, long predating Islamization, something that compelled listening and bestowed well-being.

Let us turn to the description of the last performance of music by Jayengraga and his companions (v. 144–96, the final stanza in Canto 363, and the opening stanzas of Canto 364 in *Sinom* meter, verses 1–14).[78] The musicians have brought out a sacred gamĕlan named *Alun Jaladri* (Waves of the Sea) that belongs to Jayengraga's father. When the musicians begin playing on it, everyone in the village hears (v. 157) and is "startled, their hearts pounding" (*ingkang myarsa kagyat tyasira naratab*).[79]

158. *Sinru arum angebĕki ing sadhusun* *lir natab priyangga* *pĕngrasane joganneki* *prabawane Alun-jaladri yen munya*	158. Loudly, melodiously, it filled the whole village, as if striking itself, the feeling, on its arrival, of *Alun Jaladri's* awesome power when it sounds.
159. *Kang angrungu tan pangling suwaranipun* *tan ana mĕmada* *unine Alun-jaladri* *samya berag anggugah manah gambira*	159. Those who listen make no mistake about the sound: there could be nothing like the sound of *Alun Jaladri.* Everyone is passionate, alert, jubilant.

The villagers recognize the sound, and it fills them with joy. It compels them to put on their best clothes and bring money and other offerings to bestow on Jayengraga as a token of respect for *Alun Jaladri*, the women walking in a way that "showed off their beauty," the men "their manly charms" (v. 188). Kulawirya comments that when people hear a gamĕlan like *Alun Jaladri* "the energy of pagan times returns" (v. 189), for the sound "by necessity bestows good fortune" (v. 192). At the beginning of Canto 364, which is in the *Sinom* meter, a single verse sums up *Alun Jaladri*'s power to cause involuntary responses by its listeners, with five lines stating the coming together of the various kinds of instrumental sound in a mystical unity as the *rĕbab* is played:[80]

9. Nulya mungĕl Kĕmbangmara *Alun-jaladri ngrĕrangin* *sakathahe kang miyarsa* *jalwestri tyase mong brangti* *rĕmpĕg panabuhneki* *ukur jawil ngĕnut-ĕnut* *ungĕle kang angl[e]la* *rĕbab gambang lawan suling* *sarancak ungĕle kamot jroning kawat*	9. Then it resounds, *Kĕmbang Mara*, Ocean Waves ringing soft and sweet. Everyone who is listening, men, women, is deeply lovestruck. The notes struck right together, measured touches one by one, sounds that were distinctly heard, *rĕbab*, *gambang*, and the *suling*, ensemble of sounds: contained in a single string.

The description ends with an image of the *rĕbab*, the most voice-like instrument in the gamĕlan ensemble, sounding the *yatmaka*, the "soul," of all the "inner melodies" of the *gĕndhing* being played, brought together in the melody heard on a single string of the *rĕbab*. Although the villagers are not capable of the kind of *samāʿ*-like hearing that is implied by the poetry in verse 9, they are "lovestruck" (*brangti*) by what they hear, a state described in the poetry of pre-Islamic ancient Java as far back as the ninth century CE as an expression of the most advanced kind of aesthetic sensibility.[81]

The episode of "listening" from the *Sĕrat Cĕnthini* that we have been reading in Cantos 363 and 364 presents three distinct, but ultimately intertwined and mutually referential, kinds of sound-making and hearing. In the first of these, the village awakens to the sounds that people, animals, and things make every day at dawn. These sounds are there to be heard, but no one in the poem is listening intentionally to them. Those hearing the poem sung out loud, however, would be listening with intent, religious or otherwise.[82] The poetry, as we have indicated in our annotated translation of it, is intended either to protect us from all harm and lull us into slumber, or to alert us to "earth's signs and marvels," as Ghazālī puts it in his *On Conduct in Travel*, Book XVII of the *Iḥyāʾ*.[83]

In the second phase of listening presented in these verses, when the musicians first begin making music together, the instruments being played are described in a way that recalls the minute kinds of sonic and descriptive reference to the tiny details of the everyday world that we find in the first phase of listening. It sounds at times as if the poets are drawing the entire human and natural world from the dawn-section of the episode into the "inner melodies" that the musicians hear in their hearts as they now play their different instruments. But this time the reader/listener is given to understand (v. 85) that not only are musical sounds intertwined with sounds and movements from the natural and physical world, they are also becoming intertwined, through the creative *wilĕt* of the musicians, with each other, as the mystical ensemble of sounds, "the interwoven style / seized the very soul of the *gĕndhing*." Musical sounds themselves, like the onomatopoeic words that define and imitate them, express the "inner melodies" heard in the hearts of the individual musicians, that finally meet in a gracefully adorned "coil" (*ukĕl*) to reveal the *yatmaka*, the "soul," of the *gĕndhing*. The musicians in this passage are clearly engaging in a kind of *samā*ʿ as they play and which the listener to the poetry describing the scene can overhear.

Finally, when *Alun Jaladri* engulfs the village, it is as if the pre-Islamic, heathen sounds of that orchestra are supernaturally identical with ocean waves themselves. We are compelled, like ordinary Javanese villagers, to be respectfully attentive to *Alun Jaladri*. We might even imagine that some kind of good fortune may befall us just from reading the poetic account of this music, even though we are living more than two centuries later and are probably not intentionally journeying on a mystical Sufi path toward Mystery.

Notes

1. R. M. Ng. Dr. Poerbatjaraka, *Pandji-Verhalen Onderling Vergeleken* (Bandoeng: A. C. Nix en Co, 1940), 361, note 3.

2. Translation by Nancy Florida. Here and throughout the translations and discussion that follow, we are enormously grateful to Florida for her queries, suggestions, and corrections. Our warmest thanks also go to Els Bogaerts and Edwin Wieringa for their helpful corrections and suggestions.

3. C. F. Winter, Sr., ed. T. Roorda, *Javaansche zamenspraken. Eerste deel: Zamenspraken over verschillende onderwerpen* (Amsterdam: Johannes Müller, 1862), 193.

4. Theodore G. Th. Pigeaud, *Literature of Java. Volume I. Synopsis of Javanese Literature 900–1900* (The Hague: Martinus Nijhoff, 1967), 227–229.

5. T. E. Behrend, *The Serat Jatiswara: Structure and Change in a Javanese Poem 1600–1930* (PhD dissertation, Australian National University, 1987), 325–26.

6. T. E. Behrend, "Technical Prolegomena to Any Future Centhini-Critique," n.d.

7. The better of only two full-length studies of traditional Javanese poetic meters, called *macapat* and the performance of *macapat* poetry in modern Java is Bernard Arps, *Tembang in Two Traditions. Performance and Interpretation of Javanese Literature* (London: School of Oriental and African Studies, 1992).

8. Three significant contributions are those by J. Kunst, *Music in Java: Its History, Its Theory, and Its Technique*, Volume 1 (The Hague: Martinus Nijhoff, 1973); Noriko Ishida, "The Textures of Central Javanese Gamelan Music: Pre-Notation and Its Discontents," *Bijdragen tot de Taal-, Land- en Volkenkunde* 164 (2008): 475–499; and Sumarsam, "Inner Melody in Javanese Gamelan," in *Karawitan: Source Readings in Javanese Gamelan and Vocal Music*, ed. Judith Becker and Alan H. Feinstein, volume I (Ann Arbor: University of Michigan Press, 1984), 249–304.

9. For earlier studies that quote from and discuss gamělan playing in Canto 363 of the *Sěrat Cěnthini*, see Kunst, *Music in Java* and Sumarsam, *Javanese Music and the West* (Rochester: University of Rochester Press, 2013). Other scholars have attempted to capture Javanese culture in one iconic concept or another; see Benedict R. O'G Anderson, "The Idea of Power in Javanese Culture," in *Language and Power: Exploring Political Cultures in Indonesia*, ed. Benedict R. O'G Anderson (Ithaca: Cornell University Press, 1990), 17–77; A. L. Becker, "Text Building, Epistemology, and Aesthetics in Javanese Shadow Theater," in *Beyond Translation: Essays Toward a Modern Philology*, ed. A. L. Becker (Ann Arbor: University of Michigan Press, 2000), 23–70; and Ward Keeler, *Javanese Shadow Plays, Javanese Selves* (Princeton: Princeton University Press, 1987).

10. The sung poetic meter *Pocung* consists of four lines, 12u, 6a, 8i, 12a, where the numbers designate the required number of syllables per line, followed by the vowels required for the final syllable of each line. Our translation follows Florida in adhering to the syllable counts for *Pocung*, which helps us stay close to the sentence structure and concision of the Javanese original. It also means that our English translation can be sung to *Pocung* melodies. We have not attempted to reproduce the play of assonances and alliterations, which are crucial to the meanings of the original.

11. The character Jayangraga is also known as Jengraga in the poem, especially when poetic meter requires that his name be shortened from four syllables to three.

12. For more on the mystical Sufi theme of "servant-master" (*kawula gusti*) in the *Sěrat Cěnthini* and playful parodies of the goal of bringing the two together in a mystical union often involving vivid descriptions of sex see Tony Day, "Stepping on a *Wulu*: Minor Characters and Narrative Possibilities in the *Sěrat Cěnthini*," in *Storied Island: Explorations in Javanese Literature*, ed. Ronit Ricci (Leiden and Boston: Brill, 2023), 57–60.

13. The Javanese text of the *Sěrat Cěnthini*, Canto 363, that we use for this essay, with small changes to spelling, comes from the online version of Kamajaya, ed., *Sěrat Cěnthini (Suluk Tambangraras)*, Jilid VI (Yogyakarta: Yayasan Cěnthini,1988), 124–40, accessible on sastra.org, https://www.sastra.org/kisah-cerita-dan-kronikal/serat-centhini/976-centhini-kamajaya-1985-91-761-jilid-06-pupuh-362-364. All the translations from Javanese are by Day, with interventions by Weiss and suggestions and corrections by Florida. Day claims responsibility for all remaining infelicities and errors.

14. In the last line of verse 7, *dhikir* are recitations associated with Sufi forms of worship that repeatedly intone God's name and attributes. *Slawatan* are religious songs in Arabic or Javanese, accompanied by one or more frame-drums (*těrbang*).

15. The word *bawa* means "voice, sound" but also specifically the solo opening, sung by a male, to a gamĕlan piece. Here and elsewhere in the description of the first sounds heard in the village as it awakens there are allusions to gamĕlan. For a translation and discussion of verses describing a gamĕlan performance held later on in the day in this same episode from the poem see Tony Day, "The Poetry of Minor Characters and Everyday Life in the *Sĕrat Cĕnthini*," *Wacana: Journal of the Humanities of Indonesia* 22 (2021): 707–26.

16. *Subuh* is the first of the five obligatory daily prayers in Muslim Java.

17. The word *gĕrandhung*, which is preceded by the word *ting* to indicate that it is an onomatopoeic word about sound, is not found in any of the Javanese dictionaries known to us. In a personal email (May 22, 2023), Edwin Wieringa, Professor of Indonesian Philology and Islamic Studies, University of Cologne, suggests that we translate it by thinking about words like *glundhĕng*, *glendhang*, and *glondhang* that have a similar sound. All of these words mean, according to the online *Leksikon* at sastra.org, https://www.sastra.org/leksikon, "the thudding sound that a small, hard object makes when thrown into an empty hollow space like a vat, chest, (empty) house or cart." Based on our experience grinding hard substances with a mortar and pestle, we have decided that "crunch-crunch" is a good translation for *gĕrandhung* here. Wieringa's suggestion makes a further important point about the way sound is heard and assigned meaning in Java. Words that sound alike in Javanese usually have similar meanings but are often associated with different events and experiences. Hearing the word *gĕrandhung* in verse 10 above would have stimulated memories of other, related words and life experiences for the Javanese listener in Surakarta in 1815, as it would today. Even if it is invented, which it probably is, *metri causa*, because it only appears in this place in the entire online version of the Kamajaya *Sĕrat Cĕnthini*, or in any of the hundreds of Javanese texts that have been put online at sastra.org, the word *gĕrandhung* not only fits the 12-syllable requirement of the first line in the *Pocung* meter, but also has good assonance-alliteration with the words *galĕpung* and *nglĕmbur* in the same line, better than any of the other words that sound like it and mean something like "crunch." In other words, the poet achieves a play on sound and meaning with his invented "crunch" word within the line and encourages the listener to extend that play to other contexts in the listener's experience. This kind of "synesthesia," if we can call it that, has a Sufi religious meaning in the *Sĕrat Cĕnthini*.

18. Passages and whole stanzas describing singing birds occur regularly in both Old and Modern Javanese poetry. In our translation of verses 12 and 13, we add some descriptive detail to give the reader a sense of the diversity and colorfulness of the tropical birds mentioned, each of which also has its own distinctive song. In the third line of verse 12, *cingcinggoling*, which has lost a syllable, *metri causa*, in the Javanese text, is a synonym for *sikatan*, the black wagtail, but the word sounds onomatopoetic and, grammatically, is here the sound the *sikatan* are making. Similarly, we render *aklikikan* as the song of the female night cuckoo, whose mate answers her in the first line of verse 13. The word only appears once in the entire *Sĕrat Cĕnthini* and cannot be found in the dictionaries. We have added the adjective "dark" to describe the trees (*kaywan*) from which the *kolik* is singing to suggest the nocturnal nature of the two cuckoos as well as give the scene a slightly ominous feel. Javanese people believe that when these birds sing, they are warning the listener of the presence of thieves nearby.

19. In this line the chirping of the birds is sounded by means of the assonance of the "ch" and "k" in the words themselves.

20. In the last line the bird song is communicated by means of the alliteration of "e," "a," and "o" sounds. In the last two lines of verse 13, the *kuthilang* is a thrush, *kedhasih* a swallow, *bethet* a small green parrot, *jalak* a starling, *menco* a magpie, and *kĕpodhang* a golden oriole.

The Javanese word that we have translated as "bicker" is *imbal*, which in gamělan terminology refers to the interlocking patterns, a kind of musical "conversation" or friendly "quarrel," that can be played on two *bonang* or *saron*. In the last line the word *sauran* means literally to "answer one another" or, in the case of sounds coming from all directions, to seem to do so, but the expression *saur pěksi*, "answer like birds," means for a group of people responding to a request by someone in authority to say "yes" in unison. The suggestion is perhaps that the four different species of bird are answering one another politely and, as the first line of the next stanza suggests, agreeing respectfully with those in charge in chirping unison to participate in the activities of the new day, along with the villagers.

21. In the first line of verse 14, two words have associations with gamělan. *Umyung*, translated as "resound," can describe the sound of a flock of birds echoing from the forest but also the sonorous resonance of gamělan instruments being played together. *Barung* means to "accompany" but more specifically to make a sound, with the voice or another instrument, at the same time as other gamělan instruments are playing.

22. For more on Sufism in early nineteenth-century Surakarta and the Sufi character of the *Sěrat Cěnthini*, see P. J. Zoetmulder, *Pantheism and Monism in Javanese Suluk Literature: Islamic and Indian Mysticism in an Indonesian Setting*, ed. and tr. M. C. Ricklefs (Leiden: KITLV Press, 1995); Nancy K. Florida, "Shaṭṭāriyya Sufi Scents in the Literary World of the Surakarta Palace in Nineteenth-Century Java," in *Buddhist and Islamic Orders in Southern Asia*, ed. R. Michael Feener and Anne M. Blackburn (Honolulu: University of Hawai'i Press, Amazon Kindle Edition, 2019), 226–70; and Day, "Stepping on a *Wulu*."

23. See Peter Carey, *The Power of Prophecy: Prince Dipanagara and the End of an Old Order in Java, 1785–1855* (Leiden: KITLV Press, 2008) on the Dipanagara rebellion and Eric Tagliacozzo, ed., *Southeast Asia and the Middle East: Islam, Movement and the Longue Durée* (Singapore: NUS Press, 2009) for other Southeast Asian examples of the debate between more conservative and more tolerant forms of Islam in Southeast Asia in the nineteenth century.

24. Other descriptions include: "[for use] as [an] interlude (*sěsělan*) when tension is slack (*grěgětipun kěndho*); and when [the] atmosphere is leisurely (*saseěkecanipun*)" (Arps, *Tembang*, 423).

25. Banten is a province in West Java known for its Islamic culture.

26. The word *arum*, which is used 408 times in Kamajaya's edition of the *Sěrat Cěnthini*, is given the primary meaning "fragrant" in the dictionaries of the Leksikon and is used to describe flowers and women in the poem. But it also frequently describes how people speak and how music sounds. This synesthesia of meanings between sight, sound, and smell, which we encounter repeatedly in the passage we are discussing, is very old in Java. P. J. Zoetmulder and S. O. Robson, *Old Javanese-English Dictionary*, Part II (The Hague: Martinus Nijhoff, 1982), 1569 give the meanings of the root word *rūm* in Old Javanese as "beauty, loveliness, charm, attractiveness, sweetness, gentleness, fragrance."

27. As translated in Nancy K. Florida, "Better to Touch the Heart: Performing Sufi Songs (*Suluk*) in the *Sěrat Cěnthini*," in *Storied Island: Explorations in Javanese Literature*, ed. Ronit Ricci (Leiden and Boston: Brill, 2023), 66–95.

28. The third line of this verse consists of four words that relate to playing the *gambang* or other instruments in the gamělan. *Gětěr* means a loud noise that causes you to shudder but is also the trilling motion made by the fingers playing the *rěbab* or while dancing. *Glěběg* means to make a trilling sound on a *gambang* with the round, padded ends of the *gambang* mallets. *Kur* is short for *ěkur*, the clucking sound that hens make. And *jawil* is to poke someone with your finger. All these words precisely evoke the sound and manner of playing the *gambang*.

29. As with Hindustani *bols*, Javanese drumming or *kĕndhangan* is taught and learned through spoken syllables. Each hand stroke has a spoken sonic reference that sounds reasonably like the one produced on the drums. Although these sounds can be rendered visually and written parts can be used in teaching, students typically learn using the spoken syllables, and teachers often sing the *kĕndhang* parts while playing other instruments to remind students where they are in the piece and which patterns and sounds they should be playing. When musicians perform without a drum, as depicted in this line, they often sing the *kĕndhang* stroke syllables in transitions when the drum would lead the change or when they want to "hear" the drum part. Here Kulawirya joins in with Jayengraga's *gambang* playing by speaking the drum syllables, even though he does not yet have a *kĕndhang* in hand. He receives one in verse 72, after Jayengraga tells people to bring more instruments.

30. Playing *gadhon* style refers to the performance of music with the principal, elaborating instruments—*rĕbab, gĕnder, gambang, suling*—but without the large group of instruments in the *saron* family, which play the *balungan* or skeletal (sometimes referred to as "main") melody of the piece. The term comes from the Javanese verb *anggadho*, which means eating a meal without eating rice, that is, eating the vegetables, protein, and *sambal* (hot sauce) only. Whether this means eating the most important parts of the meal (the meat and vegetables) or NOT eating the most important part of the meal (the rice) is open to debate. Nevertheless, the analogy makes a clear comparison between the melody played by the *saron* family and rice, and the music played by the elaborating instruments as the vegetables, protein, and *sambal*. In this passage, only the elaborating instruments and a voice are heard. As we argue below, following Sumarsam, "Inner Melody," this ensemble is all that is required to evoke "the inner melody" of the *gĕndhing*, the sound that matters most to musicians and that is central to the mystical meaning of the passage.

31. *Man* is short for *paman*, "uncle."

32. The musicians named in verses 71–73 appear elsewhere in the poem. The most important of them for the story is Kulawirya, who is Jayengraga's uncle, the younger brother of his father, and a constant traveling companion in later episodes in the poem.

33. Kamajaya footnotes this word, which doesn't exist as spelled, and corrects it to *thinthing* (also spelled *thentheng*), which means the sound obtained by striking a key or stroking a *rĕbab* string to test the accuracy of its pitch.

34. *Gĕndhing Kuwung-Kuwung* is today most often performed in *laras pelog pathet barang*. However, in Raden Tummenggung Warsadiningrat's "Wédha Pradangga," tr. Susan Pratt Walton, in *Karawitan: Source Readings in Javanese Gamelan and Vocal Music*, volume 2, 47, the piece is listed as a *gĕndhing rĕbab* in *pathet manyura*. In Javanese modal theory (*pathet* theory), *pathet barang* and *pathet manyura* are comparable in terms of mode hierarchy and some interpretive elements.

35. The word we have translated as "resonance," *grengek*, does not appear in the collection of dictionaries in the Leksikon of sastra.org. The word sounds onomatopoetic, like the sound the string of a *rĕbab* makes when bowed and may have been made up to convey such a sound. It only appears in this place in the *Sĕrat Cĕnthini* and nowhere else in the collection of hundreds of Javanese texts found on sastra.org, an assertion we base on a search we made on the sastra .org website using their search function, https://www.sastra.org/telusuri. There are words with similar spellings that come to mind when we hear "*grengek*" in this context: *grĕngĕng*, meaning "to make a humming, buzzing, whining" sound or *grĕnggĕng*, to sound like a buzzing bee.

36. "Sound unfurled" is our translation for the word *amiwil*, which is a variant spelling of *amiwir*, which means to spread its wings (of a bird) or flick the scarf around your waist away

from your body (a Javanese dance movement). The Javanese line says literally: "stopped, the trilling (of the wire) spreads its wings/flicks away its scarf." Bird and dance imagery (we will soon meet references to the latter) are never far from the minds of the poets as they describe musical sound in the *Sĕrat Cĕnthini*. This line is another good example of the haptic and kinesthetic nature of Javanese sounds.

37. "Caterpillar grasping" is our translation for *ngulĕr keket*. This is the only place in the *Sĕrat Cĕnthini* where this expression occurs. An *ulĕr* is a generic caterpillar; *keket* means to clamp onto something. An *ulĕr keket*, to quote from one of the dictionaries in the Leksikon of sastra.org, is "a very large, green, hairless caterpillar." Literally, *ngulĕr keket* means "to behave like an *ulĕr keket*." In the next line, which is the first line of verse 76, the word *puk* is the sound of the *rĕbab* player's finger as it pushes quickly and hard against the wire and the neck of the instrument.

38. Kamajaya amends the word *ganggang* to *nganggang*, meaning to lightly touch, brush. An *anggang-anggang* is a kind of water spider, whose skimming movements across the smooth surface of a pond are imagined here along with those of a skittish fly, to describe, in a vivid way, the touch and movement of the *rĕbab* musician's finger as he plays.

39. "Standard" is our translation of the word *pacak*, which means a pole stuck into the ground to fix something to the earth or lay claim to the spot, and hence, by extension, a rule or standard of behavior or performance that should be achieved. The listeners here, who feel that the Crĕmasana's *rĕbab* playing has achieved exactly the right quality and mood, are the musicians as well as any and all of the implied audience. Throughout the passages we are discussing, the assessment and description of sounds imply standards of excellence as well as precision of metaphoric analogy by which the quality and nature of sound production is evaluated. Florida suggested to us that *pacak* might refer to the musicians themselves, rather than their playing, and indeed the first line of v. 80 and the second line of v. 82 below support this possibility. We concur that *pacak* refers to both at the same time, fusing the sound, the manner of playing, and the appearance of the musician in performance together in a single, kinesthetic image.

40. This refers to the kind of sounds produced by the *gĕnder* as well as the rapid motion of Gandasana's hands, which gesture in a furious manner as he plays. The description is of the sounds-as-kinds-of-motion.

41. In the first line, *ngĕnuk-ĕnuk* means to form little hills or clumps of cooked rice with a cup and describes the way the *gĕnder* player's hands look while he plays. *Bĕngkak* is a term used for playing the *gĕnder* or *gambang* when the hands are spread, playing low and high notes that are wide apart together. *Pinjal* or *kĕpinjal* are fleas and *ngĕpinjal* is to jump around like a flea, but specifically in the context of playing gamĕlan, to jump into off-beat spaces and fill them up with sound.

42. In the first line of v. 80, the word *pacak* is used again to define the quality of the playing and/or the elegance of the musician, this time by comparing it/him to the masked-dance character Onĕngan. It is interesting that although here the *gĕnder* player is male, the instrument was traditionally played by women in shadow-puppet (and perhaps also masked-dance?) performances. Onĕngan is the elegant sister of Panji, the famous hero of a large cycle of stories that were performed as sung poetry or by masked dancers in Central Java, married to the proud Wukirsĕkar, or Gunungsari, to whom the style of Jayengraga's *gambang* playing is compared in verse 82 below. Both of these characters had prominent roles in masked dance performances of the Panji epic in Surakarta in the nineteenth and early twentieth centuries. In the last line, the alliteration and assonance of "t's," "u's," and "i's" imitate the sounds Tunjangkara makes as he tongues and blows

into his *suling*. The dictionaries do not tell us what kind of bird the *tilěpan* is, only that it is little; its song may well be imitated by the reiteration of "t's" and "i's" in the line, sounds that require the tongue to be pressed forward against the teeth, as does tonguing the *suling*.

43. The last two words of the verse, *aliyu-liyu*, imitate the four-note flourish the *suling* makes at the very end of a cadence.

44. "Thokkingly" is our own invented sound-word to translate Javanese *glěběg*, which is used only here and in verse 69 above in the entire *Sěrat Cěnthini*. For a discussion of how the sound "*gěblěg*," which is onomatopoetic, is made on the *gambang*, see note 29 above.

45. The *Suluk* or *Pathětan Manyura Ngělik* is the song, accompanied by *gěnder*, that a Central Javanese puppeteer sings at the conclusion of *gěndhing* played for certain scenes in the *Pathět Manyura*, or third and final section of a shadow-puppet play in the early hours of the morning before dawn. Probohardjono (see Radèn Ngabèhi Samsudjin Probohardjono, "Serat Sulukan Sléndro," tr. Susan Pratt Walton, in *Karawitan*, volume I, 476-77, 489, and Warsadiningrat, "Wèdha Pradangga," 90), say that *Pathět Manyura Ngělik* was composed by Pakukubuwana IV (1768–1820), the ruler of Surakarta at the time when this version of the *Sěrat Cěnthini* was created.

46. "Sweeps away" in the first line of v. 83 is our translation for *nganyut-anyut*, which refers specifically to the enrapturing effect of musical sounds on a listener and is constantly used in the *Sěrat Cěnthini* to describe the effect of gamělan music on an audience. The same expression is used above in v. 7. In the first line of v. 83, Widiguna's *Suluk Manyura Ngělik* calls attention to the approaching end of the *pathětan*, but gamělan music and singing can also serve as a protective lullaby that allows the listener to doze off, protecting her from harm. Florida once told us that there used to be a special group of soldiers who sang all night from poetic texts as they guarded the palace of the ruler of Surakarta. This practice was probably a variant of the nighttime performance of a poem in *Dhandhanggula* meter, the *Kidung Ruměksa ing Wěngi* (Song Guarding the Night), which can still be heard at night in Central Java to protect listeners and sleepers from a wide variety of evils. Bernard Arps, "The Song Guarding at Night," in *Towards an Anthropology of Prayer: Javanese Ethnolinguistic Studies/Vers une anthropologie de la prière: études ethnolinguistiques javanaises*, ed. Stephen C. Headley (Marseille: Publications de l'Université de Provence, 1996), 47–113.

A marvelous visual depiction of the soothingly sleep-inducing and protective effect of singing *macapat* poetry to an audience at night is the illustration in Figure 6.1, from folio page 96r of British Library MSS Jav89., a late eighteenth-century manuscript of the Damar Wulan romance written in *macapat* poetry. The drawing shows four guards in the evil Menak Jingga's palace after night has fallen, three of them asleep or listening drowsily, eyes shut, heads resting on arms, in different locations and postures, while the fourth reads from a text on which is written: *mamaca damar wulan* (reading *Damar Wulan*). The whole picture is an ironic allusion to the fact that the hero, Damar Wulan, will later kill the evil Menak Jingga and so win the hand of the Queen of Majapahit. To see this illustration in context, as well as feast your eyes on the rest of the wonderful drawings made for this manuscript, including depictions of gamělan playing and accompanying dancing, go to http://www.bl.uk/manuscripts/FullDisplay.aspx?ref=MSS_Jav_89.

47. *Sarayuda* is mentioned ten times in the *Sěrat Cěnthini*, eight of which appear at moments of music-making. The word is described by Becker and Feinstein (see Becker and Feinstein, eds., *Karawitan*, volume III, 27) as designating "the end [last note] of a *pathětan* accompanied by *kěndhang* and *gong* or *kempul* and (sometimes other colotomic instruments)." In such moments in the *Sěrat Cěnthini*, only the *kěndhang* and *gong* are mentioned.

48. In the third line of verse 85, the word *rěbut*, "seize" is a key concept for a certain kind of Javanese behavior during ritual events in which unruly crowds grab food and offerings that have been ritually sanctified in order to capture and possess the blessings that now inhere within them. This practice is associated with Sufi beliefs in Java. See John Pemberton, *On the Subject of "Java"* (Ithaca: Cornell University Press, 1994), 257 and Tony Day, "Stepping on a *Wulu*: Minor Characters and Narrative Possibilities in the *Sěrat Cěnthini*," in *Storied Island: Explorations in Javanese Literature*, ed. Ronit Ricci (Leiden and Boston: Brill, 2023), 51, fn. 30. Later in the canto, in verse 114, when the musicians are playing the *gěndhing Gambir Sawit*, they again *rěbut ngěs kěsaman*, "seize joy in the sameness" of the beauty of their interwoven playing, and then they all become *birai*, "besotted," with what they are playing and hearing, a feeling of ecstasy that brings them closer to God, if they are playing with a religious intention (*niyat*) in mind. Here and elsewhere in the *Sěrat Cěnthini* playing music together can be an exercise in Sufi worship.

49. The word *niba*, "fall," refers to the transition from the first section of a *gěndhing* or *merong* to the second section or *inggah*.

50. This line is difficult to translate. We think that it describes the way drumming patterns change when "falling" from one section of a *gěndhing* to the next. This kind of transition is characterized by an increase in subdivisions of the beat. This increase calls for the drummer to play different kinds of sound, some of which are low, like *běm* sounds, and some of which are higher and faster moving. The word *nělampet* refers to folding or doubling and *kěmpyang*, while usually referring to an instrument, also refers to the smaller, higher-sounding side of the *kěndhang* opposite to the one on which the lower *běm* sounds are played (Javanese drums are usually two-headed).

51. See note 41 above. Here *Pathět Ngělik* refers to an arhythmic, coda-like piece performed after, and in the same mode of, a *gěndhing*. *Ngělik* refers to a version of the piece beginning in a higher register of the mode.

52. See note 23 above. Jayengraga or Jengraga's name can also mean "triumphant in love"; *raga* can mean "body, physical appearance, love." Jayengraga is not only gorgeous to look at and devoted to sexual conquests of both women and men, he loves food, fine clothes, playing music, and singing. Whenever he sings, even when he is intoning religious texts or reciting the Quran in the mosque, his performance is enthralling.

53. Steven Feld, "Aesthetics as Iconicity of Style, or 'Lift-up-over Sounding': Getting into the Kaluli Groove," *Yearbook for Traditional Music* 20 (1988): 82, italics in the original.

54. Sumarsam, "Inner Melody," 303.

55. Passage quoted in Leonard Lewison, "The Sacred Music of Islam: Samāʿ in the Persian Sufi Tradition," *British Journal of Ethnomusicology* 6 (1997): 1, from Book XVIII, "Responses Proper to Listening to Music and the Experience of Ecstasy" of the *Iḥyāʾ ʿulūm al-dīn* (The revival of the religious sciences) by the famous twelfth-century Sufi scholar Abū Ḥāmidal-Ghazālī (1058–1111). The importance of al-Ghazālī's writings and ideas for understanding Javanese Islam, particularly in the literary world of early nineteenth-century Surakarta, where the influence of Shaṭṭārī thinking was very strong, is suggested by the fact that his *Iḥyāʾ* is mentioned in the *Sěrat Cěnthini*. See Tony Day and Nancy K. Florida, "A Sufi Traveler on the Road in Nineteenth-Century Java," *Philological Encounters* 9 (2024): 60–95 for a discussion of another episode in the poem where Ghazālī's ideas from the *Iḥyāʾ* (in this case, from a chapter about conduct during travel) seem to have been on the minds of the *Cěnthini* poets.

56. We take this terse, definitional phrase for *samāʿ* from J. During and R. Sellheim, "Samāʿ," in *Encyclopaedia of Islam, Second Edition*, ed. P. Bearman, T. Bianquis, C. E. Bosworth, E. van Donzel, W. P. Heinrichs. http://dx.doi.org/10.1163/1573-3912_islam_COM_0992.

57. Lewison, "The Sacred Music," 2–3.

58. For an English translation of Ghazālī's Book XVIII, see Duncan B. MacDonald, "Emotional Religion in Islam as Affected by Music and Singing," *The Journal of the Royal Asiatic Society of Great Britain and Ireland* Apr. (1901): 195–252; Oct. (1901): 705–48; Jan. (1902): 1–28.

59. Navid Kermani, *God Is Beautiful: The Aesthetic Experience of the Quran* (Cambridge: Polity Books, 2018), 321. Kermani's chapter 6, "The Sufi Listeners," 293–345, especially the section, "The Pleasure of Listening," 320–345, is the most relevant to our essay.

60. Ibid., 323.

61. Ibid., 325.

62. Ibid., 326.

63. Ibid., 327.

64. Ibid., 331–332.

65. Kamajaya, *Sěrat Cěnthini*, Jilid IX, 1990, 277–78, 281.

66. Jaap Kunst translates all of Wiracapa's comments following the performance of *Gambir Sawit*, Canto 595, v. 41–49, in which the village sage criticizes musicians for their godless infatuation with music. Kunst, *Music in Java*, 267–68.

67. We discuss Canto 596, v. 1–38, paraphrasing and making use of short quotations from the poem as we do so. Kunst translates v. 28–32 in full (Kunst, *Music in Java*, 268–69), and Judith Becker quotes and slightly revises Kunst's translations of these lines in her analysis of this passage (Judith Becker, *Gamelan Stories: Tantrism, Islam, and Aesthetics in Central Java* [Tempe: Arizona State University Press, 2004], 106–8). Becker is the only previous scholar to have suggested a connection between Wiracapa's critique of musicians and the Sufi idea of *samāʿ*.

68. Wiracapa says: "The *yatmaka* is the *ukěl* of the *gěndhing*" (Canto 596, v. 6). The basic meaning of *ukěl* is "tendril-shaped coil." It is widely used in descriptions of hand-movements in dance and terms for variation in musical ornamentation. The hands of the legendary female *gěnder* musician, Nyai Jlamprang, when playing close together, were said to resemble those of a dancer making the *ukěl pakis* (curled fern-tendril) movement with her hands (for this comparison and the legend of Nyai Jlamprang, see Warsadiningrat, "Wédha Pradongga," 115–18).

69. The dictionaries explain that Widayaka is another name for Aji Saka, the legendary bringer of culture and literacy to Java. Another meaning given for "widayaka" is *pujangga*, "poet."

70. Canto 596, v. 21 (Kamajaya, *Sěrat Cěnthini*, Jilid IX, 1990, 280).

71. Compare Kunst's translation of this verse 29 in Kunst, *Music in Java*, 268. We are indebted to Kunst for helping us arrive at our own understanding of the verse.

72. Passage from Ghazālī's *Iḥyāʾ*, Book XVIII, quoted in Lewison, "The Sacred Music," 17. The translation "heart-savour" for *dhawq*, one of Ghazālī's most important philosophical and aesthetic concepts, is Lewison's. Eric Ormsby defines *dhawq* as "to experience directly, without mediation. It is the confluence of perception and action. . . . In his Sufi writings, Ghazali uses 'taste' as a coded metaphor for experience. The deepest truth is perceptible only through experience; truth must be tasted to be known." Eric Ormsby, *Ghazali* (Oxford: Oneworld Publications, 2007), 105. Wiracapa uses the Javanese word *rasa* in a way that reminds us of Ghazālī's *dhawq*. As the *Sěrat Cěnthini*, and Sumarsam, in the passage below, describe it, feeling the "soul" of a *gěndhing* is a practice, something Javanese make happen and experience when they play music

together. See Florida's discussion of *rasa* as direct experience when the word is used in another passage in the *Cĕnthini* where musical performance enables "the presencing of God" in Florida, "Better to Touch," 76.

73. Sumarsam's stress on the primacy of the human voice in his concept of melody and melodic motion in gamĕlan is also noteworthy in the context of our discussion (Sumarsam, "Inner Melody," 262). He says: "If inner melody is the essence of gendhing, based on the inner feeling of the musicians, we may say that the range of the gendhing is actually comparable to the range of the musicians' voices" (263). Commenting on the *rĕbab*, the most voice-like of gamĕlan instruments, Sumarsam writes: "Some gamĕlan teachers believe the rebab presents the 'soul' of the gendhing, and they call it the *pamurba yatmaka* ('that which has authority in presenting the soul')," 264. Recall that it is the *yatmaka* of the *gĕndhing* that the traveling musicians in the *Sĕrat Cĕnthini* are said to seize at a certain moment in each of their consummate ensemble performances. Note also the similarity of the "inner feeling" that, according to Sumarsam, Javanese musicians activate to listen for the *yatmaka* of a *gĕndhing* when they are playing it, and the "heart-savouring" mode of mystical listening to music described by Ghazālī.

74. Kermani, *God Is Beautiful*, 334, quotes from Nietzsche, Adorno, and Schopenhauer to suggest that these German philosophers, particularly Schopenhauer, like the Sufis, thought of music not as a representation but as the "authentic medium" of the Absolute (Ibid., 339–41).

75. Ibid., 344.

76. Lewison writes that Ghazālī "enumerates seven reasons why listening to poetry is more conducive to rapture than hearing the cantillation of the Koran" (Lewison, "The Sacred Music," 19).

77. Florida, "Better to Touch," 91–92. As an alternative translation for line 6 in the first stanza in the Asmarandana meter above, Florida gives: "through the meaning, the Mystery of *suluk* alone." In her Glossary to Ronit Ricci, ed., *Storied Island: New Explorations in Javanese Literature* (Leiden and Boston: Brill, 2023), 287, Ricci defines the word *suluk* as: "Sufi discipline; name of a genre of 'classical' Javanese poetry (Sufi songs) composed in *macapat* verse; related to Arabic *sulūk* or 'pathway' in Sufism." We have italicized the words *rasa* and *suluk* in our quotation of Florida's translation.

78. Kamajaya, *Sĕrat Cĕnthini*, Jilid VI, 124–42.

79. Ibid., 137–40.

80. Ibid., 142. See Tony Day, "The Poetry of Minor Characters and Everyday Life in the *Sĕrat Cĕnthini*," *Wacana: Journal of the Humanities of Indonesia* 22 (2021): 707–726 for further discussion of this verse and the episode at the beginning of Canto 364 in which it appears, especially at 724. The translation of the seventh line published in that article has been revised here following the suggestion by Florida that a better reading for the word *anglola* is *anglela*, which appears in this place in the edition of the poem published in 1912. Serat Tjenthini, *Serat Tjenthini, Babon Asli Saking Kitha Leiden ing Nagari Nederland, Djilid I-II* (Betawi: Ruygrok, 1912), 252, meaning "audible, note by note."

81. The word *brangti*, which we translate as "lovestruck," can also mean "passionately longing" for God. It is a very old word in Javanese associated with wandering, mystical poets as described in Old Javanese poetry. For a discussion of the poet in ancient Java and the "cosmic unity" of his world, see P. J. Zoetmulder, *Kalangwan: A Survey of Old Javanese Literature* (The Hague: Martinus Nijhoff, 1974), 126–214.

82. Just as the pre-Islamic sound of the gamĕlan ensured the safety of the world in which it was sounded and heard, the reading/singing of the poem has a similar effect on the worlds

in which we listen to it. The singing of the poetic episode we have been examining, apart from its potential for inducing mystical states of mind, actualizes and activates the all-encompassing feeling of well-being described in the fictional world of the passage, so that it is transported, via the ears of our hearts, to us as we listen, encouraging us to bring that feeling to bear on our own worlds and lives.

83. The phrase is quoted in Day and Florida, "A Sufi Traveler," 77, in their discussion of the attention that the poets of the *Sĕrat Cĕnthini* pay to minute details of sight and sound, a reflection, they argue, of Ghazālī's belief that even the tiniest, flea-like detail of the physical and living world, all the "specks in the heavens and the earth," bear witness to their "Creator's glory" (79–80). Such an attention to nature in all its particularities can be found in Old Javanese poetry as well; it doesn't commence with the coming of Islam to Java or the acquaintance of early nineteenth-century Surakarta poets with the ideas of Ghazālī.

PART III

Ecstasy

CHAPTER 7

Mystics and Skeptics: Joseph Segond's Psychology of Prayer

Alexandra Kieffer

Pragmatist philosopher and psychologist William James, in the series of Gifford lectures given in Edinburgh in 1901 and 1902 that would comprise his book *The Varieties of Religious Experience*, observed that music has a special relationship to the mystical. "Music," he wrote, "is the element through which we are best spoken to by mystical truth."[1] This idea that music can "speak" a "mystical truth" smacks of a familiar nineteenth-century musical metaphysics, a vestige of Schopenhauerian claims that music is the unmediated revelation of a metaphysical Will. Musical listening transports us out of ourselves into higher regions of being. "Lyric poetry and music," James tells us,

> are alive and significant only in proportion as they fetch those vague vistas of a life continuous with our own, beckoning and inviting, yet ever eluding our pursuit. We are alive or dead to the eternal inner message of the arts according as we have kept or lost this mystical susceptibility.[2]

Thus music "gives us ontological messages which non-musical criticism is unable to contradict."[3]

James was a hugely influential, though polarizing, figure in the development of an early twentieth-century "science of religions." This essay focuses primarily on his influence on the French intellectual Joseph Segond, a graduate of the École Normale Supérieure and professor of philosophy at the Lycée in Toulon and later in Aix-en-Provence. Segond's doctoral thesis was a psychology of prayer, which he defended in 1911 and which was subsequently published by Félix Alcan. Segond was positively disposed toward James, and his work bears the clear imprint of James's influence. But Segond was also highly attuned to objections to James's work raised by French sociologists of the time, specifically Émile Durkheim and Durkheim's nephew and close associate, Marcel Mauss. Segond's attempt to insulate his account of prayer from the objections of Durkheim and Mauss drove him to theorize the psychology of religious experience in a very different way than James. Segond's account of musical attention, in particular—which is very much unlike James's paeans to "mystical truth"—provides insight into his attempt to articulate a psychological stance that could preserve aspects of a Jamesian approach to religious experience while also addressing the critiques of Durkheimian sociology.

My goal, then, is to use Segond's psychology of prayer, and specifically his discussion of musical attention as a tool to inculcate religious contemplation, as a nodal point from which to triangulate a nexus of social-scientific thought on musical attention and religious experience in the early years of the twentieth century. Implicit in this undertaking is the idea that the coalescence of the modern social sciences around the turn of the twentieth century was crucial for defining and reifying a modern understanding of "religion" itself, a claim that has been argued with particular cogency by anthropologist Talal Asad. As he observes, "It was late nineteenth-century anthropological and theological thought that rendered a variety of overlapping social usages rooted in changing and heterogeneous forms of life into a single immutable essence, and claimed it to be the object of a universal human experience called 'religious.'"[4]

Understandings of music, I argue, are tangled up in this intellectual-historical project of defining the "religious" in a variety of ways, even as the theorization of a specifically *musical* attention—that is, attention as it pertained to ostensibly "autonomous" art music—in early twentieth-century French thought differed fundamentally from Segond's attempt to formulate a psychology of prayer. The divergence between these two varieties of attention, at a formative moment in the early history of the social sciences, tells us a great deal about the ways in which the categories of "music" and "religion" emerged as starkly different kinds of epistemic objects in twentieth-century thought. It also sheds light on the complications that attended the entire project of making "experience" into a stable object of social-scientific knowledge, which exposed basic incompatibilities in understandings of both experience and attention in different disciplinary contexts. In particular, this project was from the outset beset by a methodological push-and-pull between fundamentally opposed impulses underlying the increasingly professionalized and distinct disciplines of psychology and sociology: on the one hand, the study of the inner recesses of white European subjectivity, and, on the other, the decentering of European cultural practices through the othering gaze of ethnography.

Segond and the Liberal Protestants

James's *Varieties of Religious Experience* was a watershed in the development of the "science of religions," a text that reoriented the study of religion away from intellectual argumentation and religious doctrine sanctioned by the institutions of the Christian church and toward the psychology of feeling. James understood religious experience as a fundamentally interior phenomenon; he approvingly cited the French Protestant theologian Auguste Sabatier, whose book was a touchstone of the project of so-called "liberal Protestantism" underway in the late nineteenth century, and endorsed Sabatier's contention that the essence of religion lies in interior prayer. James was especially interested in mysticism, the "root and center" of "personal religious experience":[5] The fact of mystical experience, in James's view, provided a compelling body of evidence for the supposition that the truth of our existence exceeds what we are able to discern in our ordinary and rational course of life. "No account of the universe in its totality can be final which leaves these other forms of consciousness quite disregarded. They may determine attitudes though they cannot furnish formulas, and open a region though they fail to give a map."[6] While those who have not had such experiences are not enjoined to "accept their revelations uncritically," James maintains that the fact that many people have had these experiences "break[s] down the authority of the non-mystical or rationalistic consciousness, based upon the understanding and the senses alone. They show it to be only one kind of consciousness. They open out the possibility of other orders of truth, in which, so far as anything in us vitally responds to them, we may freely continue to have faith."[7]

Thus, when James talked about music as a conduit of "mystical truth," he was speaking quite literally and deliberately about a kind of disclosure that exceeds what can be known in our everyday, rational mode of existence. James was, in fact, centrally concerned with this "truth" of the mystical, the truth of the sphere beyond ordinary knowing. "The counting of that wider world of meanings, and the serious dealing with it," he writes, "might, in spite of all the perplexity, be indispensable stages in our approach to the final fullness of the truth."[8]

The Varieties of Religious Experience was published in French translation by Alcan, Paris's renowned publisher of philosophical and psychological works, in 1906, just four years after its initial publication in English, disseminating James's ideas more widely in French intellectual circles. Segond's dissertation on prayer, published five years later by the same press, drew from this translation extensively. Soon into his introduction, Segond notes that he follows James in his understanding of religious experience itself: "I take the term [*états réligieux*] in a broad sense, as M. W. James does, without a specified relationship to the idea of the Divine";[9] Segond cites James, along with Sabatier, in support of his contention that prayer is the central fact of religious life.[10] By referencing the work of both Sabatier and James favorably at the outset of his work, Segond was positioning himself, in a broad sense, as congenial to the project of liberal Protestantism associated with Sabatier. Further, in a very long footnote, Segond undertook a defense of James responding, in part, to a critical review of *The Varieties of Religious Experience* by Mauss (to whom I will return later in this essay), who, Segond claimed, did not "understand James correctly."[11] Characterizing the experience of prayer as, most essentially, a "felt Presence," Segond reproduced precisely what Mauss had critiqued in James and Sabatier: the elision of the principles of "the most liberal Protestantism," and specifically the doctrine of the "interior God," with religion writ large.

Like James, moreover, Segond noted the special affinity between music and mystical religious experience. Indeed, he raised the phenomenon of music as a point of reference immediately after establishing the basic identity of mystical experience as reverent contemplation, or *recueillement*, and appealed to the nineteenth-century French Catholic theologian Auguste Gratry as an authority:

> It would seem that contemplation [*le recueillement*] constitutes the essential character of the activity of prayer. It is thus through the analysis of contemplation that we must begin our study. Yet, if this act [*acte*], or this state [*état*], offers a religious aspect in itself, if any disposition that responds to this term presents a mystic allure, the analysis that we must make does not constrain us to remain enclosed in mystical life, in the ordinary and "technical" [*positif*] sense of the

> word. Because it is a disposition of this type that determines melodic "invitation"; and it is not without reason that prayer so easily takes on the musical mode; it is not without reason that Gratry insists on the kinship [*parenté*] between music and prayer.[12]

Yet Gratry by no means positioned music as a conduit of "mystical truth"; in fact, his account of its purpose and effect was very different from that of James. Lauding music as a "means of true repose," Gratry had focused on the bodily effects of musical rhythm, which "regularizes movement in us, and accomplishes, for the mind and the heart, as well as the body, what sleep accomplishes for the body—it reestablishes, in its plentitude and calm, the rhythm of the beating of the heart, the circulation of the blood and the rising of the chest." Thus, he concluded: "True music is the sister of prayer, like poetry."[13] Segond's account of music, too, focused particularly on the effects of rhythm—and not only rhythm understood in a musical sense. "All rhythmic rocking," he writes, "engenders a similar 'return' [*retour*] to oneself; and all regular actions, even if they are mechanical and physical, tend to bring about, not so much concentration on the action itself as on the regularized and lulled consciousness that assists the action."[14]

Segond's account thus placed under special scrutiny the relationship between the experience of consciousness and the materiality of the body. Like James, Segond believed in the utility of thinking deeply about the testimony of mystics, but he located this utility not in any extrarational intimations of a metaphysical realm but in the physiological aspects of mystical states themselves, which presented an opportunity to investigate human capacities of being and knowing. Regularized gesture, he tells us, is an aide to contemplation because it turns consciousness toward the body: What facilitates the "absorption" of one's attention

> is not the gesture itself, mechanical and automatic (thus a woman who paints her face in a contemplative silence), but the feeling that one acts according to the rhythm of the gesture and that all functions are regulated in this rhythm, thus the consciousness of "internal rhythms."[15]

Musical rhythm, too, though "seemingly intellectual," as Segond acknowledges, acts in precisely the same way. Even leaving "auditory rhythm" and its "physical vibration of the tympanum" to the side, the "heard sounds" of music "may well occupy all of one's attention." As he continues, "one 'communes with oneself' [*se recueille*] only through inward concentration, and the content of the feeling, here again, is not so much the object of hearing itself as the collection of vital rhythms regulated by the changing harmonies."[16] As an outward stimulus that induces "inward concentration," the "heard sounds," precisely by being an object of absorbed aural attention, create a state to which they are ultimately incidental; what enables us to be "absorbed" or captivated by these experiences is not a matter of close attention to, say, the thematic unfolding of a sonata form, of actively listening for aesthetic features of a musical composition, but a simple regularity of rhythm or gesture that induces a contemplative state in the physical body. From there, Segond makes an easy leap to states of religious contemplation, which he maintains can be induced through similarly physiological means:

> If, from these "secular" states, which are only such in a "technical" sense [*d'un point de vue "positif"*], we move to truly "religious" states, the functional nature of contemplation does not change at all: the best proof of this is furnished by the method of prayer [*oraison*] that St. Ignatius of Loyola recommends in a curious passage of the "Spiritual Exercises," which consists in modeling the recitation and meditation of vocal prayer on the physiological intervals that separate inhalations.[17]

Yet, as Segond goes on to explain, even though ecstatic or mystical states are characterized by the manipulation of physiological states, the effects of such manipulation are not experienced as such by the person in the throes of an ecstatic experience. He turns, once again, to music by way of explanation:

> The "melodic" contemplation that constitutes musical enjoyment . . . is a consciousness of respiration and of visceral and circulatory rhythms; but the feeling of lightness and the ebb and flow of vital rhythms does not appear to the man who listens "ecstatically" as a respiratory, visceral, or arterial phenomenon. If the faithful who participate in the gestures of the Office and who inhale the perfume of the incense, if the religious who conform the modulations of their thought and the articulation of their vocal prayer to the rules of Saint Ignatius, determine by this "mechanism" a manner of vital being and

> a consciousness of this manner of being, the "concentration" that is also thereby established does not appear to them in the form of a state of cænesthesia.[18]

Segond's invocation of cænesthesia—the generalized coordination of bodily sensibility—was a reference to Théodule Ribot's widely influential *Les maladies de la personnalité*, published in eighteen editions between 1885 and 1921. Ribot had argued that the apparent unity or coherence of the self is not a psychological given but rather an emergent phenomenon that arises from the coordination of disparate physiological states.[19] Segond, then, was drawing on Ribot's conclusion that the foundation of the self is physiological in essence, with the implication that contemplation must also be considered a fundamentally physiological phenomenon.[20] Yet, Segond's point was that this state of contemplation is not *experienced* primarily as a physiological phenomenon, as a state of coordinated bodily sensibility, but as a change in consciousness. Consequently, Segond was not simply appealing to Ribot as a psychological authority but also issuing a subtle rebuke to his conception of the body as an opaque, static, inert materiality—the kind of bodily materiality supposed, for instance, by the physiological optics of Hermann von Helmholtz (by whom Ribot had been strongly influenced), which took as a given an unchanging mechanism of bodily sense perception. That is, Segond was contending that *recueillement* reveals the relationship between consciousness and bodily materiality to be complexly dynamic, subject to alteration simply by cultivating certain kinds of contemplative attention. Bodily gesture and musical rhythm were, in his estimation, useful as aids to contemplation because they returned attention to the bodily coordination, the "internal rhythms," that are the foundation of the self, in such a way as to change the experience and perception of one's embodied personhood.

In short, to scrutinize the psychological reality of mystical experience, for Segond, is to realize that the relationship between consciousness and bodily materiality is other than what we often take it to be (and other than what the prevailing empirical psychology of the time, as represented by Ribot in particular, took it to be). But, crucially, this awareness of the relationship between consciousness and bodily materiality that arises from using rhythm and gesture as tools of mystical contemplation does not disclose anything beyond the experience itself, *even though belief in such disclosure may be central to what the experience is.* Put another way, the physical and psychological reality of these experiences, Segond contends, does not imply a corresponding *metaphysical* reality, even though the mystic herself understands the experience in a metaphysical way. "We envision this consciousness of 'internal' rhythms and this 'interior life,'" he writes, "as the mystics themselves envision it; we adopt, as *psychologically real*, the aspect under which the life of the body appears to them, [even] if we do not adopt, as *metaphysically real*, the interpretation that conforms to this appearance."[21] Unlike James, Segond insistently bracketed the metaphysical truth of religious experience as outside the bounds of psychological inquiry. Musical sound becomes relevant to this inquiry not because it discloses anything, not because it "speaks" to us of "mystical truth," but because the effect of its rhythm, and specifically the capacity of rhythm to regulate attention, is an especially apt example of a potential contained within the human person, a coextension of consciousness and the materiality of our "vital being" that challenged common conceptions of what consciousness was.

Segond and the Bergsonians

The distinction between the "psychologically real" and the "metaphysically real" was central to Segond's project, and while this distinction has affinities with the general orientation of liberal Protestantism (as represented by Sabatier), Segond's articulation of it speaks equally clearly to the influence of Henri Bergson. In addition to citing Gratry as a theological authority on the contemplative potential of musical attention, Segond also, in passing, mentioned an account of aesthetic experience at the beginning of Bergson's 1889 *Essai sur les données de la conscience*, which comes very close to his own description of a consciousness "lulled" by rhythm. From Bergson:

> In the processes of art we shall find, in a weakened form, a refined and in some measure spiritualized version of the processes commonly used to induce the state of hypnosis. Thus, in music, the rhythm and measure suspend the normal flow of our sensations and ideas by causing our attention to swing to and fro between fixed points.[22]

Segond's conclusion also extensively discussed Bergson's more recent *L'évolution créatrice* (1907), whose ideas he attempted to align with his own. Bergson himself seems to have been persuaded by his efforts: After the publication of Segond's thesis, Bergson wrote to his younger compatriot, assessing his work favorably and analogizing Segond's analysis of the mystical state to the harmonics of a vibrating string:

> You have submitted to penetrating analysis the fundamental state of a praying individual and the states that are like *harmonics*, those that, reinforcing it, give it its psychological *timbre*. You have studied the mystics as they should be studied, with sympathy, but also submitting them to penetrating critique: for my part, I think that they gain more than they lose by being critiqued in this way, and that they can give the philosopher extremely valuable information, precisely that which intuition brings to reason. . . .
>
> I would accept the majority of views that you develop in your conclusion. You have kindly established a connection between your ideas and mine. I am pleased to see that, by different routes, we come to similar conclusions.[23]

Yet not all ostensibly "Bergsonian" thinkers of the early twentieth century took up the connection between musical rhythm and mystical contemplation that so interested Segond and to which Bergson himself seems to have been inclined; indeed, the link is largely absent from French psychological discourse in the first few decades of the century. Instead, a number of thinkers within the Bergsonian orbit drew on Schopenhauer's legacy to situate music as a means to access unconscious phenomena. Albert Bazaillas (who wrote favorably of Bergson's ideas in his 1905 book *La vie personelle*) published another study in 1908 on music and the unconscious that was essentially an elaboration of Schopenhauerian music aesthetics, purporting to reinterpret the German philosopher's metaphysics in light of "the modern psychology of feeling"—a method of which, Bazaillas claimed, Schopenhauer had "had a presentiment."[24] Bazaillas saw music as a means of access to our unconscious interiority: "The sentiment that we feel in the course of [musical] experience," he wrote, "is that of a regression, of a return to primitive simplicity."[25] This "primitive simplicity" consists in the vestiges of a consciousness that evolutionarily preceded so-called rational thought, persisting in human psychology in a kind of subaltern form that, even as it coexists with rational consciousness, is unavailable to its scrutiny.

Music thus achieves an apparently paradoxical feat in Bazaillas's account: It is able to direct conscious attention to that which is definitionally excluded from conscious attention. Though Bazaillas at times pointed to music's capacity to induce an altered state of consciousness, much as Segond did (e.g., that musical rhythm can "bring about a state comparable to hypnosis"),[26] he was also keenly aware that musical attention involved conscious cognition, including forms of recognition prepared by stylistic competence in a specific musical tradition.[27] His account, then, was constantly toggling between the idea of music as a conscious activity and claims about its "analogical" correspondence with the unconscious, which purportedly "reveal[ed]" the true nature of our deep sensibility.[28] As such—and like the human unconscious itself, which Bazaillas likened to the unconsciousness of plant life—music is incapable of deceit or subterfuge; rather, it "speaks profoundly to the soul through its great naïveté. The soul gives way to its spontaneous movements without thought of dissimulating them to itself."[29] In Schopenhauerian terms, music is a "true copy of psychological life, transposed into the register of sound."[30]

In his *Musique et inconscience*, Bazaillas did not posit any direct links between his ideas on music and Bergson's philosophy. Five years later, however, André Joussain, in an article entitled "L'expansion du Bergsonisme et la psychologie musicale," took Bazaillas as attempting a Bergsonian psychology of music:

> Without insisting here on the musicality of M. Bergson's literary forms, one must observe that, since [Bazaillas's] effort was to rediscover in pure durée the deep life of the soul [*la vie profonde de l'âme*], music is precisely the art that unfurls in pure *durée*, that expresses the deep life of the soul, the elements of which, like those of the deepest consciousness, succeed one another or penetrate one another without ever being juxtaposed. But this relationship between music and deep life was barely indicated by M. Bergson. It was inevitable that a philosopher would take up his thought on this point in order to continue, clarify, and expand upon it. This is what M. Bazaillas has attempted.[31]

Joussain held that this "Bergsonian" understanding of music, and of the intimate relationship between music and *durée*, was a recognition of music as a means of knowledge, a method of psychological investigation. "Without leaving the soul [*âme*], quite a large field of research is opened up to us. Music does not reveal to us the deepest foundation of our being; it simply expands the domain of the knowable [*le domaine des faits*]. It should be a means of psychological investigation for us."[32] The music of Beethoven and Wagner allows us to penetrate into the deepest recesses of consciousness—that is, into what is old and "primal" in human nature, something that is inaccessible to historical as well as empirical scrutiny. "Music is our final resource when reflection becomes powerless; it allows us to grasp the part of our psychic life that eludes concepts."[33] Like Bazaillas, moreover, Joussain held that in musical experience there is no duplicity, no possibility for deception or delusion:

> Music renders all of the emotional content of our subconscious activity perceptible to us. Through the effects that it produces on our soul, it uncovers for us all of the potentialities that are in us, and the innumerable energies that our being holds within itself manifest through [music] their reality and their power. This, for example, is how Wagner's music reveals to us some of our highest aspirations in all their purity and in all of their depth.[34]

As long as the "truth" to which music gives us access is imagined as subjective, inner truth (i.e., the truth of unconscious interiority), there is no reason to doubt the veracity of music's supposed disclosure of such truth. It "renders perceptible" our inner emotional life; it reveals what is otherwise hidden within our consciousness. There is no sense of a gap between the experience of musical listening and the otherwise inaccessible domain of human psychology to which it corresponds.

In stark contrast, then, to Segond's insistence on maintaining a distinction between the "psychologically real" and the "metaphysically real," there is a certain conflation here—and along with it, a credulity about the power of disclosure that (autonomous) music was imagined to have. Whereas Segond's text continually grapples with the precise nature of the psychological "reality" of mystical prayer, highly attuned to the problem of parsing what is felt in the body and how the "life of the body" enters into conscious awareness, Bazaillas and Joussain seemingly felt no need to interrogate how the physiological and cognitive processes involved in musical listening ostensibly make aspects of the unconscious mind available to conscious experience. By insisting that musical attention was a direct *means* of acquiring knowledge of the workings of the unconscious, Bazaillas and (still more explicitly) Joussain thereby disallowed this attention from becoming an *object* of knowledge, as Segond had made mystical contemplation an object of knowledge.

Segond and the Sociologists

The striking disjunction between Segond's approach to mystical contemplation and musical experience and that of other Bergson-inclined thinkers might be understood in light of another thread running through Segond's text: his intense, and contentious, engagement with sociological accounts of religion put forward by Durkheim and Mauss. Segond's notes refer frequently to arguments made by these thinkers; indeed the central issue of Segond's entire approach—the question of the "psychologically" versus "metaphysically" real—can be understood as his response to the challenge posed by Durkheim's sociological study of religion.

In particular, Segond's text attempted to answer a central question in both psychology and sociology during this period regarding the nature of human experience and how it can enter into social-scientific knowledge. The differences between Segond (as a psychologist) and the sociologists on this point were fundamental. They were made clear in Mauss's 1902 review of James's *The Varieties of Religious Experience*, where Mauss complained that James did not give sufficient weight to the distinction between "experience" and "belief":

> To the extent that M. James's formula is limited to expressing an important current of Christian religious opinion, it is quite right. The states that were thus described to us are and were actually felt [*ressentis*]. . . . But when one comes to attribute to these ideas and impressions an experimental value, an ontological import; when, instead of saying 'the conscious person feels himself [*se sent*], believes himself [*se croit*] in relationship with a superior consciousness,' one says, 'the conscious person is,' the language of 'experience' [*expérience*] does nothing but mask the much clearer ideas of faith and belief.[35]

Segond's answer to this critique attempted to show that a religious "experience," as a phenomenon of human consciousness, was more than simply a matter of "belief." Rather, the psychology of religious experience, he contended, could not be explained if it was regarded as mere delusion; that is, if the psychologist was not willing to take religious subjects as reliable witnesses to their own experience. He quoted a different portion of Mauss's review of James, in which Mauss claimed that "M. James's last stand would be to deny the unrepentant 'scientist' the right of speaking of these states that he has not experienced [*éprouvés*]. [Yet] the psychologist has no need whatever of having felt [*ressenti*] any feeling in order to speak of it with lucidity."[36] Segond rejoined:

> [Mauss] does not seem to have grasped that this is precisely the essential difficulty, and that these observations 'made by the mystics of themselves,' by the very fact that they are 'personal' and have a 'subjective' character, cannot be treated as 'institutions,' that is to say we cannot behave toward them like a pure 'sociologist,' nor take the 'mental attitude' in which we imagine them as 'things' [*choses*]. Rather it is necessary, if one wants to not only 'speak with lucidity' but 'know' them 'deeply,' to in turn 'experience' [*éprouver*] them sympathetically and 'feel' [*ressentir*] their import [*valeur*].[37]

Segond's reference to "things" was a reference not to Mauss but to Durkheim, though precisely what Durkheim meant by a "thing"—an idea explicated in the preface to the second edition of his *Règles de la méthode sociologique* (the first edition of which was published in 1895, with the second edition appearing in 1901)—is somewhat more nuanced than Segond lets on. Durkheim had claimed, famously, that a "social fact" is a thing, a *chose,* because it is external to us (as opposed to an idea, which is known internally), and therefore cannot be understood scientifically through introspection.[38] Durkheim's argument was that the appearance of psychological phenomena to consciousness—that is, the way in which these phenomena are experienced—does not disclose their nature in the way that properly "scientific" methods can. Rather, all "faits" [facts] in Durkheim's assessment

> are necessarily, from the moment at which we attempt to do science, unknown to us, unknown *things*, since the representations that we have been able to make in the ordinary course of life, having been made without method and without critique, have no scientific value and must be kept at a distance. The facts of individual psychology themselves have this character and must be considered under this aspect. In fact, though they are internal to us by definition, our consciousness of them reveals to us neither their internal nature nor their origin. It lets us know them to a certain point, but only as sensations make heat, light, sound, or electricity known to us; they give

> us confused, fleeting, subjective impressions of them, but not clear and distinct ideas or explicative concepts.[39]

The "mental attitude" that Durkheim takes towards "things" was thus a kind of skepticism that he saw as indispensable to the social-scientific enterprise—the need to set these "things," these objects of study, at a distance from our habitual ways of knowing and assume a posture of (ostensibly) self-imposed ignorance so that they can be approached anew through the steadying methods of science.

Segond, however, insisted that by taking this "attitude" Durkheim had merely deceived himself into believing he had found a solution to an epistemological problem that was far more complex than he was willing to acknowledge. The methodological divergence between the two was thrown into particular relief at Segond's doctoral defense, over which Durkheim presided (and the proceedings for which were published in the *Revue de métaphysique et de morale*). Over the course of a long and combative line of questioning, Durkheim eventually came to the heart of his critique of Segond's work:

> *M. Durkheim:* The point of my objection is this: The mystics do not know the essence of prayer; you have to look for it.
> *M. Segond:* Who knows this essence?
> *M. Durkheim:* We do, when we study things.
> *M. Segond:* Minds are not things.[40]

The disagreement between Segond and Durkheim, ultimately, was a disagreement about witnessing and the status of witness accounts vis-à-vis the scientific study of human life.

Crucially, the shape of this disagreement was determined in no small part by the specific methodological challenges presented by the study of religious experience: It was, after all, because Durkheim regarded the testimony of religious believers as "objectively" uncredible that he sought a different explanation for their experience, an explanation that located the *true* object of that experience (unbeknownst to the believer) in the phenomenon of sociality itself. As Donald Nielsen writes, "'The unanimous sentiments' of believers [in Durkheim's view] could not be based on pure illusion. There was a truth to religion. However, the object of this religious 'experience' was different from what the faithful believed."[41] In a crucial sense, this stance emerged from Durkheim's conception of "religion," which differed profoundly from Segond's understanding of the term. The foundations of Durkheim's (and Mauss's) understanding of religion were not the accounts of premodern Christian mystics but instead recently published ethnographies of Indigenous societies (especially those of Indigenous Australians). Mauss's critique of James was precisely that he had misunderstood "religion" by supposing that its essence was determined largely by historically recent incarnations of European Protestantism rather than by the broadly comparative perspective enabled by ethnography.[42]

Both Segond and James thus can be understood as engaged in an (ultimately unsuccessful) attempt to insulate European religious experience as a meaningful component of human life against what they saw as the encroachment of sociology, specifically the efforts of thinkers like Durkheim and Mauss to redefine the category of religion based on ethnographic evidence. Sociologists, so it seemed, were making religion into a "thing"—a concept fundamentally illegible to the educated, modern European. James says this explicitly:

> [The scientist's] antipathy to religion finds an echo within the very science of religions itself. The cultivator of this science has to become acquainted with so many groveling and horrible superstitions that a presumption easily arises in his mind that any belief that is religious probably is false. In the "prayerful communion" of savages with such mumbo-jumbos of deities as they acknowledge, it is hard for us to see what genuine spiritual work—even though it were work relative only to their dark savage obligations—can possibly be done.[43]

Notwithstanding the validity of Mauss's critique of James's stubbornly narrow Eurocentrism, James had a point, too: that the way in which Durkheim and Mauss were able to imagine religion as an object of social-scientific investigation was possible only because they were not theorizing about people whom they imagined to be fundamentally like themselves or whose full humanity they were willing to acknowledge. Durkheim's focus on Indigenous societies whom he encountered as abstractions in ethnographic literature was surely closely tied to his conviction that the people whose religious practices were chronicled in this literature were not doing and experiencing what they thought they were doing and experiencing, and that the modern European sociologist therefore had access to the truth of those experiences in ways that the participants themselves did not.

We might understand Segond's intense focus on the relationship between consciousness and physiology, experience and truth, as a response to the challenge represented by Durkheim and Mauss, a refutation of their skepticism regarding the nature of social-scientific evidence and its relationship to mystical experience as attested by mystics themselves. In the context of this broader project, Segond's account of musical attention lent support to a conception of consciousness and the human person that could understand mystical experience as psychologically, if not metaphysically, real. When it came to theorizing about musical experience as autonomous art form, on the other hand, as Bazaillas and Joussain demonstrate, the intellectual landscape was very different. The sociology of music was a curiously uncultivated area of inquiry in the early years of the century. Charles Lalo, a student of Durkheim, wrote on the subject, but his "scientific," sociological music aesthetics was premised on the history of European music, not on ethnographic accounts of musical practices in Indigenous societies.[44] His work did not find any particular success with other members of the Durkheim school and was met with derision from musicologists.[45] As a result, Bazaillas and Joussain did not need—as Segond did—to interrogate the nature of musical attention itself as a way of validating the reality they attributed to the experiences made possible by such attention; there was no one to question that such experience was meaningfully "real."

This fundamental, radical asymmetry between conceptions of music and conceptions of religion in human life and culture—an asymmetry through which religion, in Durkheimian thought, became the root and center of sociality, to the point that to explain religion was to explain sociality itself, while musical activities were never more than peripheral to the sociological enterprise—was of course made possible and conditioned by existing ideas about music as an "autonomous" art object forged in the crucible of nineteenth-century, largely German, music aesthetics. But it was surely no foregone conclusion that such ideas should receive so little scrutiny when they came into contact with ostensibly nonmetaphysical strains of early twentieth-century social-scientific thought. It is not that the arguments espoused by Bazaillas and Joussain can be said to characterize the whole of intellectual thought on music of the time (or that the links they posited between music and the unconscious were universally accepted); rather, simply that their ideas could be taken seriously at all speaks to the contours of so-called "psychological" debate about music and their remarkable difference from those of debates about religious experience. In a variety of ways, and in the work of a variety of thinkers, music became fodder for an especially naïve, if covert and unacknowledged, metaphysics that continued to assert itself from within ostensibly empirical psychologies. Musical attention, in this context, could only be understood as the act of a self-present and self-aware perceiving subject—not as anything resembling a Durkheimian "thing," an object of study whose true nature remains opaque to the person experiencing it.

Conclusion

One cannot draw overly broad intellectual-historical claims from the work of marginal thinkers whose influence, even in their own milieu, quickly dissipated; yet, in their insistence on music as a (more or less reliable) means of knowledge, Bazaillas and Joussain are illustrative of an emerging stance toward musical experience that has bearing on broader questions about the place of music in academic thought in the twentieth century. As Schopenhauerian claims about music's immediate disclosure of unconscious interiority faded from academic respectability, the knowledge to which musical attention grants access often became, instead, a knowledge limited to the formal structure of musical objects. What for James had been a disclosure of mystical truth, and for Bazaillas and Joussain a disclosure of psychological truth, became, in much mainstream academic thought on music, a disclosure of a purely formal truth. Accordingly, musical attention, in prominent strains of twentieth-century music-academic thought, became a matter of training and discernment to ascertain what is present *in* music as formal object, and such attentiveness was the hallmark of the successful music academic (who can thereby draw the attention of others to subtle features of the work). This approach is aptly summed up in Schoenberg's observation that "efforts to discover laws of art" can, at best, "influence the way in which the sense organ of the subject, the observer, orients itself to the attributes of the object observed," a process that can "bring closer what is too distant, thereby enlarging details, and remove to some distance what is too close, thereby gaining perspective."[46] Decades later, Allen Forte, in his 1959 explication of the work of Heinrich Schenker, noted that Schenker's achievement was "the deepening of musical understanding through the discovery of structural levels"; this "discovery," Forte contended, rested not on "abstruse speculation, or in acoustical or metaphysical formulations" but on "the organization of the music itself," and, more specifically, on "aural experiences with actual musical compositions"—aural experiences that can then be shared with the student willing to attend in the way that Schenker prescribes.[47]

We might see Schoenberg and Forte as examples (among many) of a distinctly twentieth-century attempt to regulate musical attention, most notably and emphatically within the discipline of music theory, by insisting on the importance of focusing on formal details rather than on the "subjective" images, narratives, or affective connotations that characterize so much nineteenth-century music criticism (E. T. A. Hoffmann's hearing of "radiant beams" in Beethoven's Fifth Symphony, say, or A. B. Marx's *Idee*).[48] And from this vantage, the increasingly narrowed scope of musical disclosure in certain spheres of Euro-American intellectual discourse—from Schopenhauerian metaphysics, to early twentieth-century attempts to adapt Schopenhauer to an ostensibly nonmetaphysical *psychologie*, to the purely formal relationships explicated by twentieth-century music theorists—suggests an attempt to ensure, according to ever more stringent criteria, that the posited object of study (unlike the visions of Segond's mystics) is demonstrably real.

Yet, ensuring the reality of "music" as an object of quasi-scientific inquiry meant eliding or ignoring fundamental epistemological questions of the sort that had so preoccupied Segond; that is, questions about the variegated ways in which song, rhythm, and dance are *experienced*, across and within human cultures, and how these experiences—unanchored by any objective aesthetic content—may be shaped and made palpable by beliefs that are unverifiable from the standpoint of academic inquiry.[49] Such an enlarged perspective would necessarily encompass not only those experiences that are recognizably "aesthetic" in post-Enlightenment European musical culture but also, perhaps more crucially, those that are not (including trance and other altered states of consciousness and the kinds of communal experiences that Durkheim referred to as "collective effervescence").[50]

Put another way, the transformation of a nineteenth-century metaphysics of musical disclosure into a twentieth-century aesthetics of the formally closed, autonomous musical art object went hand in hand with the constitution of music as a secular object of knowledge. The reinvention of nineteenth-century musical metaphysics within an avowedly nonmetaphysical social science in the work of Bazaillas or Joussain should thus be understood in tandem, and in implicit dialogue, with the contemporaneous social-scientific invention of religion in the work of Durkheim and his school. Musical experience was able to avoid a certain kind of "scientific" scrutiny precisely because such scrutiny had been displaced onto religious experience from which music-as-object had been definitively cleaved; yet, the price for evading this scrutiny was that "music" in this sense, requiring a narrowly conceived, aesthetic mode of attention, could have only a fraught and undertheorized relationship to beliefs about what music is and does (an exclusion that occasionally surfaced in music discourse as questions about "presence" or the "ineffable"). Building on arguments made by Gary Tomlinson about the entire category of "music" in post-Enlightenment European intellectual culture, we might posit that a fundamental condition of autonomy in twentieth-century European musical thought—an autonomy that was conceived quite differently from its nineteenth-century progenitor—was that this "music" exclude anything that it could possibly have in common with Indigenous religious ritual.[51]

Notes

1. William James, *The Varieties of Religious Experience* (London and Bombay: Longman, Green, and Co., 1903), 420–421. First published 1902.
2. James, *The Varieties of Religious Experience*, 383.
3. James, *The Varieties of Religious Experience*, 421.
4. Talal Asad, *Formations of the Secular: Christianity, Islam, Modernity* (Stanford, CA: Stanford University Press, 2003), 31.
5. James, *The Varieties of Religious Experience*, 379.
6. James, *The Varieties of Religious Experience*, 388.
7. James, *The Varieties of Religious Experience*, 423.
8. James, *The Varieties of Religious Experience*, 428.
9. Joseph Segond, *La prière: étude de psychologie religieuse* (Paris: Félix Alcan, 1911), 43.
10. Segond, *La prière*, 52.
11. Segond contends that Mauss does not appreciate the extent to which James is aware, at the end of his text, that he is presenting "overbeliefs," and that James does not take the "immediate" experiences of mystics as objectively true statements of fact but simply as real, felt experiences that have import as experiences.
12. Segond, *La prière*, 69.
13. Auguste Graty, *Les Sources* (Paris: Librairie P. Téqui, 1876), 34.
14. Segond, *La prière*, 69.
15. Segond, *La prière*, 71.
16. Segond, *La prière*, 71.
17. Segond, *La prière*, 71–72.
18. Segond, *La prière*, 77.
19. "The unity of the self is thus not the single entity of the spiritualists that is dispersed in multiple phenomena, but the coordination of a certain number of continually renewing states, having as their only point of reference the vague feeling of our body. This unity is not 'top-down,' but rather 'bottom-up'; it is not a starting point, but an endpoint." Théodule Ribot, *Les maladies de la personnalité*, 7th ed. (Paris: Félix Alcan, 1897), 170–171. First published 1885.

20. Segond observes that "because caenesthesia, as the psychologists have proved, is the very basis of individual consciousness, the fact of contemplating [*se recueillir*], of returning to oneself, of taking consciousness of oneself, must have the same essential nature as caenesthesia," that is, the consolidation and coordination of the material body. Segond, *La prière*, 71.

21. Segond, *La prière*, 76.

22. Henri Bergson, *Essai sur les données immédiates de la conscience* (Paris: Félix Alcan, 1889), 11; Bergson, *Time and Free Will*, trans. F. L. Pogson (London: G. Allen, 1913), 14–15. Cited in Segond, *La prière*, 79.

23. Henri Bergson, letter to Joseph Segond (18 January 1911); in Bergson, "Henri Bergson (1859–1941) Lettres à Joseph Segond (1872–1954)," *Revue philosophique de la France et de l'étranger* 205, no. 4 (October–December 2015): 543–560, at 543.

24. Albert Bazaillas, *Musique et inconscience* (Paris: Félix Alcan, 1908), 85.

25. Bazaillas, *Musique et inconscience*, 297.

26. Bazaillas, *Musique et inconscience*, 155–156.

27. "The purest melodies, the richest sonorities are lost in the vacuum of an unprepared consciousness: they find no echo, lacking a sufficiently prompt adaptation of auditory images to our emotional life." Bazaillas, *Musique et inconscience*, 154.

28. "Instinctively, music . . . intuits that each of [our feelings] is an individual organization and that it has a deep, dynamic constitution. It is this dynamism of sensibility that [music] reveals to us and of which, in some way, it causes us to touch the foundation." Bazaillas, *Musique et inconscience*, 160.

29. Bazaillas, *Musique et inconscience*, 169.

30. Bazaillas, *Musique et inconscience*, 149.

31. André Joussain, "L'expansion du Bergsonisme et la psychologie musicale," *La Revue bleue* 50, no. 24 (15 June 1912): 758–763, at 759.

32. André Joussain, "L'expansion du Bergsonisme," 759.

33. Joussain, "L'expansion du Bergsonisme," 759.

34. Joussain, "L'expansion de Bergsonisme," 759.

35. Marcel Mauss, "William James: *The Varieties of Religious Experience*," *L'Année sociologique* 7 (1902–1903): 204–212, at 205.

36. Mauss, "William James," 207.

37. Segond, *La prière*, 62, n. 2

38. This was precisely the argument advanced by Ribot for psychology as an "objective" science two decades earlier; see Ribot, *La philosophie anglaise contemporaine* (Paris: Librairie philosophique de Ladrange, 1870).

39. Émile Durkheim, *Les Règles de la méthode sociologique*, 3rd ed. (Paris: Félix Alcan, 1904), xi–xii.

40. "Thèses de doctorat," *Revue de métaphysique et de morale* 19, no. 1 (January 1911) (supplément): 26–35, at 33.

41. Donald A. Nielsen, "Auguste Sabatier and the Durkheimians on the Scientific Study of Religion," *Sociological Analysis* 47, no. 4 (1987): 283–301, at 294.

42. Mauss, "William James," 204.

43. James, *The Varieties of Religious Experience*, 490.

44. Charles Lalo, *Esquisse d'une ésthetique musicale scientifique* (Paris: Félix Alcan, 1908).

45. See J. C. [Jules Combarieu], "*Esquisse d'une Esthétique musicale scientifique*, par Charles Lalo," *La Revue musicale* 8 (1908): 235.

46. Arnold Schoenberg, *Theory of Harmony*, 3rd ed., trans. Roy E. Carter (Berkeley and Los Angeles: University of California Press: Faber, 1978), 11. Third edition first published 1922; first edition published 1911.

47. Allen Forte, "Schenker's Conception of Musical Structure," *The Journal of Music Theory* 3, no. 1 (1959): 4.

48. See, e.g., Mark Evan Bonds, *Music as Thought: Listening to the Symphony in the Age of Beethoven* (Princeton: Princeton University Press, 2006); and Scott Burnham, "Criticism, Faith, and the *Idee*: A.B. Marx's Early Reception of Beethoven," *19th-Century Music* 13, no. 3 (1990): 183–192.

49. The entire question of belief is, to be sure, an enormously complex one to which I cannot do justice here. Far from disputing important critiques of the idea of belief by Rodney Needham and Talal Asad (among others) as it tended to figure in anthropological writing of the mid- to late twentieth century—in which, to quote Asad's critique of Clifford Geertz, the anthropological conception of religion is "a modern, privatized Christian one because and to the extent that it emphasizes the priority of belief as a state of mind rather than as constituting activity in the world"—my suggestion here is that this anthropological critique has considerable utility, in a musicological context, for revisiting the role of belief as it relates to the experience of musical effect. Along lines that historian Ethan Shagan has recently advocated, such a project would historicize, rather than take as given, the contours of "belief" in a specifically Euro-American secular modernity: "Modern belief" for Shagan is "not a particular epistemic regime" but "a space in which epistemic controversy is managed," and this space itself vitally depends upon not subjecting the coherence of the (inherently incoherent) category of "belief" to scrutiny. Talal Asad, *Genealogies of Religion: Discipline and Power in Christianity and Islam* (Baltimore: Johns Hopkins University Press, 1993), 47; and Ethan H. Shagan, *The Birth of Modern Belief: Faith and Judgment from the Middle Ages to the Enlightenment* (Princeton and Oxford: Princeton University Press, 2017), 21. See also Rodney Needham, *Belief, Language, and Experience* (Chicago: University of Chicago Press, 1972); and Bruno Latour, "'Thou Shalt Not Take the Lord's Name in Vain': Being a Sort of Sermon on the Hesitations of Religious Speech," *RES: Anthropology and Aesthetics* 39 (2001): 215–234.

50. Durkheim, *Les formes élémentaires de la vie religieuse* (Paris: Félix Alcan, 1912).

51. Gary Tomlinson, "Musicology, Anthropology, History," *Il Saggiatore musicale* 8, no. 1 (2001): 21–37.

CHAPTER 8

Among the Aissawa: Music, Hypnosis, and Disruptions of Attention

Céline Frigau Manning

In *Les Aïssâoua à Tlemcen* (1900), the French sociologist Edmond Doutté describes how a young initiate of the Aissawa, a mystical Muslim order, endured hefty saber blows on his arms and stomach without apparent pain. Later, this same young man

> returns and places himself in the midst of the music which redoubles in vigor: it is a terrible racket. There, he begins to dance. . . . His eyes seem to pop out of their sockets, he reaches the pinnacle of exaltation and falls like a lifeless mass, limbs stiffened.[1]

Doutté concludes, "There may be some pretense in this; but there is also sincerity and certainly a great deal of the phenomena of hypnosis." For him, both the multisensory stimulation and the powerful violence associated with Aissawa ritual are linked to a trance-like state that the sociologist can only compare to behaviors familiar to him from the (European) realms of hypnosis and ecstasy.[2]

The Aissawa brotherhood[3] was founded in late fifteenth-century Morocco by Sheikh al-Kamil al-Hadi ben Issa and spread throughout Northern Africa in the eighteenth century. By the nineteenth century, the group had notable branches in Algeria, Tunisia, Libya, and Egypt. During their religious ceremonies, disciples famously performed physical ordeals that European observers interpreted as self-inflicted tortures. These trials, performed to ritual music, were inspired by the legend of their founder Sheikh al-Kamil, who was reputed to have tamed ferocious beasts and venomous snakes. Colonial accounts—by settlers, explorers, and tourists—often described Aissawa ceremonies viewed either in the desert or during street processions celebrating the birth of the prophet Muhammed (the *Mawlid*). Such testimonies, tending toward Orientalism, focused primarily on the Aissawa of Algeria, who formed part of a recent and restive French colony.[4] But beyond enjoying the exotic *frisson* associated with foreign religious demonstrations, French writers were astounded at being unable to ascertain what was actually going on. How could the Aissawa be apparently insensitive to poison, iron, and fire? By what means did they achieve such profound states of trance? Were their rituals simply a performance involving music, dance, and acting—a staged deception, not unlike the marvels one might experience at a bourgeois hypnosis show in France?

Among those who attempted to answer these questions was Hippolyte Bernheim, founder of the "Nancy school" of hypnosis, the main rival of Charcot's Salpêtrière school. In his 1891 study *Hypnotisme, suggestion, psychothérapie,* he described the power of Aissawa music and the multisensory environment of their trancing ritual:

Figure 8.1. "Les aïssaoua.—Feschta religieuse," engraving, *L'Illustration*, April 27, 1850.

> The Aissawa . . . plunge into an ecstatic and anesthetic hypnosis. . . . To the sounds of the Arabian drum and iron castanets, intoxicated by the scent of aromatics whose vapors drift beneath their noses, they prepare themselves by rhythmic movements of the head and of the torso, by guttural sounds modulated according to the musical rhythm, and by violent and disordered contortions. They thus become insensible, swallow crushed glass, pierce their cheeks with sharp blades, walk on red-hot bars, etc.[5]

For Bernheim, the rhythmic sounds and motions of the ritual were gateways to hypnosis, which produced (among other results) a physical insensibility akin to anesthesia. His appeal to hypnosis to explain Aissawa ritual was unsurprising given the importance of the practice in European medical and popular discourses during the long nineteenth century.[6] Hypnosis—as scholars including Clara Gallini and Jacqueline Carroy have shown—was a powerful culture in itself.[7] Often associated with music and widely popular across all social classes, it was invoked within the context of medical research but also had strong ties to magic, charlatanism, and entertainment. All of these resonances are present in European discussions of the Aissawa.[8]

Here I am particularly interested in the role of attention in nineteenth-century accounts of hypnosis. As we shall see, fixed forms of focus were considered key to anesthetized or hypnotic states, especially the self-hypnosis associated with European religious ecstasy. In Christian contexts, ecstatic experience could be marked as divine, conceived as a deep form of prayer with revelatory potential. However, when associated with non-Western traditions, attentional absorption was generally not understood to be uplifting or ennobling but instead regarded as problematic, a "morbid state of the attention" as Théodule Ribot put it, or a form of "mystical degeneracy" in Emmanuel Régis's formulation.[9] These assessments are particularly important in the context of French reports on Aissawa religious trance, many of which rely strongly on the trope of the animalistic other in order to make sense of the participants' evident lack of pain in the face of gruesome self-mutilation. Rather than a sacred attention, in other words, Muslim forms of religious focus were linked with an involuntary cessation of mental activity and thus a degradation of the human into quasi-mechanical or nonvolitional states.

Authenticity and Simulation

In colonial accounts of Aissawa rituals, emotional contagion is key and is almost always linked to a specific and totalizing soundscape that dominates the attention of participants and observers alike. The generation of this sound environment followed prescribed stages: After several hours of prayers recited by the disciples and the audience, the mystics then began to chant alone, swaying back and forth and from one side to the other, repeating at first various phrases and finally only the name Allah. Musicians then initiated a rhythmic, repetitive song accompanied on wooden instruments and tambourines that lasted until the conclusion of the ceremony. The final preparation corresponded to the moment when the Aissawa themselves would get up, partially disrobe, and lean over stoves that gave forth odorous vapors. Then they undertook a series of ordeals frequently involving fire, poison, or iron, ranging from touching hot coals or licking scalding iron rods to swallowing cactus leaves, snakes, or scorpions and even piercing their faces with needles or their bodies with knives.[10] We can get a sense of what this might have involved from a drawing that the traveling illustrator Adrien Marie published in *Le Monde illustré* in 1889 (Figure 8.2): The central image, framed by the geometric decoration of an Oriental salon, shows seven drummers seated around a man who dances in front of an incense burner; this spectacle, visually dominated by the drums, is surrounded by vignettes representing different practices of self-harm, as in a gallery of curiosities.

For nineteenth-century Europeans, the apparently supernatural physical insensitivity of the Assaiwa was a point of great concern. Historical reports on the group's rituals frequently presented the participants as nonhuman, their minds fixed in such an intense state of irrational attentive absorption that they were deprived of their reason and of any restraint imposed by social rules. In his 1904 account of an Aissawa spirit possession ceremony, for example, the career diplomat Eugène Aubin observed that

> the venerable person who is performing [the initiation] first spits in [the initiate's] mouth, and then, in certain cases, dubs him with the name of an animal to which he considers him to possess a resemblance. The lions, tigers, panthers, and jackals will have to demonstrate the ability of the Aissawa to eat raw sheep . . . others will be content to be modest *gzouliyin* [gazelles], whose contortions and

Figure 8.2. Adrien Marie, "Exposition universelle.—Les Aïssaouas à l'Esplanade des Invalides," engraving from the Société générale d'applications photographiques (SGAP), *Le Monde illustré*, October 12, 1889.

> howlings are performed to the accompaniment of tambourines, big drums, and bagpipes.[11]

Here we have something akin to the relationship between magnetizer and magnetized: The initiator induces the initiate to act according to his command. A similar depiction of Aissawa ritual as a form of animal possession can be found in the writings of the military doctor Auguste Warnier, who reported seeing hordes of Aissawa "rush like hungry beasts on a living sheep, still with its fleece, and devour it whole, except for the bones, as hyenas or jackals might do, eating wool, skin, flesh, fat, intestines with everything they contain."[12] Warnier emphasized his feeling of humiliation in seeing "men, my fellow men, making such great efforts to get as close to the brute as possible."[13] This repertoire of dehumanizing graphic images—a kind of lexicon of radical otherness—distanced North-African bodies from European ones by drawing on obvious racial stereotypes.

Indeed, French observers took pains to clarify that the Aissawa's altered states were related to degraded conditions familiar from the bourgeois European salon. Their accounts link the violent trancing of the brotherhood to intimations of pathology or mental alienation that had woven through medical texts on ecstasy since the early nineteenth century. Frédéric Bérard, for instance, in the *Dictionnaire des sciences médicales* (1812–22), had described ecstasy as a pathological state "consist[ing] in a lively exaltation of certain ideas, which absorb the attention so much that the external sensations are suspended, the voluntary movements stopped."[14] A decade later, Alexandre Bertrand identified the deep link that unites ecstasy with religious beliefs, but emphasized that the state did not necessarily entail any mystical content, arguing that ecstasy occurred more often among women or in strongly religious societies because imagination in such cases prevailed over reason; the progress of civilization therefore tended to reverse this relationship.[15] In contrast, when associated with celebrated intellects, such as Archimedes, Newton, or Pascal, or with enlightened Christian mystics such as St. Teresa, Théodule Ribot claimed that states of "intellectual rapture" and exaltation could be ennobling and even transcendent.[16] However, as the case of the Aissawa indicates, these forms of uplift and noble concentration were rarely afforded to other religious traditions.

Colonial writers typically ascribed the apparent state of absorption characteristic of the Aissawa as involving trickery or training, rather than pure mental concentration. French diplomat Eugène Aubin, in his travelogue

Morocco of To-day, argued that demonstrations such as those made by the Aissawa should be understood not as spiritual activities but as empty displays meant to exploit "popular credulity" in order to "live in idleness on the alms brought by the poor to the Koubba of their ancestor."[17] And Baudel, the author of *Un an à Alger, excursions et souvenirs* (1887) had made much the same point, cautioning that the Aissawa were "fanatical sectarians and vulgar magicians, whose trickery and violent acts the Arabs love and admire with a superstitious terror."[18] According to Baudel, such ritual could interest Europeans only as a test of their critical faculties: To believe in such phenomena would be to allow oneself to be fooled, and hence to become more like the local population. The physician Jules Davasse similarly urged: "Scholars and travelers who have the fortune to attend such spectacles, please remove all our doubts, if possible. Beware the mirages of credulity."[19]

These references to credulity invoke a broader nineteenth-century debate presented by the English philosopher William Clifford in an influential article entitle "The Ethics of Belief" (1877) and developed in a subsequent exchange with the American philosopher William James. Clifford argued that it was not only "wrong to believe on insufficient evidence, or to nourish belief by suppressing doubts and avoiding investigation," but that credulity itself created lies and deceit, rendering society "credulous, [so that it will] lose the habit of testing things and inquiring into them; for then it must sink back into savagery," ultimately resulting in a "sin against mankind."[20]

This is the position adopted by the aforementioned Warnier in his report on the Aissawa at the 1867 Universal Exhibition. Warnier writes:

> Algerian displays, whose international theater gives an exhibition to astonished Parisians, passes in the Muslim world of North Africa for real miracles, and, by special privilege of the Prophet Mohammed—may the blessing of God be upon him!—these miracles can only be performed by members of the religious brotherhood of Aissawa, which is today widespread throughout Morocco, Algeria and Tunisia.[21]

By pastiching the eulogy that pious Muslims pronounce at each occurrence of the name of Mohammad or of any prophet, the doctor expressed irony with regard to the feats of the Aissawa. Yet Warnier also takes note of the clear-cut facts presented by his own senses: "It is indisputable," he admits, that "they crunch scorpions between their teeth; that they lick white-hot iron blades with their tongues; that they bite into thorny cactuses."[22] The odor of burning flesh, he continues, is unambiguous: "One can feel the warmth of the red plates and the smell soon warns those present that a scent of burnt flesh or horn has spread throughout the room. So there is burning, but burning without pain. Ordinarily it is either the tongue or the sole of the foot that is subjected to this ordeal."[23] He goes on, however, to explain these ordeals as resulting from mere physical conditioning (rather than any spiritual preparation), noting that "the Aissawa never approach their miraculous exercises without having prepared for them with a sort of bewilderment and daze which makes them look more like animals than humans."[24] Warnier thus suggests that the experience of pain could be largely overcome through training and practice.[25]

Training to Become a Machine

The question of veracity versus mere performance runs strongly through nineteenth-century European debates on hypnosis. These practices involved subjects coached by magnetizers whose bodies would then testify to the reality of mesmeric or hypnotic power. Whether hypnotic phenomena really took place or were instances of charlatanism often remained unclear; indeed, rather than pitting one against the other, hypnotic displays involved a curious entanglement of simulation and reality. Simulation often occurred after the induction of authentic altered states and was intended to offer to an audience the spectacle of a learned "magnetic" performance.[26]

Some European writers mapped this same complexity onto their interpretation of Aissawa rituals. Doutté emphasized that these partly simulated practices, when combined with dance and music, made it possible to "bring those who apply themselves into states more or less close to hypnotic sleep," allowing for the performance of spectacular feats of pain resistance or endurance for the benefit of audiences. As in mesmeric displays, a suspension of awareness could paradoxically be created from the states of intense focus that occur during preparations for the rituals. As Doutté points out, Aissawa practices "are thus called *idjdzâb* in Arabic, that is to say 'trainings in ecstasy.'"[27]

Warnier, still striving to distinguish the true from the false, also focused on the training involved in attaining these absorbed states. In his account, musical stimulation, combined with the lack of air and water, provoked and prolonged states of unconsciousness. The disciples spent "hours at a time, in a narrow and airless room":

> Soon . . . we can no longer breathe, all lungs are panting, the audience is overwhelmed by thirst; the moment is coming when man will become an unconscious machine, obeying the orders of the drum and the voice. . . . Only those who have mounted an assault or charged with a bayonet, driven by the sound of hundreds of throbbing drums beating the pace of the charge, will understand the effect that two hours of iron cymbals and drums, combined with a generalized howling, can produce on the brains of such unfortunates.[28]

It might seem ironic for Warnier to compare the trance of colonized Algerian subjects to the experiences of members of the French military at the front. And yet European military training was comparable to the ritualized preparation of the Aissawa, in that it too involved a complete submission of the subject's consciousness and the absorption of the individual moral and physical body into a collective.[29]

Warnier also appeals to the notion of "ivresse," or intoxication, which he disambiguates from the consumption of alcohol: "Drunk with thirst, with movement, with sounds—especially with sounds—[the Aissawa] do not know what they are doing, they do not feel anything."[30] The notion that states of trance led to the suspension of conscious actions and perceptions, and were thus a form of "ivresse," also appears in other accounts of Aissawa sessions.[31] In 1883, the science journalist G. Kerlus reported that the Aissawa relied upon "a kind of hypnosis, what might be called a mechanical intoxication . . . to support their trials." He concluded, "The dances and songs that precede [their trials] merely aim at bringing the patient to a state of cerebral arousal, a dizziness, a real drunkenness that removes the perception of pain from the brain."[32] Those individuals undergoing trance, whom Kerlus deems "voluntary torture victims," are thus able to induce altered states in themselves as a result of these focused preparations, a kind of self-hypnosis.[33]

In addition to accounting for the onset of the trance and outlining the necessary preparations for the subject's altered mental and physical state, European theories of hypnosis also provided a framework within which medical writers could comprehend the absorptive experience of the trance ritual itself. This is evident in Dr. Crouigneau's report in *Promenades d'un médecin à travers l'Exposition* (1889). Crouigneau emphasized the repetitive, industrialized character of the performance and the consequent link between physical and musical forms of repetition leading to loss of individual subjectivity:

> Before beginning their exercises, all the Khouans [initiates, here of the Aissawa Brotherhood] sit on the ground, legs crossed, in a semicircle, their chorfâ, or descendant of the Prophet, placed in their midst. They recite or rather chant their prayer taken from the Quran, sometimes spreading their arms, sometimes hiding their faces in their hands. Then they accompany themselves on *bendios* or large tambourines, whose skin they stretch from time to time by turning them over a stove of hot coals. They strike with a rhythmic and monotonous movement, rather similar to that of a thresher, alternately with the tip of the fingers in the center of the drum and with the palm of the hand on the peripheral region.[34]

Comparing the movement of musicians to mechanical threshing—a process invented at the end of the eighteenth century, which produced an immense increase in agricultural productivity—renders these Aissawa, in a sense, laborers of performance.[35]

An Aissawa advances, Crouigneau continues, "according to the measure of the battery of seven or eight tambourines which surround him; he jumps upwards with both feet in the air and at the same time swings his head back and forth, at full speed."[36] Timing the duration of this portion of the ritual, the doctor estimates it at seven or eight minutes. As it comes to a close, the accompanying music accelerates, activating then deactivating the performers by shifting them from a state of entranced attention/action to a condition of insensibility:

> When each of them has finished their work, the songs accelerate, the beat of the drums quickens more and more. The Khouan [initiate] resumes jumping frantically and bobbing his head until he finally falls to the ground lifeless. Immediately one of the assistants throws himself on him, sits him up and bends him in two until his head has reached down between his knees, then, straightening him, he lifts him up under his arms and puts him on his feet. The hypnotic sleep is then dissipated. . . . These strange sessions leave one wondering about the real state of the nervous system of these fanatics, who almost every day indulge in similar practices.[37]

Crouigneau expresses fears about the toll of repeated hypnotic performance on the Aissawa, declaring himself perplexed by the health of these men, at the mercy of the mechanical repetition of "hypnotic" processes.[38]

A similar concern, couched in Western medical terms, was voiced by the writer Théophile Gautier. In his 1851 article on the brotherhood for *La Revue de Paris*, Gautier remarked on an Aissawa member who responded to the sound of the drums by "start[ing] and ris[ing] to galvanic jolts like a dead frog, at the shock of the Volta battery."[39] This comparison associates the automatic nature of the mystic's submission to music to that of an amphibian's corpse reacting to an electric current. It recurs when Gautier describes how another Aissawa's "nostrils quivered, his lips were blue, his eyes popped out of his head, the muscles tensed on his thin neck like violin strings on their bridge; nervous tremors shook his whole body up and down."[40] As a result of the violent convulsions of his trance state, Gautier continues, "his arms moved like the springs of a broken machine, with movements which no longer started from a common center and in which the will had no part."[41] Here, the Aissawa appears not just as an involuntary actor but as a broken machine. Gautier's clinical vocabulary both feeds on stereotyped images familiar from discourses on hypnosis—where hypnotized subjects were compared to dysfunctional machines or automata—but also exaggerates it.[42] For him, Muslim mystics are dehumanized as victims of bestial spasms.

Figure 8.3. Octave-Édouard-Jean Jahyer, "Les Aïssaoua, convulsionnaires musulmans," engraving, *Le Monde illustré*, January 16, 1858.

The relationship between French medical description and reports on the Aissawa went further, however. Gautier also used the example of the stages of a nervous crisis, widely known at this time from the work of Charcot and his circle, to portray the brotherhood's rituals. He made notable recourse to coded phrases that would have reminded his French readers of medical terms and behaviors already familiar from reports on mental asylums. Thus, in his staging of the ritual's climax, Gautier writes:

> The fervor of prayer was increasing; the faces of the Khouans were beginning to become distorted. . . . Foam was coming to their lips; their eyes were becoming bloodshot, their upturned pupils fled beneath their eyelid, showing only the cornea beneath. . . . The rhythm of the drums became more and more compelling. . . . The nodding, which had at first been performed by only a few, was now generalized; the oscillations however had taken on such violence that occiputs struck against shoulders and forehead against chests. Soon even this was not enough. The swinging began from the waist up, and the body arced out a frightening semicircle; it was seizures, epilepsy.[43]

This description culminates in the figures of a circle's arc, typical of the first, "epileptoid," stage of hysteria; then proceeding to the violent movements of the second, "clonic," stage, and finally the melancholy or ecstasy of the third, "terminal" stage. By employing the imagery of the body bent into a semicircle, Gautier obliquely announces what is to come. The descent into a convulsive crisis makes it possible to exhibit the unspeakable, to turn the religious narrative into one of horror. This culminates in the ultimate act, involving a living sheep:

> The Aissawa threw themselves on the poor animal, knocked it down, and, while some held its legs, in spite of its tremors and its weak kicks of agony, the others tore at its belly with their teeth, chewing its entrails among the tufts of wool. Some drew towards them, as carnivorous birds do with carrion, a long thread of gut, which they swallowed as they went; others plunged their heads into the collapsed carcass, biting the heart, the liver or the lungs.—The sheep was soon nothing more than bloody mud, a shapeless shred that these ferocious beasts fought over among themselves, with a fury greater than that of hyenas and wolves.[44]

Gautier here highlights the Aissawa's animalian state and the deep repulsion they arouse in the viewer. His sensationalist narrative follows the fashion of contemporary French horror writing, known as "frenetic" literature, which combined Gothic tropes with melodramatic scenes of shocking violence.[45]

Gautier's account further implies that observing the Aissawa entailed a risk of contagion, even to Western audiences:

> My eye clouded over and my reason was confounded before this dizzying scene. The singular imitative sympathy which makes your jaws relax when you yawn caused me involuntary jolts on my carpet; I shook my head mechanically and I, too, felt mad urges to scream. A horseman from the Maghzen, seated not far from me, could resist it no longer and rolled in the dust with laughter and nervous sobs, heaving himself to the hurried, jerky rhythm, panting in time to the snoring tarboukas in an ever-increasing fury of percussion.[46]

The effects of this contagion—not an emotional response, but rather a mechanical sympathy relating to the phenomena of hypnosis—even induces mild symptoms in the narrator himself. And while he manages to withstand the power of the scene, a North African near him is unable to resist and submits by joining the fray. Soon, Gautier reports,

> Through the persistence of singing, drumming and oscillation, the Aissawa had reached the degree of climax necessary for the celebration of their rites; delirium, catalepsy, magnetic ecstasy, cerebral congestion, all the nervous disorders translated into sobs, contortions, tetanic stiffness which convulsed these dislocated limbs and faces which no longer had anything human about them.[47]

Here the phenomenon of imitation produces a deep malaise of reason. As Gautier's story climaxes, it features an accumulation of medical terms alongside a dehumanization of the participants into a mere assemblage of body parts.

Gautier's portrayal invokes a familiar trope that the contemporary anthropologist Johannes Fabian has critiqued in his book *Out of Our Minds* (2000): the figure of the European explorer as an intrepid emissary of science, presented as the polar opposite of the native who is chronically intoxicated by alcohol or drugs.[48] As Fabian notes, European travelers to exotic lands were often found "out of their minds," overwhelmed by "extreme fatigue, fear, delusions of grandeur, and feelings ranging from anger to contempt, under the influence of tropical diseases, alcohol, opiates, and other ingredients from the expedition's medicine chest."[49] The forms of reason attached to Western colonizers was revealed as a sham; indeed, the altered physical and mental states these figures associated with the Aissawa could easily become their own. Reports of the Aissawa, then, can be seen as attempts to grapple with the liminal state of ecstasy itself and the attentional extreme it represented, which promised both to deliver metaphysical revelation and catapult its practitioners into states of exotic degradation.

* * *

The accounts of the Aissawsa I have analyzed in this essay describe fixed attention as a state that simultaneously enables supernatural abilities and leads to self-automation, violence, and degradation. European Orientalizing reports are not just tools in the imperialist arsenal, forms of degradation and subjugation that justify the colonial endeavor, but also evidence of discomfort with mysticism itself, especially the forms of radical absorption it entailed. As an escape from the rigid bounds of reason, hypnotic states were enticing, and even (in some accounts) compatible with religious transcendence. But they were also a threat to order and to the sharp distinctions between colonizer and colonized, and thus the site of an irreconcilable tension. Even firsthand accounts of the Aissawa were dangerous, liable to draw reasoned observers into states of attentional transfixion at odds with rational volition and virile forms of control. Music and sound played an inescapable role in this cognitive slippage, raising old fears around its dissociative power—its ability to shut down or hypnotize the mind, acting solely on the body and therefore reducing the individual to mere physicality. Attention (auditory, religious, journalistic) was key to familiar forms of European transcendence and yet when taken to excess could also undo that power. Where was the boundary? And where did "proof" become involuntary "participation"? These were among the questions that galvanized French discourse on the Aissawa, and which raised as-yet unanswered historical questions around attention, audition, power, and the limits of reason.

Notes

A previous version of this chapter was published in French as *Ce que la musique fait à l'hypnose. Une relation spectaculaire au XIXe siècle* (Paris: les presses du réel, 2021), 99–174.

1. Edmond Doutté, *Les Aïssâoua à Tlemcen* (Châlons-sur-Marne: Imprimerie de Martin frères, 1900), 11. All translations are my own unless otherwise indicated.

2. Ibid. I translate the French word "extase" as the English "ecstasy," in spite of the critique raised by Ian Hacking, who observes that "the French medical name for trance was *extase*—which in that context did not strictly mean 'ecstasy,' as some translators of old medical texts have supposed. But of course it still connotes a more exalted state than the comparatively neutral English word 'trance.'" Ian Hacking, *Rewriting the Soul: Multiple Personality and the Sciences of Memory* (Princeton: Princeton University Press, 1995), 142.

3. Both Aissawa (*ʿIsawa*) and Aissawiyya (*ʿIsawiyya*) come from the name of the order's original founder; the former refers to the disciples, the latter to the order itself.

4. Such accounts supported the colonizing ideology of France at the time; indeed, many justified their texts from this perspective. An 1845 account of the religious orders of Algeria by the career soldier Édouard de Neveu, for example, underlined the danger of Muslim brotherhoods and illustrated their supposed political activities. In a dedication to the Duc d'Aumale, Neveu declared that "to expose the beliefs of the populations of Algeria . . . shows the civilizing task that France has taken on, and perhaps, in some cases, will help to dominate them." Édouard de Neveu, *Les Khouan, ordres religieux chez les musulmans de l'Algérie* (Paris: Imprimerie de A. Guyot, 1846), 8; for the section of this study dedicated to the Aissawa in particular, see pp. 67–110. See also Saaïdia Oissila, "Islam et ordre colonial dans les empires britannique et français: entre collaboration et contestation," *Histoire, monde et cultures religieuses*, 1, no. 25 (2013), 75–105.

5. Hippolyte Bernheim, *Hypnotisme, suggestion, psychothérapie: études nouvelles* (Paris: Octave Doin, 1891), 64–65.

6. See also Dr. Georges Crouigneau's report on Aissawa performances given in a wider account of the 1899 Exposition universelle, which framed North African mystical ritual in more familiar European terms: "We are forced to recognize that some [practices] are not possible except under the influence of a quite particular modification of the nervous system, rather similar to that which we observe in certain states of hypnotism." Georges Crouigneau, "Une visite chez les Aïssaoua," in *Promenades d'un médecin à travers l'Exposition (souvenirs de 1889)* (Paris: Société d'éditions scientifiques, 1890), 19–23.

7. Jacqueline Carroy, *Hypnose, suggestion et psychologie. L'invention de sujets* (Paris: Presses Universitaires de France, 1991), 19. See also Clara Gallini, *La sonnambula meravigliosa. Magnetismo e ipnotismo nell'Ottocento italiano* (Rome: Feltrinelli, 1983, republished by L'Asino d'oro, 2003); Alan Gauld, *A History of Hypnotism* (Cambridge: Cambridge University Press, 1992); Alison Winter, *Mesmerized: Powers of Mind in Victorian Britain* (Chicago–London: University of Chicago Press, 1998). In English, the word "mesmerism" remained a generic term over the course of the nineteenth century, applied to various practices associated with animal magnetism and hypnotism.

8. See Céline Frigau Manning, *Ce que la musique fait à l'hypnose*, 99–174.

9. Théodule Ribot, Psychologie de l'attention, trans. J. Fitzgerald as *The Psychology of Attention* (New York: Humboldt Publishing Company, 1889), 72–103; Emmanuel Régis, *Des*

Hallucinations oniriques des dégénérés mystiques (Paris: C. R. du Congrès des méd. aliénistes, etc., 1895), 260.

10. See, e.g., Théophile Gautier, "Les Aïssaoua," *Loin de Paris* (Paris: M. Levy Frères, 1865), 78–106, at 96–106; Émile Masqueray, *Souvenirs et visions d'Afrique* (Paris: E. Dentu, 1894), 154–159; Émile Laurent, "Les Aïssaouas," *Annales de la santé* 2 (February 1909), 18–19.

11. Eugène Aubin, *Morocco of To-Day* (London: J. M. Dent & Company, 1906), 347. The French original travelogue was published two years earlier; see Eugène Aubin, *Le Maroc d'aujourd'hui* (Paris: Armand Colin, 1904), 435–436.

12. The description is reminiscent of the scenes of divine possession in Euripides's *Bacchae.* See, for example, Auguste Warnier, "Les Arabes ou les Aïssaoua au théâtre international," *L'Exposition universelle de 1867 illustrée* (26 August 1867), 37–39. Warnier was the author of *L'Algérie devant l'Empereur* (Paris: Challamel aîné, 1865). His functions are listed on the title page: "officer of the Legion of Honor, retired military doctor, member of the Scientific Commission of Algeria, former director of civil affairs of the province of Oran, former member of the Council of the Government of Algeria." On Warnier, see Vincent Wright, *Les Préfets de Gambetta* (Paris: Presses Paris Sorbonne, 2007), 420–421.

13. Auguste Warnier, "Les Arabes ou les Aïssaoua," 38.

14. Frédéric Bérard, "Extase," in Marie Joseph Louis Alard, Jean-Louis Alibert et al., eds., *Dictionnaire des sciences médicales par une société de médecins et de chirurgiens* vol. 14 (Paris: Panckoucke, 1812), 294–298, at 296.

15. Alexandre Bertrand, *Du magnétisme animal en France et des jugements qu'en ont porté les sociétés savantes, suivi de Considérations sur l'apparition de l'extase dans les traitements magnétiques* (Paris: J.-B. Baillière, 1826, then Paris: L'Harmattan, 2004, Serge Nicolas, ed.), 473–474. As Jan Goldstein observes, "The term 'ecstasy' as used by Bertrand had an entirely idiosyncratic meaning and . . . included both the extreme religious experience usually associated with it *and* altered states of consciousness utterly lacking in religious content." See Goldstein, *Hysteria Complicated by Ecstasy: The Case of Nanette Leroux* (Princeton: Princeton University Press, 2010), 70. Goldstein refers to Bertrand's *Du magnétisme animal en France* as well as to his articles published in *Le Globe* from 1824 until his death in 1831.

16. Ribot, *The Psychology of Attention*, 93–100.

17. Aubin, *Morocco of To-day*, 363; Aubin, *Le Maroc d'aujourd'hui*, 433–434.

18. [M.-J.] Baudel, *Un an à Alger, excursions et souvenirs. Illustrations de Fromentin, Guillaume, Gérard, etc.* (Paris: Ch. Delagrave, 1887), 130.

19. Jules Davasse, *Les Aïssaoua ou les Charmeurs de serpents (Nouvelle édition revue, corrigée et augmentée)* (Paris: E. Dentu, 1862), 83.

20. William K. Clifford, "The Ethics of Belief," *Contemporary Review* (January 1877); reprinted in William K. Clifford, *Lectures and Essays*, ed. Leslie Stephen and Frederick Pollock (London: Macmillan and Co., 1886), 169, 174–75.

21. Auguste Warnier, "Les Arabes ou les Aïssaoua," 37.

22. Ibid., 38.

23. Ibid., 38.

24. Ibid., 39.

25. Warnier's explanation of the Aissawa's insensibility is reminiscent of French accounts of the so-called "needle test," a contemporaneous method of ensuring that a hypnotic state had been achieved that involved pricking the subject and demonstrating the lack of apparent reaction. Indeed, an apparent muting of pain receptors was one of the most commonly identified

effects of hypnosis, a phenomenon that had raised hopes in the field of surgery since at least the 1830s, especially in the absence of safe anesthetic substances. See Charles Ozanam, *L'anesthésie. Histoire de la douleur*, excerpt from *Le Correspondant* (Paris: Douniol, 1857), 1–5. Tracing recent efforts to Anton Mesmer, Ozanam bemoans their scientific instability: "Mesmer and animal magnetism, in 1776, revived the hope of overcoming pain; but all these attempts were so surrounded by mystery and bravura that serious minds gave up trying to unravel what was indeed true." On the wider history of anesthesia in nineteenth-century France, see Marguerite Zimmer, *Histoire de l'anesthésie: méthodes et techniques au XIXe siècle* (Les Ulis, EDP sciences, 2008), and the special issue, *Pioneers in Academic Anesthesiology: Pitié-Salpêtrière Hospital, Paris, France* in *Anesthesiology* 117, no. 6 (Dec. 2012).

26. The question of simulation is key in hypnosis and constitutes a crucial point of debates until the end of the nineteenth century. See for instance Henri Bergson, "De la simulation inconsciente dans l'état d'hypnotisme," *Revue Philosophique de la France et de l'Étranger* 22 (1886), 525–531; and Carroy, *Hypnose, suggestion et psychologie. L'invention de sujets*, 18: "Unlike sleep, the altered state of consciousness created by hypnosis remains largely a biological enigma. It has a contradictory character. Indeed, the subject relives original or archaic associations in the mode of an intense game. The duplication which is created evokes both a state of complete unconsciousness and the theatrical stage. The hypnotized person is an actor automaton, a paradox with which the scientists of the nineteenth century were confronted, and with which we are still confronted today. . . . even in the eyes of an experienced hypnotist, a truly hypnotized subject and a simulating subject are almost indistinguishable."

27. Doutté, *Les Aïssâoua à Tlemcen*, 9.

28. Warnier, "Les Arabes ou les Aïssaoua," 39.

29. See Eugène Azam, *Hypnotisme et double conscience: Origine de leur étude et divers travaux sur des sujets analogues* (Paris: Ancienne Librairie Germer Baillière et Cie., 1893), 327: "Certain conditions favor and excite courage, especially that of men of lively character: the drums, military music, good weather, the honor of the flag, the uniform, the self-esteem of the body in which the soldier serves, patriotic songs." To this list are added, "among peoples who are little or not at all civilized, the cries of war." See also Marie-Léon Coste, *L'Inconscient: étude sur l'hypnotisme* (Paris: J.-B. Baillière et fils, 1889), 46–47: the conscious being is "suppressed, as during hypnosis," it is unable to react, "exhausted, annihilated; the unconscious being remains alone, being able to undergo the various impressions. The phenomenon reminds many observers of the whirling dervishes of Turkey."

30. Warnier, "Les Arabes ou les Aïssaoua," 37.

31. See notably Laurent, "Les Aïssaouas," 18–19, at 19, and Davasse, *Les Aïssaoua ou les Charmeurs de serpents*, 52.

32. [G.] Kerlus, "Les anesthésiques des jongleurs," *La Nature* 11 (6 Nov. 1883), 245–248.

33. Ibid., 245.

34. Crouigneau, "Une visite chez les Aïssaoua," 19–23, at 20–21.

35. As Emmanuel Reibel has shown, the technologies resulting from the Industrial Revolution—from the metronome to the gramophone, or the first telephone broadcasts of music—affected performers, listeners, and composers, and changed thinking about the nature of music itself. See Emmanuel Reibel, *Du métronome au gramophone. Musique et révolution industrielle* (Paris, Fayard, 2023).

36. Crouigneau, "Une visite chez les Aïssaoua," 21.

37. Ibid., 22.

38. Ibid., 22. The Algerian poet Qaddūr bin 'Umar bin Banīnah, similarly reflected on the misadventures of the Aissawa in Paris, deploring their conditions of travel and housing and critiquing their exploitation, which he blamed on the composer Francisco Salvador-Daniel, who organized the Aissawas' trip and stay in Paris. See Qaddūr bin 'Umar bin Banīnah, "Les Aïssaoua à Paris," in *Chants arabes du Maghreb: étude sur le dialecte et la poésie populaire de l'Afrique du Nord*, ed. Constantin Sonneck (Paris: E. Guilmoto, 1904), 310–319, at 310.

39. Gautier, "Les Aïssaoua," 78–106, at 98.

40. Ibid., 98.

41. Ibid., 98–99.

42. See, e.g., Albert de Rochas, *Les Sentiments, la musique et le geste* (Grenoble: Librairie Dauphine, 1900), 13 (in the first stage of hypnosis occurs a "suspension of all personal will in the subject, who becomes a true automaton"); see also ibid., 49, 112, and 197. On the automaticity of the mind under hypnosis see Carroy, *Hypnose, suggestion et psychologie*, op. cit., 8–18, in particular, p. 18: "The hypnotized subject is an acting automaton, a paradox that nineteenth-century scholars struggled with and that we continue to grapple with today."

43. Gautier, "les Aïssaoua," 97–98.

44. Ibid., 104–105.

45. Gautier's depiction is reminiscent of the frenetic visions authored by authors such as Charles Nodier, Pétrus Borel, Gérard de Nerval, Charles Baudelaire, and J. K. Huysmans; see, e.g., the anthology of French texts published in *The Dedalus Book of French Horror: The 19th Century*, edited by Terry Hale, translated by Terry Hale and Liz Heron (Sawtry: Dedalus, 1998). On frenetic literature, see Anthony Glinoer, *La Littérature frénétique* (Paris: Presses Universitaires de France, 2009).

46. Ibid., 99–100.

47. Ibid., 100.

48. Johannes Fabian, *Out of Our Minds: Reason and Madness in the Exploration of Central Africa* (Berkeley: University of California Press, 2000), 1–6, at 3.

49. Ibid.

CHAPTER 9

Franz, Stumpf, Riemann, Kafka: Toward an Auditory Unconscious ca. 1883

David Trippett

> Beethoven's extraordinary will power and his imagination that accommodated sensation apparently even brought about a temporary improvement in his hearing.
>
> —Carl Stumpf, 1883[1]

Franz

Not paying attention proved unusually costly for the Lied composer and baroque arranger Robert Franz (1815–1892). In 1848, while standing on the platform of a newly built railway station at Dessau, his train—bound for Leipzig—was slowly entering the station when it let off a steam whistle not a meter from his head. The acoustic shock was life-changing, as figures from Liszt to Franz's American friend, the writer and critic William Foster Apthorp, noted. Reflecting in 1903, Apthorp relates the consequences:

> He immediately felt as if something had given way in his ear, and could hear nothing but a mighty internal buzzing. The trip to Leipzig was given up. In a few days the buzzing in his ears stopped, and he could hear again; but it was as if the higher tones of the scale were cut off—above a certain acuteness of pitch he was deaf. From this time forward, note after note dropped out of the range of his hearing, until at last his deafness became complete.[2]

This story of a descending scale of deafness, and variations on it, appeared in articles and commentaries following the composer's death in 1892. It laid the blame squarely at the door of the noise caused by the fierce expulsion of steam and water from locomotive cylinders, something Richard Pohl categorized—without irony—as "physical music" in 1852, and about which practical handbooks had begun cautioning users of the railways even by the late 1830s.[3] With little experience of locomotives, Franz simply hadn't anticipated the fierce blast. Though a quintessentially modern scene, the principle was ancient. If movement set up by an object is too strong for the organ of sense, as Aristotle has it in *De Anima*, "the form which is its sensory power is disturbed; it is precisely as concord and tone are destroyed by too violently twanging the strings of a lyre."[4] The physiological conditions for facilitating accuracy of sight and hearing were a constant for Aristotle (color of the eyes, consistency of the water surrounding them, "purity" of the ear membrane, etc.), so depended on the material health of sense mechanism—that is, lyre strings' density and elasticity—rather than conscious attention.[5] By extension, unguarded attention broke the tonality of Franz's lyre. As Liszt observed poetically in 1872: "He has set his lyre aside."[6]

Against commentaries on a life shrouded in anecdote, his correspondence revealed a more textured story: that his hearing in fact began to deteriorate as early as 1841, that his real name was "Knauth" (rather than the musically auspicious telescoping of *Robert* Schumann and *Franz* Schubert), and that as early as 1867, he had been forced to give up his organist, conductor, and teaching positions in Halle owing to an accumulation of physical incapacities. This did little to hinder the circulation of tales of impairment. With a novelist's eye for ironic coincidence, the American critic Henry Finck reported that Franz's audible tones vanished progressively, from highest to lowest, "until 1876—the year of the first Bayreuth Festival—found him totally deaf," linking the beginning of the world in Wagner's *Ring* (or: E-flat at 18 Hz) with the end of Franz's acoustic world. For good measure, he added a faddish biological cause, ascribing Franz's physical condition to his father's age (60) at the time of his conception: "Physiologists tell us that the children of aged parents are peculiarly liable to all sorts of degenerate nervous conditions," he explains, "Franz's fate did not disprove this doctrine."[7]

All this might appear an intriguing but arguably irrelevant datum of music history, but for the extravagant regard in which Franz's ca. 279 Lieder were held by certain German contemporaries, and—as we shall see—his post-traumatic auditory hallucinations. Let us take these in order. First, Franz's reception as a curate's egg. Beyond the early encomium he received from Schumann ("he seeks more than pleasant- or unpleasant-sounding music, he seeks to recreate the poem in all its vital depth"), still cited in Hanslick's sympathetic obituary,[8] and Wagner's sense of rivalry during the late 1850s,[9] the critic Wilhelm Heinrich Ambros ascribed to Franz's Lieder the "extreme 'depth and inwardness' of Schumann, bound to that "naiveté" of Schubert." When, during the summer of 1871, Ambros and Franz strolled together in the garden of the early Lied composer Johann Friedrich Reichardt (1752–1814), smelling the roses first planted by Goethe that now encircled the house, Ambros later ventriloquized for his readers the words he imagined Reichardt would have said to Franz: "You have *completed* what we once *started*!"[10]

Such genealogical stylizing, along with the organicist imagery and name-as-Frankenword, is almost too obvious a pedestal, too forced a platform for one whose career had fossilized from the perspective of genre and enjoyed only localized acclaim. As a composer of Lieder, almost exclusively so, Franz tacitly rejected Schumann's public caution to "try his hand at new forms" for "success in small forms often leads to one-sidedness, to mannerisms."[11] Just a year after his garden walk with Ambros, Franz offered bitter self-justification for his narrow output and resulting professional isolation. "From the outset," he told the biographer Ida Marie Lipsius in 1872,

> my Lieder found a very small circle of admirers, and a very large, fierce circle of adversaries. . . . The critics of our specialist journals . . . preferred to keep the Lieder in dead silence rather than bump into a hornet's nest that might eventually become uncomfortable for them. But when they opened their big mouths, you could be quite certain that much stupidity would emerge. For one, I had too little melody, for another my accompaniments were too difficult; a third labelled me a weak follower of Schumann, a fourth mocked [my] Bachian turns etc.[12]

In light of these shades of bitterness, Ambros's words might appear excessively sympathetic—consolation toward a living colleague with acute hearing loss, partially paralyzed, experiencing professional decline, and in need of benevolence. In short, the case of Franz all but *inverted* the Beethoven paradigm, as a musical mind defeated by deafness, creating a problem for commentators seeking to articulate his genius. "Unlike Beethoven," Finck confesses, "Franz did not write his best music after he had become deaf."[13] Moreover, his turn from original composition to arrangement (of J. S. Bach and Handel) offers a positively anti-Beethoven model for so-called "disablist music," for it robs the creative subject of originality (i.e., "exemplary originality" constituting the primary criterion of the Kantian genius), hastily widening the aesthetic remit ascribable to a "musical practice that rejects the normal performance body."[14]

Yet aspects of the Beethoven paradigm linger in Franz's reception. One obituary notes how he "swiftly withdrew from the world, which largely alienated him with its constant search for novelty, and in which he hardly felt comfortable."[15] And among his staunchest defenders, it was Liszt who first presented a stratagem of *attentive* listening as poetic straining against deafness, a trope also applied to Beethoven. Here, then, we turn to the topic of auditory hallucinations as the flip side to inner hearing. Liszt's public endorsement from 1855—seven years post–train whistle—carries the claim that Franz

> feels the urge *to hear*, to immerse himself, as it were, in waves of tones, to rock himself on her infinite element, to navigate her ether, to let her fragrant breezes caress her unfolding wings / vibrations [*Schwingen*], to surround himself with her miraculously scenic cloud-images, to eavesdrop on their tragic or poignant dialogue, to transport himself in its world of more expressive atoms, glowing and sparkling like the magic formulas of a heavenly language. Franz wants to *hear music making* and make *music himself*.[16]

This "urge *to hear*" implied hallucination. Writing in 1855, he sidesteps the label of impairment by evoking a quasi-magical science ("more expressive atoms") in language more typical of Wackenroder's generation. Accordingly, he remains elliptical regarding the elephant in the room, the loss of hearing itself as an inward turn in Franz's destiny: "He dreamt of happiness . . . his grand piano [*Flügel*] touched him softly . . . and then it fled! This catastrophe of his inner destiny decided his entire path."[17] Buried in metaphor and the agency of ambiguity, the wordplay on *Flügel* (wing, grand piano, pinna) lets the reader decide. It solicits an aesthetic idiolect that makes the ellipses within a system of meaning mutually functional. The reader must intervene in the text to fill in semantic gaps, in other words, making Liszt's tiptoeing around the issue of deafness a prototype for reader response theory in which "each reading . . . is an actualization of something that by its very nature is a potential, which can never be totally actualized."[18] Significantly, these unrealized potential readings were not primarily the outcome of a poetic flight, but of diplomatic ambiguity.

Liszt's article was benevolent propaganda, after all. He and Franz had discussed it on a two-hour train journey a few weeks prior to its publication in the *Neue Zeitschrift*; "naturally" Liszt tells his partner, Princess Carolyn zu Sayn-Wittgenstein, Franz "will be glad to fill in the blank spaces on the little handwritten manuscript I left in his hands."[19] This helps to contextualize the positive spin placed on "catastrophe," as a speech-act within the sphere of illocution, channeling a reading of openness and possibility rather than a destiny stymied:

> He broke with all temptations of wavering desires, wavering hopes; the pain radiated and concentrated his spirit and gave it that consecration, that energy which allows the soul all its freedom, in order to affirm this freedom with all its might. With these renewed powers he felt called to take his place among men of action, to speak his own language in the name of his inner zeal.[20]

But times change. When this article was reprinted in book form in 1872 (accompanying a fundraising concert tour to secure a pension of 30,000 thaler for the composer),[21] Liszt included a frank postscript that dropped the diplomatic ambiguity. He now confirms the "hearing impairment . . . [that] has spread," making all music making impossible, with conversation only forcible via a raised voice at his ear.[22] This moment of *anagnorisis* brought an optimistic shift toward the belief that medical science might ameliorate Franz's condition and recast the composer's encroaching auditory hallucinations as quasi-supernatural,[23] borrowing a personification familiar to Germanic legend: "In front of the aging master's door lies the gray woman [*das graue Weib*] . . . who whispers all kinds of frightening sayings into his ear and heart."[24] Franz's acousmas—occurring "especially at night . . . [such] that his friends feared he might become insane"[25]—figure as the flip side of a creative imagination lodged between inner hearing and aural infirmity.

Not three years after the train whistle incident, Franz's attraction to the lines of the young poet Max Waldau (Georg von Hausenschild) about inaudible music might encourage a frail life-works reading of this predicament:

An ancient song resounds in the air . . .	Es klingt in der Luft uralter Sang . . .
It is like a forgotten tone	Ihm ist wie einen vergessnen Ton
That nobody can recognise.	Den Niemand wieder erkennen mag.

Taken from the final stanza of the poem "Es klingt in der Luft," which Franz set strophically as Op. 13, No. 2 (1851), it is tempting to suppose the memory of sounds once heard resonated, for him, with Liszt's sense of inner hearing as a creative amulet. If this has a wishful valence as inner hearing on the side of the angels, negative readings are also possible. Auditory hallucinations were not just invitations to musical invention; they also served to actualize the negative claims of critics in Franz's aural imagination, especially in his later years of near-total deafness. He complained privately in 1889, for example, that unflattering reports from those "who knew in earlier years how to judge my little tunes" led him to imagine in sound various musical failings at a concert by the soprano Amalie Joachim—a lost duet-like melody in the tenor ("only the clatter of the demi-semiquavers!"), amid a wholly misconceived character ("the lady showed not a trace of understanding for the tender reticence which is absolutely necessary"). Here the distinction between interior and exterior sound (and between actual and imagined composition) became moot: "I, a deaf man, could of course not attend to the concert."[26] This, it seems, was inner hearing on the side of the devil.

By contrast, Liszt only ever understood inner hearing as a spur to creativity (of the gray lady's many associations, she supposedly knew where the Niebelung's treasure was hidden and carried a set of magic keys that opened mountain vaults: "At her touch a mysterious door opened," one popular account reports, "glimpses of gold and jewels flashed within; but no living man had ever been brave enough to follow and seize the treasure").[27] Beyond such allegories, it was an underexplored link between physiology, attention, and the creative imagination that would draw the interest of empirical researchers.

In this chapter, Franz's experience of his declining aural sense and the personal uncertainties it instilled in him opens the door to a discussion of three modernists from radically different fields: Carl Stumpf (empirical psychology), Hugo Riemann (music theory), and Franz Kafka (fiction). All engage the discourse of aural uncertainty, hallucination, and conscious attention, preparing the ground for a concept of the auditory unconscious—what our reasoned mind tells us we hear when there is nothing to hear, and vice versa, a deafness to rationally present sonic stimuli that appear insensible. Their voices speak indirectly to one another, but collectively broach the uncertainties of aural deception as they dovetail with the solipsist's torment of a possible self-deception in what is heard.

Stumpf

Our first empirical researcher, the prominent philosopher and psychologist Carl Stumpf, declared in 1883 that "attention is paramount" for all sensory judgment. As an expression of interest (that is, a feeling), attention constitutes "a *theoretical* feeling," he explained, "since it stands in an especially close causal relation to perception."[28] But attending to the intensity of any sound impression—heard or hallucinated—immediately presents a paradox:

> The intensity to be observed would indeed become *eo ipso* greater through observation. Hence I could never observe a faint tone, but only a maximally forceful one or at least one that steadily grows during observation, where I can, as a matter of fact, even follow a diminuendo with growing attention.[29]

His swift qualification—"the tone does not become louder, but we become more intensely conscious of it"—foregrounded the role of conscious attention. It was picked up by reviewers who noted that if attention did strengthen a sonic impression it would only hinder an accurate judgment of its intensity.[30] This meant that attention had a Heisenberg problem: Sensory phenomena could become *intensified* for listeners through attention while the stimulus itself remained constant: "We merely become more intensely conscious of it."[31] This splitting of measurable amplitude from consciousness of intensity defined the agency of attention within an empirical domain and influenced the research for volume one of *Tone Psychology* ca. 1875–83. What the split accomplished, for Stumpf, was nothing less than the claim that the experience of sensation itself is real as an autonomous phenomenon, that is, a thing in Heidegger's catchall sense of "a something and not nothing," independent of consciousness.[32] What it did not do was distinguish between attention to inner and "outer" hearing, distantly forecast by the case of Franz.

Shortly after being appointed full professor at Würzburg University in 1873, Stumpf began working on what would become the first volume of his *Tone Psychology* (1883). The project's empirical methods offered an invigorating break from what Stumpf saw as the fruitlessness of recent work in philosophy: "It has always been reassuring and a kind of solace to switch from theories to observation, from reflection to facts, from the writing desk to the laboratory."[33] He set out from the principle that, rather than an ontology of pure sensation (a philosophical rabbit hole), it is only "sensory *judgment*" that can be scrutinized. This required a comparative process, for "every sensation is necessarily related to others; there are simply no pure sensations."[34] Judgment, in this context, means the reaction of the mind to sensations, its apprehension of their intensity and quality, and the relations of these to those of other sensations. A comparative approach is paramount because it sets like against like, avoids a reflexive stage, and prevents judgments from being fixed in language: "Sensation is itself something relative," he explains in an extreme formulation; "we feel not absolute contents but only relations, difference, changes."[35]

What could not be scrutinized within this comparative schema was the phenomenon of aural hallucinations—in Stumpf's words: "those presentations that are subjective in origin and arise with the full character of powerful sensations."[36] He gave these phenomena scant treatment. Yet they raised the possibility of two complementary problem cases: a phantasmal sound that originates in the body or mind to which only one listener can ever be sensitive (i.e., unverifiable by another mind); and a sound that exists in the world but is initially inaudible to the unpracticed listener, becoming sensate only via an effortful process of mental arousal. Stumpf spelled out the theoretical implications of these cases:

> If weak sensations that were already noticed can be intensified by increased attention, we must indeed conclude that, whenever a tone is attentively presented, central or centrifugal arousals in the brain occur and accumulate with the ones on which they are based. . . . [Hence] not merely unnoticed sensations can be made noticeable, but also *arousals of the acousticus [Akusticus] that are not at all given in sensation can be raised above the threshold of sensation.*[37]

In short, attention can make the inaudible audible by bringing about a process of recalibration. The prospect of raising the threshold of one's sensate perception thus implied that thresholds of perception are not fixed (a claim long advanced by researchers interested in mesmerism, spiritualism, and the like). Yet Stumpf's claim was biological rather than miraculous, as evidenced by his postulate of an evolutionary function whereby, if a neural process in the brain accompanies attentiveness, "gradually a special disposition, perhaps one that increases *from generation to generation*, is produced . . . for a faster crossing over the threshold of sensation."[38] Rather than arising from degeneracy, then (as Finck ascribed to Franz), inner or extrapolated hearing might result from inherited attentive acuity.

Behind this thought experiment in breeding hypersensate humans stands the ever-malleable model of Beethoven, whose sheer willpower and imagination coupled to sensation, Stumpf avers, allegedly brought about "a temporary improvement in his hearing."[39] This logic linked attention (will) to acuity based on a process of "centrifugal arousals in the brain," distantly modeled on a reinforcing overtone series. In the case of Franz, such acuity was absent, the ethical implication being that the composer lacked the will to address the challenges of his own condition (*pace* Liszt and the "urge *to hear*"). Such a claim contradicted Karl Marx's historical materialism half a century earlier, which expressly denied the capacity of individuals to swap one kind of conscious perception for another: "This demand to change consciousness amounts to a demand to interpret reality in another way, i.e. to recognize it by means of another interpretation."[40] In short, emerging from Stumpf's scheme was the extraordinary possibility that, by degree, a disciplined practice of sense judgment achieved the opposite of hearing loss.

The possibility of confusion over phantasms of the imagination (that is, hallucinations and sensations that are weak but real), presented an epistemic tangle, for both were to be defined by the variable perception of a listening subject. Adopting an autoethnographic mode, Stumpf briefly relates his experience of passing a wind harp each day where at a certain distance he became unable to tell if its chord still sounded "in the most extreme *pianissimo*" or if he imagined it out of habit. This precedes a related anecdote, borrowed from Karl von Schafhäutl, professor of geology at the University of Munich, concerning an acoustic version of Hans Christian Andersen's tale "The Emperor's New Clothes" (1837): A conductor continually instructed the horn section that they were not playing a certain passage quietly enough; exasperated, they finally placed the mouthpieces to their lips without blowing at all. (In Stumpf's telling, the conductor was satisfied: "That's the right way, gentlemen! Keep doing it like that!" In Schafhäutl's, not quite: "Now it is right; but if possible, still more *piano*.")[41]

The abiding lesson for Schafhäutl, who had also trained in acoustics and organ building, was that inner hearing led to musical experiences. Beethoven's ability to orchestrate scores while deaf, again, serves as the telling detail, predicated on his—assumed—lifelong habit of reading scores silently. Yet this ability represented a danger to acoustic experiments, Schafhäutl concluded soberly. On the border of musical audibility, listeners were prone to hear whatever they wished to hear: "We know only too well how easily one thinks one hears music tones that lie on the threshold of musical audibility, if one wants to hear them."[42]

Like a sleeve pulled inside out, this situation was inverted in cases where auditory sensations could be objectively real in their physical effect on a listening body, yet independent of a listening consciousness. The physiologist and philosopher Gustav Fechner coined the term "negative sensation" (*die negative Empfindungswerth*) for the nonexistence of a sensation where there is a corresponding neural process.[43] In his words: "If the magnitude of the stimulus *ß* or of the psychophysical activity elicited by the stimulus falls below the stimulus threshold *b*, then, according to the measurement formula, the sensation *y* assumes negative values."[44] This concept of negative sensation is not mathematically negative, merely an expression for a form of subconscious perception, and for this reason a few paragraphs later, Fechner also termed it "*unconscious* sensations" (*unbewusste Empfindungen*), acknowledging the role psychophysical activities below such a threshold play in one's "so-called unconscious life."[45]

For Stumpf, who praised Fechner as a "true scientific researcher," this results in the alarming prospect of a ubiquitous auscultation that must be ongoing below the threshold of perception, that is, "*the total mass of undistinguished sensations* which at all times make up the background [noise] of mental life."[46] This notion of an *auditory unconscious* emerges as a specter looming over the scene of German psychophysics in the final third of the century and ties together the four protagonists of this chapter. To trace the mechanisms for attenuating vibration or neural stimuli and the points at which this no-longer-heard content of the body's neural processes proves insufficient to arouse the ganglia or cerebral cortex is simply "impossible," Stumpf protests, "without going into metaphysics."[47] For Fechner, such negative sensations are autonomous phenomena and only become metaphysical with the claim that sensations exist independently of the mind. Ostensibly, unconsciously perceived stimuli always have the capacity to be perceived consciously when attention is applied, but with one qualification:

> In fact, if with fixed attention on an area of sensation I can perceive absolutely nothing of a sensation (or of a perceived difference), in spite of its physical cause being there, which is the *actual* case of the threshold in my sense, then I don't understand—as one says—that there is after all a small sensation (simple or differential) which one

simply cannot compare with others; these seem completely mystical concepts to me.[48]

A sensation that cannot be discerned by any means does not exist in his account partly, we may surmise, because it doesn't fit within the logarithmic law of sensation he had based on Weber's law (famously, Fechner, who studied with Ernst Heinrich Weber, published two laws, Weber's and his own, in his *Elements of Psychophysics* [1860]; these claimed that the intensity of sensation varies as a logarithm of the sensation's stimulus).[49] In *Tone Psychology* Stumpf retorts that there are often means for "*inferring* the existence of sensations" in subsensible cases. This led his most prominent English reviewer (and test subject), James Sully, to break ranks and side with Fechner. Stumpf's argument presented a new, unpersuasive kind of psychological dualism, he felt, where sensations were thought to exist as real entities with a defined content independent of the mind seeking to observe them: "An impression which I cannot detect with a full act of attention," Sully concluded, "cannot surely have any individual existence in my consciousness, however wide we make the region of consciousness."[50]

Stumpf's argument that an auditory unconscious exists (that sonic sensations below perceptual thresholds could be rendered accessible only by the agency of attention) stems in part from the view, current during the 1870s, that *un*consciousness implied both subperceptual phenomena (cf. Fechner's "negative sensation") and those not directed by the conscious mind (namely, automatic or reflex actions). Indeed, the British philosopher George Henry Lewes drew attention to this ambiguity in the second issue of the journal *Mind*, noting that "reflex actions may be unaccompanied by Consciousness—in one meaning of the term—without ceasing to be sentient."[51] A simple explanation of Lewes's account might be the behavior of vibrating hairs within the cochlea duct during the process of sympathetic resonance. Stumpf's insight was that the mind, too, was subject to reflex actions (cf. his accounts above of inaudible orchestral horns or his own daily hearing of an Aeolian harp), so the boundaries of auscultation below the threshold of perception, which I am calling an auditory unconscious, could be established with confidence neither by physiological experiment nor by conscious cognition. Here attentive agency becomes the dynamic element perpetuating this ambiguity, a mediating force between conscious and unconscious.

A musical example can serve to illustrate Stumpf's position. Hermann von Helmholtz had argued in *Sensations of Tone* (1863) that the perception of the phenomenon of consonance was directly related to the existence of each tone's upper partials, whose coincidences determined the nature of consonance for the ear in consecutive pitches (his theory of "melodic affinity") as well as in simultaneous intervals. Untrained listeners could not easily hear the upper partials, he explained, but "success depends more upon a peculiar power of mental abstraction or a peculiar mastery over attention, than on musical training."[52] Half a century later, Stumpf disputed the subliminal value of overtones in this regard. In the detailed endnotes to his 1909 lecture on *The Origins of Music*, he argued against the notion that overblown notes on wind instruments might have been a principal source for the origins of consonance. The makers of panpipes, he hypothesizes, would tune a second pipe so that its third overtone formed the double octave with the fundamental of the first pipe, creating a (large) perfect fourth between pipes. By continuing this pattern to the third pipe, impure consonant intervals arise (e.g., the fifth overtone of the first pipe is equal to the fundamental of the seventh pipe, "which is possible because the overtones are all a little too low"), and in-between notes would be added by a process of measured gap-filling.

Stumpf's hypothetical panpipes show that "overtones are certainly utilized" but the transfer of overtone tuning to the pitch of a new pipe "did not lead to the pure intervals." Prominent uses of the raised fourth—corresponding to the eleventh overtone—for example, in the *ranz des vaches* and associated alpine horn music, only draws attention to the *selective* use of overtones within theories of consonance, raising the prospect that some of these overtones might constitute negative sensations in Fechner's sense, at least for unpracticed listeners. In his *Tone Psychology* of 1883, we read that "a distinction must be made between the existence of a currently unnoticed sensation and the nonexistence of a sensation as such."[53] By 1909, Stumpf had developed the consequences of this into a theory:

> The fact that we do not use the seventh, eleventh, and thirteenth overtones . . . is indeed already evidence that overtones as such are not the crucial factor for our sense of hearing. We occasionally hear those overtones while overblowing, but they do not impress us, save for enthusiasts who worship natural products as such: we cannot use them in connection with our well-grounded tonal system.[54]

In Figure 9.1, reproduced from *The Origins of Music*, the raised fourth in the horn theme at the opening of the finale of Brahms's First Symphony serves as an explicit reminder of this fact: "I, at least, regard it as very probable that

Figure 9.1. Carl Stumpf's example of the influence of the 11th overtone on the Finale of Brahms's First Symphony, ascribed to the practice of overblown wind instruments. Stumpf, *The Origins of Music* (1911), 95.

the horn melody . . . arose as a consequence of such Swiss notes."[55] Disagreements over the relative prominence of different overtones belied the truism that "practice does not change sensitivity" among listeners—leading to the marginalization of certain notes from the aural habitus of Western tonality during the late nineteenth century. At an extreme, Stumpf's logic implies that listeners can revise the relative prominence and separability of different overtones in their hearing through attention and training.

Finally, within a materialist discourse in which imbrications between the physical conditions of life and mind yielded philosophical tracts such as Lewes's *The Physical Basis of Mind* (1877), research into auditory perception was directly implicated.[56] For Stumpf, this material entanglement raised the startling possibility that neural processes connected with attentive comparisons of sensation might alter physical aspects of listeners' bodies. As a violinist, his instinctive comparison was to the physical entrainment of its frequency to bodily response: "There would be molecular changes in [the ear], as we suspect, by analogy, changes in the structure of a violin when it keeps 'responding' better and better during the use of it without there being any visible changes."[57] Any changes to the body of the player would be minimal, he reassures skeptical readers, "for it is not credible that the [auditory] organ could essentially transform itself anatomically in a few hours and become more acute by half." But a few months? A few years perhaps? If physiological sensitivity could reside in the ear and attention could modify the boundaries of an auditory unconscious (and thus the workings of the brain), this established a material undergirding to disagreements over hearing component parts of tones.

Riemann and Kafka: Mobilizing the Auditory Unconscious

In the final section of this chapter, we turn to fiction and return to the interiority of hearing, its unverifiability, and consequent freedoms within a solipsistic approach to acoustic testing. First, the acoustics. Among the fabled disagreements in the history of acoustics, the moment when the twenty-four-year-old theorist Hugo Riemann claimed—and continued to argue for years—that he could hear undertones emanating from his Ernst Imler piano, arguably counts among the boldest. His work could not be replicated by others, so his response to critics relied on an insistence on the unassailable acuity of his own perception: "However this may be, and if all the authorities in the world appear and say: 'We hear nothing,' I must still say to you: 'I hear something, and indeed something very significant.'"[58] Who could disagree? Undertones, we recall, are the mirror image of overtones, their harmonic reciprocals; for believers, they appeared to offer a natural basis for the minor triad, just as overtones had done so for the major triad since the early seventeenth century. In the context of a debatable psychophysics where attentive listening could drive physical change in the auditory system, where thresholds of perception were seemingly variable based on practice, and where cases of impairment appeared to confirm the value of (a Beethovenian) will in improving acuity, the grounds for disputing Riemann's claim might seem dangerously flawed.

In fact, the phalanx of fellow theorists who disputed and finally defeated Riemann's claim that undertones had an objective existence is a matter of historical record.[59] Like Stumpf's argument against an overtone-derived consonance, Riemann initially supported the theory in his doctoral dissertation by including a melodic illustration of undertones taken from a familiar folk song ("Ach wenn du wärst mein eigen"). It parallels Stumpf's strategy to the extent a well-known melody is made unfamiliar by demonstrating the possibility of new ways of hearing, and by extension, new auditory realities, albeit the auditory realities of what cannot be heard.

Figure 9.2. Hugo Riemann's early demonstration of melodic undertones (lower stave) in *Musikalische Logik* (Leipzig: Kahnt, 1873), 46.

A number of questions arise from Riemann's claim in connection with the discourse we have been following thus far: Was the young theorist experiencing a procrustean moment, as Karl von Schafhäutl—a prominent critic—would suggest,[60] finding the sound he so keenly sought as a kind of auditory hallucination? Could his experience be an instance of sustained attention toward sounds ordinarily classed by Fechner as an "unconscious sensation," namely real stimuli coaxed into consciousness from under the threshold of conscious perception? Or might it have resulted from a previously subliminal auscultation, the unnoticed by-product of a listening body that—according to Stumpf's rather radical hypothesis—had undergone "molecular changes" resulting from months of practice in discerning differences in the sensations of tone?

Sensing arrogance, critics such as Schafhäutl had little patience for sympathetic theorizing. "Dr *Hugo Riemann* tells us that he and nobody else, except for Aristotle two thousand years ago, has heard these undertones."[61] Even within the breadth of psychophysical theories, it seemed implausible that Riemann, like some antique violin, was the only body whose resonant response had been entrained to an undertone series, penetrating a threshold of consciousness with no other tool than his Imler piano.

Within the discourse of an auditory unconscious, what was being debated was not the honesty of researchers (for the most part), but rather the mechanism by which the mind's response to auditory sensation was to be measured—something Stumpf defined in all cases as a comparative judgment, as we saw earlier. He might have had Riemann in mind when he declared that any errors in sense judgment fall into two classes: "Either we see or hear falsely, or we interpret falsely what is seen or heard."[62] That is, objectively false sensory judgments arise when the mind either imagines or misinterprets a stimulus. The crossing-point in this division, between cognition and physiology, was made unstable by the possibility that powers of attention could change the status of sounds previous thought to be imagined. As is well known, the outcome was significant for Riemann's dualist theory of harmony because undertones promised finally to reveal the natural basis of the minor triad. In the third edition of *Sensations of Tone* (1877) Helmholtz had described harmonic undertones as a property of the glass resonators he cupped to his ears during experiments to isolate over/undertone frequencies, but in a later footnote he confessed to being unable to demonstrate them directly via sympathetic resonance on the piano.[63]

As scholars from Alexander Rehding to Alexandra Hui have shown, physicists continued to dispute the reality of undertones until 1891, when

Riemann retreated into a stalemate, arguing that because a state of perfect inference existed between the sound waves of over- and undertones, the latter were rendered inaudible, in other words, they were a matter of interpreting falsely "what is seen or heard." It is striking, historiographically, that such a claim can now be read as the analogue to Stumpf's belief that sensations can exist autonomously in the mind even when not perceived consciously.

Just as the line between imagined undertones and sounds that were deemed real but below the threshold of normative conscious perception blurred, so too did the hard distinction between sounds heard in the world and "phantasmal" sounds (those created or reproduced in the mind). This loss of certainty is indicative of the collapsing boundaries that for several decades gave credence to Riemann's claims. Indeed, amid the currents of phenomenology eddying within Germany, Edmund Husserl used musical sounds to illustrate the reality he ascribed to the contents of consciousness. He defined phantasy in 1905 as consciousness "characterised by re-presentation (reproduction)," that is, a consciousness incapable of setting forth some kind of objectivity: "Not to give the object itself is the very essence of phantasy."[64] His test case for distinguishing direct perception from phantasy was hearing several tones in succession, then imagining the last sound in one's head. In terms of temporal succession, these were categorically distinct, the one occurring after the other. In terms of consciousness, he acknowledged that the consciousness of sound that remains after the directly sounding event ends "presents itself in an entirely different way from the way in which the sounding presents itself." However, it is with acts of mental reproduction that the two begin to blend:

> We never refer to a continuous transition of perception into phantasy, of impression into reproduction . . . but reproduction, phantasy-consciousness also requires precisely the same shadings, only reproductively modified. . . . Sensation which functions presentatively for the object, forms a seamless continuum; and the phantasm likewise forms a continuum for the re-presentation of a phantasy object.[65]

This claim is based on the argument that all perception arises from sensation, and in the "now-moment" of our attention perception need not distinguish between the contents of sensations originating outside the mind and those of phantasms originating inside it. A counterintuitive conclusion finally dissolves the distinction entirely: "Whoever assumes an essential difference between sensations and phantasms naturally may *not* claim that the apprehension-contents for the just-past phases of time are phantasms."[66] Here, then, to circle back to our earlier figures, the difference between Franz's inner hearing and Riemann's illusive undertones becomes moot, for as contents of a conscious mind attentive to inaudible sound, both would count as equally real for listeners by comparison to fully heard sounds. This, at least, is the phantom auditory door opened by Husserl's phenomenology ca. 1905.

Given the strained credulity of Riemann we saw above and the gradual intertwining of "actual" and "fantastical" theories of perception in Husserl and others, it is perhaps unsurprising that the contestable discourse around sonic interiority would find expression in fiction itself. Franz Kafka's last short story, *Der Bau* (*The Burrow*, 1923), remained unfinished at his death. What survives fictionalizes the controversial psychology of an auditory unconscious with uncanny precision. Its first-person narrative sees a solitary mole-like creature discover "an almost inaudible whistling noise" in his burrow.[67] He tries to determine its origin—intruder or hallucination—and fails, despite being a practiced engineer ("I have a good deal of experience in investigations of this kind"). Flashes of double-meaning alert readers to the underlying mind-ear nexus through anatomical wordplay: "I must have silence in my passages" (the plaint of a tinnitus-sufferer?) or "I stray so far that I find myself at the labyrinth,"[68] as though a giant ear were being explored. As a result, the distinction between inner thought and outer reality collapses in the creature's underground isolation, itself a metaphor for the interiority of mind, for the noise "is, so to speak, audible only to the ear of the householder."[69] As well as erasing distinctions between outer world and metaphysical solipsism, this in turn leads to improbable conclusions: "The noise can be heard everywhere and always at the same strength, and moreover uniformly, both by day and night. At first, therefore, one cannot but incline to the hypothesis of a great number of little [intruder] animals."[70] The anthropomorphic creature displays some acoustic refinement ("my ear can distinguish [the noise] in all its finest shades"[71]), as well as rendering several of Stumpf's concepts from *Tone Psychology* in narrative prose: from comparative sense judgments ("There still remains the possibility that there are two noises"[72]) to the affecting powers of continuous attention ("the noise seems to have become louder"[73]) and discernment resulting from training ("My sensitiveness to disturbances in the burrow has perhaps become greater with years, yet my hearing has by no means grown keener"[74]). By the end, Kafka's creature arguably becomes truer as an encapsulation of discourse on the auditory unconscious than the individual historical listening subjects we have traced in Franz, Stumpf, and Riemann.

It is telling that, biographically, Kafka not only confessed to being "unmusical" (a view borne out by his friend Max Brod, who soon gave up taking Kafka to concerts "since the impression they made on him were of a purely visual nature"),[75] but for him, listening to musical sound was mentally trapping, a means of demarking off the normal world with a sonic one: "The natural effect of music on me is to circumscribe me with a wall," he scribbled privately after an all-Brahms concert given by the Vienna Singverein in 1911, "and its only constant influence on me is that, confined in this way, I am different from what I am when free."[76] Rather than initiate an autobiographical reading of being confined within *Der Bau*'s subterranean environment, this raises the prospect that all aural worlds can be both world-defining and yet "ultimately unfathomable," as one reader of Kafka recently put it.[77]

Epilogue

In an aside, the French philosopher Paul Ricoeur once observed that "the relationship between fiction and history is assuredly more complex than we will ever be able to put into words."[78] This rings true for the above attempt to thread a discourse of auditory perception through the needle-eye of Kafka's unfinished story. Even the stability of the ear as a material object ran into difficulty with Franz's advancing deafness and Stumpf's "molecular changes." They point to conditions of uncertainty and hypothetical questing that were susceptible to exploitation and help to explain why false authority in audiometry could flourish. During this period a fad for "Patent Artificial Eardrums" gained prominence, including commercial fraud by figures such as J. H. Nicholson (Figure 9.3), whereby aural realities could be redefined as much by belief as by empiricism, "throughout the civilised world."[79]

Here aural fraud counts as little more than applied fiction vis-à-vis Kafka's tale. It also explains why, from the outset of his doctoral work, Riemann's belief that acts of listening are always analytical—"*so-called simple tones are not simple*"[80]—was also a statement about the inscrutability of private cognition, projected as such through Kafka's subterranean creature who discerns noise in "all its finest shades." Not only do acts of listening comprise

Figure 9.3. An advertisement for J. H. Nicholson's artificial eardrums, *English Mechanic and World of Science* (11 November 1887), vii.

composite partial tones, but they reflect the identity of the mind attending to them. Accordingly, before articulating what is meant by the analysis *of musical works*, the first edition of Riemann's *Musik-lexikon* sets out a prior sense of "analysis" as an act of cognitive discernment: "*Analysis of sounds by the ear* is an end point of the new acoustics, it means the description of partial tones contained within a single tone (sound) of our musical instruments."[81]

With this skill, it is possible to discern the different partial tones within a single sound, even if *how* this happens continues to remain "insufficiently explained" (*noch nicht hinlänglich erklärt*).

This insufficiency explains why debates over impairment and musical composition, the thresholds of sensation, the role of consciousness in acts of perception, and the prospect of an acoustic unconscious nest together as interwoven contingencies during the 1870s. In this sense, their emergence as a single entanglement, in the sense of a Foucauldian unity of discourse, is not the story of a historical concept's progressive refinement or rationalization but of its "various fields of constitution and validity . . . its successive rules of use [and] the many theoretical contexts in which it developed and matured."[82] Above all, this entanglement speaks to the centrality of *sound* in guiding theories of sensation, perception, consciousness, and imagination in the final third of the century. Within the strengthening field of empirical psychology, moreover, the role of attention *to* sound sits across these discourses, both structuring their relations and, at times, threatening to dissolve the very boundaries between them.

Notes

1. Carl Stumpf, *Tonpsychologie I* (1883), Eng. trans. *Tone Psychology: Volume 1—The Sensation of Successive Single Tones*, trans. Robin D. Rollinger (Abingdon: Routledge, 2020), 241.

2. William Foster Apthorp (ed.), *Fifty Songs by Robert Franz* (Boston: Oliver Ditson, 1903), xiv (fn).

3. "ein Geräusch hervorgrbracht wird, welches . . . auf heerstraßen jedoch gewiß bedenklich seyn würde." Carl Hartmann, *Praktisches Handbuch über die Anlage von Eisenbahnen* (Augsburg: Jenisch & Stage, 1837), 421–22. The first passenger line, the Bavarian Ludwig Railway, had opened a mere thirteen years earlier, on 7 December, 1835. See also Richard Pohl, *Akustische Briefe* [1851–52] (Leipzig: Bruno Hinze, 1853), 95–101.

4. Aristotle, *De Anima*, Book 2, 424a, ll. 28–32; Eng. trans. W. S. Hett, *On the Soul* (Cambridge, MA: Harvard University Press, 1957).

5. Aristotle, *De Generatione Animalium*, Book 5, 1–2; Eng. trans. A. L. Peck, *Generation of Animals* (Cambridge, MA: Harvard University Press 1942).

6. "er [hat] seine Leyer bei Seite gelegt." Franz Liszt, *Robert Franz* (Leipzig: F. E. C. Leuckart, 1872), 56.

7. Henry Finck, "Robert Franz: Anecdotes and Opinion," *The Looker-On* (1 October 1896): 288.

8. "er will mehr als wohl- oder übelklingende Musik, er will uns das Gedicht in seiner leibhaftigen Tiefe wiedergeben." Robert Schumann, "Rezension der Lieder op. 1," *Neue Zeitschrift für Musik* 19 (1843): 35. Hanslick's guarded appraisal speaks of "the last of that rare band that . . .

gathered about the standard bearers of musical romanticism, Schumann and Mendelssohn." Hanslick, "Dr Hanslick's Feuilleton," *Musical Herald* (1 December 1892): 53.

9. Richard Wagner, *My Life*, trans. Andrew Gray (Cambridge: Cambridge University Press, 1983), 552. See also John Deathridge, *Wagner Beyond Good and Evil* (Berkeley and Los Angeles: University of California Press, 2008), 262; and Werner Wolf, "Die Beziehungen zwischen Robert Franz und Richard Wagner," in *Robert Franz (1815–1892): Bericht über die wissenschaftliche Konferenz anläßlich seines 100. Todestages am 23. und 24. Oktober 1992 in Halle*, ed. Konstanze Musketa (Halle: Händel-Haus, 1993), 152–62.

10. "Da war wir, als trete uns Reichart entgegen und spreche zu dem lebenden Meister: 'Du hast *vollendet*, was wir einst *begonnen*!'" W. H. Ambros, *Bunte Blätter: Skizzen und Studien* (Leipzig: Leuckart, 1872), 325.

11. Henry Pleasants (ed. and trans.), *Schumann on Music: A Selection from the Writings* (New York: Dover, 1965), 190.

12. "Meine Lieder haben vom Anbeginn einen sehr kleinen Kreis begeisterter Verehrer und sehr großen heftiger Widersacher gefunden. . . . Die Kritik unserer Fachblätter . . . suchte die Lieder lieber tot zu schweigen, als daß sie in ein Wespennest stieß, daß ihr vielleicht mit der Zeit unbequem warden konnte. Wenn sie aber den breiten Mund aufthat, ließ sich mit voller Bestimmtheit darauf rechnen, daß eine dicke Dummheit zum Vorschein kam. Dem einen hatte ich zu wenig Melodie, dem anderen waren meine Begleitungen zu schwierig; der dritte nannte mich einen schwachen Nachtreter Schumanns, der vierte mokierte sich über Bachsche Wendungen u. s. w." La Mara [Ida Marie Lipsius], "Zur Erinnerung an Robert Franz: Briefliches und Mündliches von ihm." *Westermanns Illustrierte Deutsche Monatsheft* 75 (1894): 707–18, at 710–11.

13. Finck, "Robert Franz," 289. Emphasis added.

14. Blake Howe, "Disabling Musical Performance," in Howe, Stephanie Jenson-Moulton, Neil Lerner, and Joseph Straus, eds., *The Oxford Handbook of Music and Disability Studies* (New York: Oxford University Press, 2016), 191–209, at 200–201. Cf. Franz's public statement on musical arrangement as method, *Offener Brief an Eduard Hanslick Ueber Bearbeitungen älterer Tonwerke* (Leipzig: F. E. C. Leuckart, 1871).

15. "Und doch mußte er scheinbar schnell Abschied nehmen von der Welt, die ihm mit ihrer rastlosen Sucht nach Neuem fast entfremdet war, und in der er sich kaum mehr behaglich fühlte." Richard Winker, "Erinnerungen an Robert Franz," *Neue Musikzeitung* 13 (1892): 257–58, at 257.

16. "den Trieb in sich fühlt, *zu hören*, sich gleichsam in Wellen von Tönen zu tauchen, sich auf ihrem unendlichen Element zu wiegen, ihren Aether zu durchschiffen, ihre duftenden Hauche um seine entfalteten Schwingen kosen zu lassen, sich mit ihren wunderlandschaftlichen Wolkenbildungen zu umgeben, ihren tragischen oder ergreifenden Dialogen zu lauschen, sich in ihre Welt ausdrucksvoller, gleich den Zauberformeln einer himmlischen Sprache glühender und funkelnder Atome zu versetzen. Franz wollte *musiciren* hören und selbst *musiciren*." Liszt, "Robert Franz," *Neue Zeitschrift für Musik* 23 (1855): 242.

17. "Er träumte von einem Glück . . . leise berührte ihn sein Flügel . . . und dann entfloh es! Diese Katastrophe seines innern Geschicks entschied seine völlige Reise." Ibid., 246.

18. Wolfgang Iser's concise account of his own theory from *How to Do Theory* (Oxford: Blackwell, 2007), 77.

19. Liszt to Carolyn zu Sayn-Wittgenstein, 15 October 1855, in La Mara (ed.), *Franz Liszt's Briefe*, 8 vols. (Leipzig: 1893–1905), 4:273.

20. "Er liebte mit aller Hingebung, wie sie nur in seiner reinen edlen Natur keimen konnte. Er träumte von einem Glück . . . leise berührte ihn sein Flügel . . . und dann entfloh es! Diese Katastrophe seines innern Geschicks entschied seine völlige Reise. Er brach mit allen Lockungen schwankender Wünsche, schwankender Hoffnungen; der Schmerz strählte und concentrirte seinen Geist und gab ihm jene Weihe, jene Energie, die der Seele ihre ganze Freiheit lassen, um mit allen Kräften diese Freiheit zu bestätigen. Mit diesen neuerweckten Kräften fühlte er sich berufen, seinen Platz unter den Männern der That einzunahmen, im Namen seiner innern Begeisterung seine eigene Sprache zu sprechen." Liszt, "Robert Franz," 246.

21. Details recorded in Hugo Riemann's *Musik-Lexikon*, 5th ed. (Leipzig: Max Hesse, 1900), 340.

22. "[T]rotz seines Gehörleidens war der Verkehr mit ihm nicht schwer. Das mit etwas erhabener Stimme in sein Ohr gesprochene Wort blieb ihm ohne Schwierigkeit vernehmbar." La Mara, "Zur Erinnerung an Robert Franz," 707.

23. "Perhaps some fortunate turn of events in his story will bring him back his lost health, revive his creative drive, and bring us a new springtime of songs." ("Vielleicht, daß durch irgend eine glückliche Wendung seines Geschictes [*sic*] zum Bessern ihm die verlorene Gesundheit wiederkehrt, sein Schaffenstrieb von neuem erweckt wird, und uns einen neuen Liederfrühling bringt.") Ibid., 55–56.

24. "Das Gehörleiden . . . hat in den letzten Jahren in einer Weise um sich gegriffen. . . . vor des alternden Meisters Thüre lagert das graue Weib, die Sorge 'stets gefunden nie gesucht,' und raunt ihm allerlei ängstigende Sprüche ins Ohr und Herz." Liszt, *Robert Franz*, 54.

25. Finck, "Robert Franz," 288.

26. Franz to Carl Ambruster, 13 December 1889, Halle, in William Barclay Squire, "Letters of Robert Franz," *The Musical Quarterly* 7 (1921): 278–83, at 282.

27. See Emma Huntington Nason, "The White Lady of German Legend," *Munsey's Magazine* 9 (1893): 178–86, here 179.

28. Stumpf, *Tone Psychology*, 39. Emphasis added.

29. Ibid., 41.

30. James Sully, "*Tonpsychologie* by Carl Stumpf," *Mind* 9 (1884): 593–602, at 596.

31. Stumpf, *Tone Psychology*, 41.

32. Martin Heidegger, *What Is a Thing?*, trans. W. B. Barton Jr. and Vera Deutsch (Chicago: Henry Regnery, 1967), 6.

33. Stumpf, *The Origins of Music*, ed. and trans. David Trippett (New York: Oxford University Press, 2012), 199.

34. Stumpf, *Tone Psychology*, 8.

35. Ibid.

36. Ibid., 179.

37. Ibid., 240.

38. Ibid., 241.

39. Ibid.

40. Karl Marx, "The German Ideology" (1845), *Collected Works*, 50 vols. (London: Lawrence & Wishart, 1975–2004), 5:39.

41. Stumpf, *Tone Psychology*, 240. Schafhäutl relates the tale: "Der Director war zufrieden: 'So ist es recht; aber wenn möglich, noch etwas mehr *piano*.'" In Karl von Schafhäutl, "Moll und Dur in der Natur, und in der Geschichte der neuern und neuesten Harmonielehre," *Allgemeine Musikalische Zeitung* (6 February, 1878), 90, in a footnote.

42. "Allein man weiss wohl zu gut, wie leicht man in der Musik Töne, die an der Grenze der musikalischen Hörbarkeit liegen, zu hören glaubt, wenn man sie hören will." In Schafhäutl, "Moll und Dur in der Natur," 91.

43. Gustav Theodor Fechner, *In Sachen der Psychophysik* (Leipzig: Breitkopf & Härtel, 1877), 95–6.

44. "Wenn die Grösse des Reizes ß oder der psychophysichen Thätigkeit, die durch den Reiz ausgelöst wird, unter den Schwellenwerth b fällt, so nimmt nach der Massformel die Empfindung y negative Werthe an." Ibid., 88.

45. "im sogen. unbewussten Leben." Ibid., 91.

46. Stumpf, *Origins of Music*, 196; Stumpf, *Tone Psychology*, 247.

47. Stumpf, *Tone Psychology*, 246.

48. "In der That, wenn ich bei fixirter Aufmerksamkeit auf ein Empfindungsgebiet von einer Empfindung (oder einem empfundenen Unterschiede) durchaus nichts erkennen kann, trotz dem, das seine physische Ursache dazu da ist, was der *factische* Fall der Schwelle in meinen Sinne ist, zo verstahe ich nicht, wie man sagen kann, dass doch eine kleine (einfache oder Unterschieds-) Empfindung da sei, die man nur mit andern nicht vergleichen kann; es scheinen mir das völlig mystische Begriffe." Fechner, *In Sachen der Psychophysik*, 96. On Fechner's concept of the unconscious, see David Romand, "Fechner as a Pioneering Theorist of Unconscious Cognition," *Consciousness and Cognition* 21 (2012): 562–572.

49. Fechner's classic formulation of his laws reads: "Simple differential sensitivity is inversely proportional to the size of the components of the difference; relative differential sensitivity remains the same regardless of size." Fechner, *Elements of Psychophysics* [1860], trans. Helmut E. Adler (New York and London: Holt, Rinehart & Winston, 1966), 1:112.

50. Sully, "*Tonpsychologie* by Carl Stumpf," 599.

51. George Henry Lewes, "Consciousness and Unconsciousness," *Mind* 2 (1877): 156–67, at 159.

52. Hermann von Helmholtz, *On the Sensations of Tone* (1863), 3rd ed., trans. Alexander J. Ellis (New York: Dover, 1954), 49. Translation modified.

53. Stumpf, *Tone Psychology*, 246.

54. Stumpf, *Origins of Music*, 92.

55. Ibid., 95.

56. George Henry Lewes, *The Physical Basis of Mind* (London: Trübner & Co., 1877).

57. Stumpf, *Tone Psychology*, 48.

58. "Wie dem auch sei und wenn alle Autoritäten der Welt auftreten und sagen 'wir hören nichts,' so muss ich ihnen doch sagen: ich höre etwas und zwar etwas sehr deutliches." Riemann, *Musikalische Syntaxis. Grundriß einer harmonischen Satzbildungslehre* (Leipzig: Breitkopf & Härtel, 1877), 121.

59. See Alexander Rehding, *Hugo Riemann and the Birth of Modern Musical Thought* (Cambridge: Cambridge University Press, 2003); Alexandra Hui, *The Psychophysical Ear* (Cambridge, MA: MIT Press, 2013).

60. "It is quite reckless to build his theory on the appearance of tones that no one has heard or can hear, even if our latest theorist believes he really has heard them. But we know all too well how easy it is for one to believe he hears musical tones which lie on the border of musical audibility, if one wants to hear them." ("Es ist überhaupt gewagt, seine Theorie auf Erscheinung von Tönen zu bauen, die kein Mensch gehört hat oder hören kann, wenn auch unser neuester Theoretiker glaubt, sie wirklich gehört zu haben. Allein man weiss wohl zu gut, wie leicht man in

der Musik Töne, die an der Grenze der musikalischen Hörbarkein liegen, zu hören glaubt, wenn man sie hören will.") Schafhäutl, "Moll und Dur in der Natur," 90.

61. "Dr. *Hugo Riemann* erzählt uns, dass er und sonst Niemand, als vielleicht Aristoteles allein vor 2000 Jahren, diese Untertöne . . . gehört habe." Ibid.

62. Stumpf, *Tone Psychology*, 22.

63. Helmholtz, *Sensations of Tone*, 44–45, 365n.

64. Edmund Husserl, *On the Phenomenology of the Consciousness of Internal Time*, trans. John Barnett Brough (Dodrecht: Springer, 1991), 47.

65. Ibid., 48–49.

66. Ibid., 49.

67. Franz Kafka, "The Burrow," in *The Complete Stories*, ed. Nahum N. Glatzer (London: Vintage, 2018), 368.

68. Ibid., 378.

69. Ibid., 369.

70. Ibid., 379.

71. Ibid., 369.

72. Ibid., 370.

73. Ibid., 377.

74. Ibid., 373.

75. Max Brod, *Über Franz Kafka* (Frankfurt am Main: Fischer, 1974), 103.

76. "Die gehörte Musik zieht natürlich eine Mauer um mich, und meine einzig dauernde musikalische Beeinflussung ist die, daß ich, so eingesperrt, anders bin als frei." Kafka, *Tagebücher 1910–1923*, ed. Max Brod (Frankfurt am Main: S. Fischer, 1967), 135.

77. Thomas Martinec, "Music," *Kafka in Context*, ed. Carolin Duttlinger (Cambridge: Cambridge University Press, 2017), 128–36, at 135. Cf. Tyler Whitney, "Inside the Ear: Silence, Self-Observation, and Embodied Spaces in Kafka's 'Der Bau,'" *The Germanic Review: Literature, Culture, Theory* 92 (2017): 301–19.

78. Paul Ricoeur, *Time and Narrative*, 3 vols., trans. Kathleen McLaughlin and David Pellauer (Chicago: University of Chicago Press, 1985), 3:154.

79. On the shift of artificial eardrums from a surgical technology to a domestic medical good, see Jaipreet Virdi, "Between Cure and Prosthesis: 'Good Fit' in Artificial Ear Drums," in Clair L. Jones, *Rethinking Modern Prostheses in Anglo-American Commodity Cultures, 1820–1939* (Manchester: Manchester University Press, 2017), 48–69, at 65.

80. "diese *sogenannten einfachen Töne sind nicht einfach*." Riemann, *Musikalische Logik: Hauptzüge der physiologischen und psychologischen Begründung unseres Musiksystems* (Leipzig: Kahnt, 1873), 8. On this point, see Trevor Pearce, "Tonal Function and Active Synthesis: Hugo Riemann, German Psychology and Kantian Epistemology," *Intégral* 22 (2008): 81–116.

81. "*Analyse der Klänge durchs Ohr* ist ein Terminus der neuern Akustik und bedeutet die Unterscheidung der in dem einzelnen Ton (Klang) unsere Musikinstrumente enthaltenen Partialtöne." Riemann, *Musik-Lexikon* (Leipzig: Bibliographisches Institut, 1882), 27.

82. Michel Foucault, *Archaeology of Knowledge*, trans. A. M. Sheridan Smith (New York: Random House, 2010), 4.

PART IV

Measurement

CHAPTER 10

Salih Bey's Sonometer: Synchronizing Audition in Ottoman Istanbul

Jacob Olley

Introduction

Are ways of listening shared across different societies and musical traditions, or are they culturally specific? How are they connected to models and practices of time measurement? What role does attention play in conceptualizations and experiences of sound and temporality? This chapter approaches these questions through a discussion of experimental acoustics in the late Ottoman Empire.[1] It focuses on the activities of Salih Zeki Bey (1864–1921), a mathematician, physicist, and historian of science who also published the first substantial works on acoustics in Ottoman Turkish. Beginning in 1898, in collaboration with the musicologist Rauf Yekta (1871–1935), Salih conducted experiments with a differential sonometer to measure the intervals used in Ottoman music. This led to the formulation of a new theory of pitch that combined concepts from early Islamicate music treatises with the principles of modern acoustics, and which is the basis of the standard theory used in Turkish music today.[2] In this chapter, I situate Salih's experiments within the context of his scientific career and contemporary debates about the relationship between Europe and the Ottoman Empire in order to illustrate how developments in music theory and acoustics were connected to broader social, intellectual, and political currents. As I will argue, Salih's conceptualization of pitch can be fruitfully understood in relation to his efforts, as director of the Imperial Observatory, to implement international standards of time measurement in the Ottoman Empire. In this way, the chapter shows how global experiences of attentive listening were transformed by the adoption of new technologies and models of scientific measurement that traversed multiple domains including music theory, acoustics, and chronometry.

Attentive listening takes different forms and may serve many different purposes. Before the late nineteenth century, the most extensively theorized forms of listening in Ottoman and other Islamicate contexts were those associated with Sufi rituals involving music, vocalization, and movement, generally referred to as *samāʿ* (audition). Practices of attentive listening were also integral to Ottoman court music, which drew heavily on philosophical and aesthetic elements of Sufism.[3] Proper appreciation of such music required close attention to modal (and rhythmic) structures, including nuances of pitch that distinguished one *makam* from another, demonstrated the skill and knowledge of the performer, and intensified the affective power of a text or composition. In theoretical discourse, the apprehension of musical modes and their correspondences with cosmological and natural phenomena such as planets, humors, and elements was often portrayed as an act of extraordinary auditory perception.[4] The notion that attentive listening has an ethical effect and mediates between mundane and supramundane realms was also a key aspect of Quranic recitation, which was institutionalized through the systematic training of reciters and the construction of mosques with highly resonant interiors.[5] Furthermore, attending to the call to prayer (which coincided with the stages of the diurnal cycle) constituted the principal means of time reckoning in the Ottoman lands before the mid-nineteenth century, and thus was another way in which the supramundane structures that governed human societies were rendered audible.[6]

While the forms of auditory attention associated with Sufism, courtly music, Quranic recitation, and the call to prayer should not be regarded as homogenous or unchanging, they do represent a broad set of interrelated and socially structured listening practices that were also the subject of intellectual reflection among Muslim authors. In the long nineteenth century, however, the adoption of modern scientific technologies and concepts in conjunction with a shifting geopolitical relationship between the Ottoman Empire and Europe contributed to the emergence of new practices of listening, which I argue were also entangled with changing understandings of temporality.[7] In the context of the acoustic experiments conducted by Salih and his collaborators, the purpose of attentive listening was not to produce a heightened emotional or spiritual state, but to determine the exact frequency of a pitch. Rather than revealing the cosmological or divine basis of the manifest world, the science of acoustics aimed to demonstrate universal principles of sound as a purely material phenomenon by means of standardized technologies and mathematical models. Likewise, the standardization of time involved the adoption of technologies and calculative procedures that desacralized the experience of temporality and reconfigured it as an abstract series of linear points.[8]

As a variation on the anthropologist Talal Asad's notion of "calculative reason," I use the term *calculative listening* to describe the kinds of auditory attention that are employed in acoustic experiments, and more generally in relation to the perception of sound and time as quantifiable physical phenomena.[9] The advent of calculative listening coincided with the adoption of international standards of scientific measurement that were essential to the integration of the Ottoman Empire into global networks of governance, economics, and knowledge. This process can also be described as a form of *synchronization*, which applied to various domains from telegraphic communication and railway timetables to the homogenization of social practices and aesthetic values (including, for example, perceptions of pitch).[10] The advocates of synchronization in the Ottoman Empire—including Salih Zeki—were closely associated with the Young Turk movement, which came to power in the constitutional revolution of 1908 and was driven by a materialist philosophy that promoted belief in the supremacy of modern science.[11] The concepts of calculative listening and synchronization thus provide ways to understand the interactions between large-scale geopolitical processes, localized scientific debates and social struggles, and collective practices and subjective experiences of listening.

Yet although there was a general shift in modes of auditory attention in the final decades of the Ottoman Empire, this was by no means uncomplicated or uncontested. Salih's sonometric experiments were intended to provide empirical confirmation of precisely defined intervals in Ottoman music. However, he found that most performers and listeners were dismissive of his calculative methods and continued to use less mathematically exact means of articulating pitch relations. At the same time, Salih argued that the perception of fine distinctions in pitch and the appreciation of their emotional significance depended on culturally specific modes of listening that distinguished Ottoman from European music. Both practically and rhetorically, then, auditory attention could be deployed in diverse and apparently contradictory ways: as a means of calculating exact frequencies or synchronizing experiences of daily time, and thus as a kind of audile technique that was essential to the ideology of scientific progress, but also as an enculturated listening practice that demonstrated the sophistication and subtlety of Ottoman music in contrast to European musical aesthetics. Moreover, forms of auditory attention and temporal consciousness associated with Sufism continued to hold significance for late Ottoman and early Republican intellectuals and were reinvigorated by engagement with contemporary spiritual and philosophical movements, including the thought of Henri Bergson. In the final section of this chapter, I therefore discuss how Sufi and Bergsonian ideas of time and listening were used to conceptualize the heterogenous modes of auditory attention that coexisted in late Ottoman modernity.

An Ottoman Man of Science and Letters

Before discussing Salih Zeki's interventions in acoustics, an outline of his broader professional and intellectual trajectory can give a sense of the interconnections between scientific, technological, and governmental domains in the late Ottoman Empire.[12] Born in Istanbul in 1864, Salih began working in the Science Department of the Ministry of Post and Telegraphy (Posta ve Telgraf Nezareti Fen Kalemi) in 1882. He was sent the following year to study at the École supérieure de Télégraphie in Paris, where he also pursued courses at the École des Ponts et Chaussées and the Collège de France. After returning to Istanbul in 1885, he worked again in the Science Department of the Ministry of Post and Telegraphy while also teaching physics and chemistry at the Imperial School of Administration (Mekteb-i Mülkiye-i Şahane).

In the early 1890s, Salih worked as an assistant to Aristide Coumbary (1828–1896), an Ottoman Greek astronomer who had founded the Imperial Observatory (Rasadhane-i Amire) in 1868.[13] The primary purpose of the observatory was to collect meteorological data from areas within or adjacent to the Ottoman Empire, which were transmitted via telegraph to the Paris Observatory, thus helping to maintain a global information network. A further and related function of the observatory was to regulate time measurement, ensuring that the various calendrical and diurnal time systems used in the Ottoman Empire were standardized and synchronized with those used in Europe. This was essential not only for the management of commerce, labor, transport, communication, and administrative schedules, but also for the comprehensive mapping of space through the establishment of precise geographic coordinates, as stipulated by the International Meridian Conference held in Washington, DC in 1884.[14]

After Coumbary died in 1896, Salih became the director of the observatory and was thus responsible for meteorological data collection and time regulation throughout the empire. In 1908, the Young Turks led a coup that established a new constitutional government led by the Committee of Union and Progress (İttihad ve Terakki Cemiyeti, CUP). Given his alignment with the philosophical ideals of the Young Turks, this proved favorable for Salih's career, and over the following years he occupied a series of important posts in education and government. In 1913 Salih was the Ottoman delegate to the second International Time Conference in Paris convened by the International Time Bureau, which had been founded the previous year and was headquartered at the Paris Observatory.[15] Although he resigned from the rectorship of Darülfünun (now Istanbul University) in 1917, he continued to teach and serve as dean of the Faculty of Science until his premature death four years later.[16]

Salih was a prolific author, producing scores of newspaper and journal articles, monographs, textbooks, and translations alongside his official duties.[17] At the observatory, Salih compiled almanacs in collaboration with Émile Lacoine (1835–99), a French advisor to the Ministry of Post and Telegraphy who had facilitated his studies in Paris. These provided granular information on lunar and solar cycles, altitudes, longitudes and latitudes, rainfall, air pressure, and temperatures for local and global locations.[18] Relatedly, he wrote a number of newspaper articles advocating the reform of timekeeping systems and the adoption of Greenwich Mean Time in the Ottoman Empire.[19] Salih published educational and reference works on various areas of science, mathematics, and engineering as well as articles that sought to communicate scientific knowledge and the latest technological discoveries to the reading public. He was actively involved in print journalism, serving as the director of *Resimli Gazete* and writing regularly for periodicals such as *Tanin*, which was closely aligned with the CUP. In addition, he published books and articles on the methodology and philosophy of science, mathematics, and logic, including translations and commentaries on works by figures such as Jules Henri Poincaré, George Boole, and August Comte.[20]

Another important area of Salih's work was the history of science and mathematics in the Islamicate world, which was typically referred to during this period (by European and Muslim scholars alike) as "the Orient" (*şark*). In the late 1880s, Salih had begun to study treatises on mathematics and astronomy in Arabic, Persian, and Turkish that were held in Istanbul's libraries, aiming to understand how the adoption of ancient Greek concepts by Muslim scholars had contributed to the development of mathematical sciences in

Europe. At the same time, he read the work of Orientalists and historians of mathematics such as Franz Woepcke, J. H. C. Kern, and Paul Tannery.[21] This research provided the basis for articles on the history of the mathematical sciences that were published in both Ottoman and French journals and culminated in a projected four-volume history of mathematics in the Islamicate world, titled *Asar-i bakiye* (Eternal works, 1911), of which only the first two volumes were published in Salih's lifetime.[22]

In sum, then, Salih was a cosmopolitan intellectual who believed in the benefits of modern science and the integration of the Ottoman Empire into the global order. At the same time, he was a fellow traveler of the nationalist movement led by the Young Turks, as well as of the reformist trends that sought to revive early Islamic learning in order to modernize the Muslim world and demonstrate its parity with European civilization. These complexities are representative of many Ottoman intellectuals around the turn of the twentieth century and are essential for understanding Salih's approach to the science of sound, which wove together several different ideological threads. However, as I will show in the following section, there were diverse views about the nature and practical benefits of acoustics among Ottoman intellectuals and music practitioners, views that intersected with polemical debates about science, universalism, and cultural difference.

Debating the Science of Sound

Salih Zeki's contributions to the science of sound began with the publication of his textbook *Hikmet-i tabiiye* (Physics) in 1894. This work included a substantial section on acoustics, explaining basic principles of vibration, frequency, hearing, and resonance, complete with diagrams of instruments such as tuning forks, sirens, and a sonometer.[23] In this respect, it was preceded by a number of other generalist physics textbooks, most of which were based on French sources.[24] It was not until the appearance of Salih's *Mebhas-i savt* (Acoustics) in 1910 that a work devoted exclusively to acoustics was published in Turkish.[25] *Mebhas-i savt* is also notable for its exposition of the Ottoman pitch system and comparison with other tuning systems, which had not been discussed in previous treatises on physics. This was an outcome of the experiments Salih and his collaborators began conducting with the sonometer in the late 1890s, experiments whose implications were vigorously debated in the Ottoman press at the time.

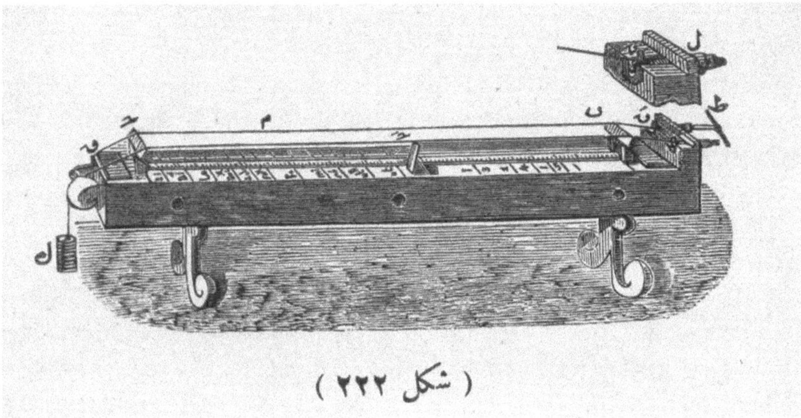

Figure 10.1. Differential sonometer from Ṣāliḥ Ẕekī, *Ḥikmet-i ṭabīʿīye*, 565. Image courtesy of Türk Tarih Kurumu.

The differential sonometer (Figure 10.1) had been developed in the mid-nineteenth century by Albert Marloye (1795–1874), an early specialist maker of acoustic instruments.[26] It comprised a wooden resonator box across which two or more metal strings were mounted, supported by two fixed bridges a meter apart, and whose vibrating length and tension could be adjusted by manipulating sliding bridges, weights, pulleys, and a tuning key. The surface of the resonator box was usually marked with different scale divisions, including the "chromatic tempered scale" and the "chromatic physicists' scale" (based on harmonic ratios), as well as a meter scale divided into millimeters, though string lengths could also be measured with a beam compass. By adjusting these parameters, it was possible to acoustically compare pitches produced by different string lengths (as well as degrees of tension, string diameter, or density) in relation to different mathematical scales.[27] In comparison to other acoustic instruments such as the siren or the tonometer, the sonometer was relatively inexpensive and simple to construct. Although it did not measure frequencies directly, it could be used in conjunction with other instruments such as pitch pipes or tuning forks to provide an accurate indication of pitch. This involved tuning one string of the sonometer to a fixed pitch (e.g., 435 Hertz), which allowed for the calculation of other frequencies with reference to the mathematical scales marked on the body of the instrument.[28]

In the section on acoustics in *Hikmet-i tabiiye*, which was based entirely on European sources, Salih had been unable to provide any information about the scientific basis of Ottoman or "Oriental" music, and his curiosity about this topic brought him into contact with Rauf Yekta, today regarded as the founder of modern Turkish music theory. Like Salih, Yekta had been collecting and studying treatises in Arabic and Persian since the 1880s, in his case with the aim of reviving mathematical approaches to pitch and mode that were no longer used or well understood by Ottoman musicians and theorists.[29] Yekta's main teacher and collaborator in this regard was the Mevlevi Sufi sheikh Ataullah Efendi (1842–1910), a highly proficient musical performer who was conversant with older theoretical works and at the same time deeply interested in contemporary scientific and intellectual developments. Yekta had read the section on acoustics in *Hikmet-i tabiiye* closely, and he and Ataullah both had sonometers built in order to demonstrate the intervals documented in earlier treatises and measure the pitches used in contemporary Ottoman music.[30]

Based on his historical research and practical experiments with Ataullah, Yekta began publishing articles on Ottoman music theory in periodicals such as *Resimli Gazete* and *İkdam* in January 1897, and the following year he personally demonstrated his findings to Salih using the sonometer.[31] Yekta's publications attracted the attention of other musicians and intellectuals, including Ahmed Midhat (1844–1912), perhaps the most influential man of letters of the late Ottoman period. Yekta and Ahmed Midhat had a series of testy exchanges in the press beginning in October 1898, in which Ahmed Midhat argued that there was a single, universal music that had reached the most advanced state in both practice and theory in Europe. By contrast, Yekta maintained that there was an essential difference between Occidental and Oriental music, and that the latter was governed by distinct theoretical rules that had been meticulously formulated by al-Fārābī (d. 950) and other Muslim scholars.[32]

In response to Yekta's articles, an unidentified writer using the pseudonym "İsmail" questioned not just the distinction between Eastern and Western music, but the value of applying mathematical models to Ottoman music. In attempting to calculate the number of vibrations in sounds, according to this critic, Yekta and his collaborators were pointlessly "counting hairs on a sheepskin" ("*pösteki saymak*") and posing riddles like those that elementary school children ask one another.[33] Yekta's response was dismissively terse, but Salih Zeki was now motivated to intervene.[34] In an article published in *İkdam* on January 7, 1899, Salih stated that he had met Yekta only once, and emphasized that he was not a musician but a scientist with an interest in acoustics. If the current century was the "age of progress" ("*asr-ı terakki*"), as the critic had asserted, it was necessary to respect the achievements of illustrious scholars such as al-Fārābī and to have a firm grasp of the latest scientific principles. Moreover, the respected European scientists who had carried out acoustic experiments, such as Helmholtz, could hardly be compared to elementary school children.[35]

Salih went on to explain in more detail the concept of frequency, asserting that it is the only possible means to distinguish between different pitches. Indeed, it was the basis for the advanced development of music theory in Europe, to the extent that the composer had become a kind of "music accountant" ("*musiki muhasibi*").[36] Yet when Salih had attempted to discuss acoustic principles with practicing Ottoman musicians, his inquiries were often met with incomprehension. Salih reported that he was astonished to meet musicians who, "although they play an instrument superbly, are . . . not convinced that the phenomenon called sound is the result of vibration."[37] Upon hearing that the air column in a pitch pipe tuned to *A* vibrates 870 times per second (i.e., 435 Hertz), one interlocutor responded: "Be reasonable! Does the air in such a tiny pipe have time to vibrate so many times?"[38] According to Salih, whereas in European music the relations between pitches had been accounted for mathematically, the musicians he spoke to confirmed that knowledge of pitch relations in Ottoman music could still be learned only from a master teacher. For these reasons, he was overjoyed to read about the progress of Yekta's research, which would establish a rational theory for Ottoman music rooted in the principles of acoustics, and thus demonstrate scientifically its differences from European music.[39]

Yekta responded a few days later to Salih's article, grateful to have the public support of a respected scientist who understood his methods and goals.[40] Ahmed Midhat also responded, expressing admiration for Salih but reiterating his opinion that "what we call Western (*alafranga*) music is scientific music. What is opposed to that is opposed to science."[41] Over the following weeks a flurry of articles by Salih, Ahmed Midhat, Yekta, and others appeared in the Ottoman press, addressing a range of questions connected to acoustics, music theory, science, aesthetics, cultural difference, and modes of listening. Salih announced that he was now engaged in further experiments with the sonometer in collaboration with Yekta and other musically knowledgeable people.[42] Like Yekta, he invoked the authority of European musicologists, Orientalists, and acousticians to support his argument that Eastern music was based on theoretical principles that differed from those of Western music, and that these had been systematically worked out by earlier

Muslim scholars. According to Salih, different civilizations and ethnic groups had distinctive musical languages, which could only be appreciated if the listener had some degree of theoretical understanding and previous experience of the music being heard. While Salih concurred that science was universal rather than particular, he maintained that this related to *acoustics* rather than *music*.[43] Hence, like science and mathematics in general, acoustics was not "European" but transcended cultural differences, while at the same time providing a practical and intellectual framework for demonstrating these differences in quantifiable and testable ways.

A Subtle Art: Attending to Ottoman Music

At the heart of these debates was the relationship between the mathematical representation of musical intervals and their auditory perception. For Salih, attentive listening mediated by the sonometer and other acoustic technologies could prove the existence of certain intervals in Ottoman music as well as their scientific validity. As a specific variety of auditory attention, we can refer to this as *calculative listening*. While the broader social and political implications of this mode of listening will be discussed in the following sections, for now I will use it mainly with reference to experiments in pitch perception. The primary purpose of calculative listening for Salih and his allies was to demonstrate that intervals described in early Islamicate treatises corresponded to those used in contemporary Ottoman music, and that they were grounded in universal acoustic laws. This offered a powerful riposte to critics such as Ahmed Midhat, who believed that pitches that fell outside of the recognized tonal framework of Western music were either incorrect or superfluous. As Midhat asserted, when closely related *makams* with subtle intonational differences were played on the piano, "we cannot detect these differences with our ears."[44] Yet if such intervals could be precisely calculated and perceived by means of the sonometer, this meant that the inability to hear pitches beyond the chromatic notes of the piano was due to a lack of proper auditory attention—meaning a mode of focused listening that was properly informed by scientific knowledge. At the same time, calculative listening offered a superior audile technique in comparison to the seemingly intuitive and unquantifiable modes of auditory attention that were employed by practicing Ottoman musicians, as illustrated by Salih's fruitless efforts to explain the physics of sound to his interlocutors.

Salih's acoustic experiments supposedly revealed the scientific basis of musical intervals that lay at the limits of auditory perception and yet constituted the essential difference between Ottoman and European music. According to Salih, the measurement of frequency using the sonometer made it possible to distinguish precisely between the major tone (9:8) and minor tone (10:9), "a difference the ear will be unable to sense" ("*kulağın his edemeyeceği farkı*").[45] Likewise, Yekta argued that while tetrachords based on either major or minor tones (as described by the thirteenth-century theorist al-Urmawī) can be realized exactly on an instrument such as the sonometer, "a violinist or a singer cannot, however, execute them one after the other in all their precision, because these different forms of the perfect fourth confuse the hearing."[46] The difference between a major and minor tone is equivalent to a syntonic comma (81:80 or 21.51 cents, referred to as *koma*). The same discrepancy is found, according to Salih, between the European note *mi* (in just intonation) and the Ottoman note *hüseyni* in the equivalent scale position.[47]

Although various types of comma were discussed by early Islamicate theorists based on their engagement with ancient Greek sources, the concept had been obsolete in Ottoman music since the early sixteenth century and was reintroduced by Yekta partly on the basis of modern European sources.[48] The reintroduction of the comma was integral to the project of establishing a rational, accurate, and mathematically defined pitch system for Ottoman music. In isolation or in close juxtaposition with other intervals, the difference of a comma might be virtually imperceptible or easily misinterpreted by the untrained listener. Yet it now offered a conceptual means to explain and represent the subtle inflections of pitch and the diversity of intervallic and modal structures that distinguished Ottoman from European music. Furthermore, by empirically verifying the existence of the comma in Ottoman music, it became possible for an attentive listener to *hear* the difference between two closely adjacent pitches *as* a comma, rather than as "out of tune" (*falso*) in relation to the piano, or according to the mathematically imprecise and variable terms that were used prior to the establishment of the new theory.

As other acoustic experiments and debates during this period demonstrate, the relationship between the physical properties of sound and its auditory perception is rarely straightforward. Sounds that are not immediately audible to the ordinary listener (such as upper partials) require special apparatus and the training of auditory attention to be accurately perceived.[49] Moreover, the inherently subjective nature of listening as a psychophysical process as well as the influence of conceptual or metaphysical frameworks may produce peculiar auditory effects including the perception of nonexistent sounds, as in Riemann's undertones.[50] More generally, experiences of pitch relations are determined to a large degree by learned habits and cultural values, so that even if the laws of acoustics are universal, the manner in which they are deployed, heard, and conceptualized within different musical systems is highly variable.[51]

Salih was evidently aware of some of these epistemological problems, as reflected in his distinction between the universality of "acoustics" and the particularity of "music." It followed that if an interval such as the comma was explainable according to the universal laws of physics, its status as a distinguishing feature of Ottoman music had to be accounted for in cultural rather than purely scientific terms. The comma therefore not only demonstrated the scientific validity of Islamicate music theory, but also offered a ready explanation for the aesthetic values, modes of listening, and emotional effects that were believed to differentiate Ottoman from European music. Salih conceptualized this distinction in terms of the relative importance of subtle intonational nuances signified by the comma, which he explained by means of a spatial and sensorial analogy:

> One of them [i.e., European music] resembles a very large and expansive park. The other one [i.e., Ottoman music] brings to mind the image of a pleasure garden that has been planted and adorned with perfumed flowers. While it is true that in such a large park the presence or absence of, for example, a carnation is of no importance whatsoever, in a blooming garden like Ottoman music, its lack will be immediately sensed.[52]

This analogy suggests that although the differences between European and Ottoman music can be scientifically measured, they are also determined by culturally specific aesthetic values and modes of attention. The image of the pleasure garden evokes the multisensorial and socially interactive context in which Ottoman music was ideally performed and enjoyed, and which was thought necessary to produce the desired emotional effects on the listener. Salih emphasized that he did not intend to disparage European music, only to point out that Ottoman music was an independent art with specific characteristics, which demanded a particular attention to subtle details. As he wrote, "I wish to explain that Ottoman music is not a part of Western music, and that while the latter is a great art (*sanat-i aliye*), the former is a subtle art (*sanat-i dakika*), and thus they are different things."[53]

Although calculative listening was the dominant mode of audition in Salih's acoustic experiments, his interpretive arguments suggest that other—more subtle—forms of auditory attention are required to fully understand and appreciate Ottoman music. The ambiguous relationship between physical sound, auditory perception, and aesthetic judgement is likewise encapsulated in the epigraph to Salih's treatise on acoustics, which is taken from a speech given by the British mathematician and physicist Lord Kelvin in Birmingham in 1883: "There [giving a sudden clap of the hands]—that is a sound. There is no question about it—nobody will ever ask: Is it a sound or not? It is sound if you hear it. If you do not hear it, it is not to you a sound."[54] This choice of epigraph signals Salih's commitment to scientific empiricism, in which objective knowledge can be obtained only through sensorial perception. At the same time, however, Salih sought to demonstrate in his treatise the existence of pitch relations specific to Ottoman music, the correct hearing and enjoyment of which depend on learned modes of auditory attention. Thus, the statement "If you do not hear it, it is not to you a sound" ("*Eğer işitmez iseniz . . . sizin için sada değildir*") can simultaneously be understood as an acknowledgment that what constitutes a sound or pitch is at least partly determined by the culturally specific and subjective perception of the listener.

Calculative Listening, Synchronization, and Difference

By the early Republican period, the comma-based theory of pitch relations formulated by Salih and Yekta had been established as the definitive theory of Ottoman-Turkish music, which it remains (with some modifications) to this day. Its genesis in experimental acoustics and mathematics indicates a connection to the ideals of scientific progress and philosophical materialism that were widespread among Ottoman intellectuals in the 1890s, and which laid the ideological groundwork for the Young Turk revolution and eventually the founding of the Republic. The technique of calculative listening is thus directly related to the concept of calculative reason, which, as Asad shows, is central to the formation of the modern secular state.[55] Calculative listening involved a disenchantment of audition based on the quantification of sound (in the form of frequencies, beats, intervals, ratios, and so on), which was conceived as an essentially material phenomenon whose nature could be objectively measured through instruments such as the sonometer, pitch pipe, and tuning fork. The cognitive, perceptual, and philosophical reorientations that this entailed were aligned with the project of creating rational subjects who were conversant with modern science and integrated into the social, bureaucratic, and technological structures of both the centralized state and the international order.

Calculative listening was an aspect of a broader shift in modes of auditory attention precipitated by the introduction of standardized systems of time measurement. Before the nineteenth century, daily time had been marked sonically in the Ottoman Empire primarily by the call to prayer, which was coordinated with the passage of the sun.[56] Since the diurnal cycle depended on the time of the year and the location of the observer, the hours of the day were correspondingly variable in their onset and duration. The implementation of centralized technologies and systems of measurement was intended to eliminate such local variations and inaccuracies and demanded a new kind of auditory attention to the universal intervals of time regulated by the Imperial Observatory and signaled by the ticking of a clock, the clicking of a telegraphic message, or the firing of the noon cannon.[57] Likewise, while the intervals of Ottoman music had previously been subject to local variation and flexibility, the introduction of modern acoustic instruments and systems of measurement required a focused attention to objective and invariable standards rather than a sensitivity to ephemeral moments of aesthetic pleasure or affective relationality. Such calculative modes of aural attention also engendered a desacralization of time, which became a neutral and objective series of numbers relating to the physical world rather than an expression of religious practice and cosmology.

Salih's advocacy of Greenwich Mean Time was part of a larger effort to regulate the diverse calendrical and temporal systems that were used in the Ottoman Empire and to integrate them with international standards of time measurement. By this process, local perceptions of time would be synchronized both with the Imperial Observatory in Istanbul and with other metropoles such as Paris and London. This synchronization was achieved through technologies such as the chronometer and telescope as well as telegraphy and print media (including periodicals and almanacs), which depended on the international compatibility of quantitative data. Similarly, the standardization of pitch relations involved a process of temporal quantization using acoustic instruments, rendering the sounds of Ottoman music into measurable frequencies that adhered to the universal logic of numbers and were thus made legible and translatable across geographical and cultural boundaries. Furthermore, the institutionalization of an empirical theory of pitch relations would ensure that the variable and potentially conflicting intervals that occurred among music practitioners would be synchronized with one another (that is, all musicians would now consistently play the same intervals, precisely defined by their frequency ratios) and with the universal norms of modern science.

However, the ideal of universal synchronization could not be directly and instantaneously applied and had to contend with a multiplicity of material realities, everyday practices, and alternative subjectivities. Moreover, the materialist frameworks of modern science could not eradicate uncertainty and ambiguity. In 1909, Salih had argued that the noon cannon should be replaced by a time ball, since the time it took for the sound of the cannon to reach international ships in the Bosporus strait (not to mention the inevitable delay in telegraphic transmission between more distant locations) created an unacceptable margin of error.[58] However, although he proposed that a time ball should be mounted on top of the Galata tower (the highest point in Istanbul), he discovered that vibrations caused by the nearby funicular railway meant that a telescope placed within the structure could not be completely stable.[59] Hence, while modern technologies could in principle facilitate precise and objective measurement, in practice they were not infallible and could even create distortions and delays of perception. Furthermore, technologically mediated forms of attention nonetheless required a human observer or listener whose perception and interpretation of phenomena were inevitably conditioned by a variety of unpredictable factors. The sound of a pitch pipe tuned to 435 Hertz might signify a universal scientific value, or it might be interpreted as an absurd proposition about the nature of sound that was as irrelevant to musical practice as counting hairs on a sheepskin. There was no guarantee that an interval of duration or frequency would be attended to in the same way by different subjects, and standardized systems of time and pitch measurement therefore remained open to contestation and misinterpretation.

Salih's description of Ottoman music as a "subtle art" encapsulates many of the tensions that existed within the project of scientific universalism and its acoustic, temporal, and psychological dimensions. The word *dakik* (*dakika* in its feminine adjectival form) connotes exactness or precision, and also supplies the noun for "minute" (*dakika*) in both its temporal and geometrical senses. A closely related word, *dikkat*, signifies careful attention. This cluster of meanings suggests that the objective grasp of phenomena such as time, space, or pitch involves a process of quantization into small, regular values, which is achieved by a narrowly focused and technologically mediated attention to detail. Yet this also entails the detection of intervals, gaps, or differences that are essential to calculation. The differential sonometer was designed not to measure single, isolated pitches, but the intervals between them; hence, the Ottoman-Turkish term for sonometer, *mikyas-i savt*, implies both "measurement" and "comparison" of sound. As we have seen, the difference between two pitches may be exceedingly subtle—another of the meanings of *dakik*—and indeed almost imperceptible for listeners who are unaccustomed to valuing such fine distinctions. It is therefore at risk of being misheard and may require a variety of attention that is less like comparing two notes on a piano than noticing the pleasing smell of a carnation in a garden in full bloom.

While differences between pitches could occur at the limits of perceptibility, the measurement and integration of such intervals into a larger system of pitch relations became the basis for another kind of differentiation: that between Ottoman (or "Eastern") and European (or "Western") music. The notion of cultural difference was as central to Salih's experiments in acoustics as it was to his historical scholarship, which sought to document the existence of a distinctive Islamic or "Oriental" intellectual tradition even as it reaffirmed the universal validity of modern science and a related belief in civilization and progress. Such apparent tensions are also found in the careers of Yekta and his teacher Ataullah Efendi, both of whom continued to be active members of the Mevlevi Sufi order while also promoting calculative methods and materialist ontologies of sound. In order to better understand these complexities, it will be useful to consider how Sufi ideas of listening, attention, and temporality intersected with debates about modern science and psychology in the late Ottoman Empire, particularly in relation to the philosophy of Henri Bergson.

Auditory Temporalities: Sufism and Bergsonism

Sufi practices of listening are closely bound up with conceptions of temporality and attention, which is understood as inseparable from consciousness.[60] As Ines Weinrich writes, audition (*samāʿ*) can be defined as "listening with spiritual intent in Sufi ritual practice," and was central to the traditions of the Mevlevis as well as other orders active in the Ottoman Empire.[61] Indeed, Sufism provided arguably the most important conceptual and ethical framework for practices of audition across the Islamicate world until the end of the nineteenth century. For the philosopher al-Ghazālī (d. 1111), whose tract on listening remains one of the most widely influential texts on the subject, "there is no entrance to the heart save by the ante-chamber of the ear."[62] Thus, attentive listening provides access to the innermost recesses of the soul, in which both the true state of the auditor and the presence of God can be felt. However, this can be achieved only with the right conditions of time, place, and company. Listeners in a ritual gathering should not be distracted by worldly concerns or by the behavior of those who are less spiritually developed, and must be calmly attentive and "present in heart."[63] Furthermore, the effects of music depend on the listener's stage of spiritual attainment. The most adept listener enters a state of ecstasy (*wajd*) in which the self is dissolved into the oneness of being (*waḥdat al-wujūd*). Proper auditory attention is therefore possible only in the social context of ritual practice, and although it is determined by the subjective state of the listener it also produces a mode of consciousness in which there is no autonomous self.

These spiritual practices are closely connected to Sufi ideas of time. The concept of *waqt* (time, hour, moment) refers to an intimate experience of divine presence in which the practitioner perceives the eternity of the past and the future in each moment.[64] According to al-Sarrāj (d. 988), *waqt* is "the breath between two other breaths, or the thought that lies between two other thoughts."[65] The adept may experience the timelessness of *waqt* through audition and the closely related practice of *dhikr* (remembrance), in which attention is directed to the repetition of divine names as a means of constantly recalling God's presence. In the Mevlevi *ayin*, a complex musical structure orchestrates a series of coordinated cyclic movements whose choreography mirrors the revolutions of the heavenly spheres.[66] While these ritual practices might be understood as forms of synchronization, they are better described as involving entrainment, or the spontaneous emergence of embodied coordination under specific conditions. Relatedly, although Sufi practices of auditory attention require the exercise of will and intention, their goal is neither to gain precise knowledge of the physical world nor to establish a sovereign subject, but to create conditions for the "passing away" or "annihilation" (*fanā'*) of the individual self through collective ritual, and thereby to become conscious of the fundamental unity of existence.

In some ways, there may seem to be a clear opposition between Sufi understandings of auditory attention and what I have referred to as calculative listening. In Sufi and Islamic cosmologies, neither physical phenomena, time, nor individual subjectivity exist independently of the will of God. Experiences of audition and temporality in this framework appear fundamentally distinct from the varieties of attention produced by modern technological, scientific, and social structures, which are predicated on the essentially material nature of the world and its perception by an autonomous subject within empty homogenous time.[67] Furthermore, this ontological and epistemological divide seems to fit neatly with broader historiographical plots concerning the relationship between the Ottoman Empire and Europe, in which Islamic religious tradition is overcome or suppressed by secular Western modernity. However, while I argue that new modes of auditory perception did indeed emerge in conjunction with modern scientific technologies and discourses, and that these developments were profoundly shaped by Ottoman relations with Europe, more complicated realities lie beneath this schematic narrative.

As a number of scholars have shown, the implementation of standardized time measurement in Ottoman territories did not immediately or wholly erase existing temporal practices and subjectivities, and in fact provided conditions for the articulation of a variety of "counter-temporalities."[68] Relatedly, if modernity promoted calculative and disenchanted modes of attention, it also led to a proliferation of interest in more elusive forms of subjectivity such as hypnosis, trance, and the unconscious.[69] This was equally true in the Ottoman Empire, where intellectuals of diverse social backgrounds and political orientations engaged with a range of contemporary discourses, including positivism and materialism as well as spiritualism, psychology, and occultism, while also critically reevaluating older scientific and philosophical traditions.[70] Whether in terms of institutions and personalities or concepts and ideologies, these engagements were typically characterized by complex syntheses rather than strict oppositions between different elements.

An apt example is the reception of Bergsonian philosophy by Ottoman and other non-European intellectuals. Bergson likened aesthetic experience, and attending to music and rhythm in particular, to a receptive state of hypnosis.[71] This was part of a broader argument about the nature of perception and subjectivity in which Bergson rejected the physical determinism of experimental psychologists and scientists such as Ribot or Helmholtz and instead posited a model of consciousness defined by the continuous interpenetration of heterogenous temporal and psychic states, which constituted "duration" (*durée*).[72] As Martin Jay argues, rather than the kinds of abstract spatialization associated with vision, Bergson was interested in multimodal forms of sensorial perception that produce affective states shaped by the interplay of presence and memory, such as the smell of a rose.[73] For Bergson, intuition—as opposed to analysis, which separates heterogenous experiences into spatialized and differentiated representations—offers a means to grasp the multitudinous singularity of duration as the fusion of embodied perception and memory. This is connected to a distinction between the qualitative multiplicity of duration as lived experience and the quantitative multiplicity of time as numerical calculation.[74] Accordingly, in Bergsonian terms the cognitive representation of relative differences in pitch involves a quantitative spatialization that is at one remove from the indeterminate and affective experience of pure sonic quality.[75]

There are significant points of contact between Bergson's ideas and Sufi philosophies of audition and temporality. The concept of *durée* bears a notable resemblance to *waqt*, while Bergson's portrayal of music and dance as forms of embodied receptivity that allow access to deeper unities of experience might easily be applied to Sufi ritual practices. More generally, his refutations of physical determinism, mechanization, and calculative reason appealed to moderate Islamists who wished to critique the materialism of European modernity as well as the secularism and scientism of the CUP and the early Republican state.[76] But although Bergsonism offered emancipatory possibilities for some non-European intellectuals, it was also interpreted in conflicting ways and pressed into the service of diverse political ideologies including fascism and ethnic nationalism.[77]

Furthermore, Bergson himself was susceptible to racist discourses of cognitive difference between civilized Europeans and primitive non-Europeans, and readily justified the French colonization of North Africa on the grounds of Muslims' supposed indolence and indifference to change and progress.[78] Although few maintained that Islam itself was a cause of psychological passivity or backwardness, the idea that Muslim civilization had declined due to a lack of scientific progress was commonplace among late Ottoman intellectuals. Thus, while some scholars argued that Bergson's concept of intuition had been anticipated by al-Ghazālī, others believed that the latter's approach to philosophy had overshadowed more rationalistic and scientific currents of Islamic thought.[79] In reformist discourses, Sufism was often associated with states of idleness and mental dissipation, and with an irrational, timeless mysticism that was out of step with the mechanized, scientific, and economically productive modes of attention that were necessary to keep pace with modern life.[80]

These discourses were a product of ideological battles fought both locally and globally over the meanings of modernity in the early twentieth century, rather than an accurate description of the complexities of individual subjects and social groups. The interest of Rauf Yekta and Ataullah Efendi in acoustic instruments such as the sonometer should be understood both in the context of the historical role of Sufis as timekeepers, astronomers, clockmakers, and music theorists in the Ottoman Empire, and in relation to their long-standing philosophical, ethical, and practical commitment to audition as a medium of knowledge.[81] Hence, the utilization of analytical concepts and technologies did not preclude an intuitive grasp of the immediacy of experience, or of varieties of consciousness that lie beyond the measurable domain of material phenomena. Conversely, although Salih Zeki's philosophical outlook was generally oriented toward positivism, he also translated works by Henri Poincaré, whose ideas about intuition and indeterminacy (together with those of his brother-in-law Émile Boutroux) were invoked by reformist Muslim theologians to critique the vulgar materialism of more radical Ottoman thinkers.[82] There was, then no clear divide between secular European modernity and Islamic religious tradition, but a plurality of positions that attempted in various ways to articulate the complex subjectivities that emerged through the entanglement of local and global forces.

Conclusion

Forms of auditory and temporal attention in the late Ottoman Empire, whether defined by the intellectual frameworks of Sufism, Bergsonism, or scientific materialism, were inescapably social and political—in terms of how they were experienced and mediated through technological and social structures, how they were conceptualized by practitioners and observers, and what they signified in the context of broader philosophical debates and political formations. They were also multiple, heterogenous, and indeterminate, in contrast to the idea that certain cultures are confined to a single, unchanging mode of consciousness, or that the advent of modernity precipitated a universal progression from one mode to another. The mapping of oppositions between intuitive and analytical states of consciousness onto the geographies of East and West—whether by Orientalists, colonialists, radical secularists, or religious conservatives—tends to overlook the fact that calculative reason was not a uniquely European invention, but enjoyed a prominent place within the history of Islamicate science and mathematics, as the scholarship of Salih Zeki demonstrated. It was only the instrumentalization of calculative methods by modern states, institutions, and economic systems that made it seem as if there was an unbridgeable divide between analysis and intuition, or between different forms of social organization and cultural production based on these varieties of consciousness. Thus, if the imposition of modern scientific discourses and technologies produced modes of auditory attention that were increasingly shaped by calculative reason and the synchronization of global networks, it also offered means to enunciate the value of cultural difference, to revitalize older aesthetic and intellectual traditions, and to formulate alternative ways of being in the modern world.

Notes

1. Research for this chapter was supported by the project Ottoman Auralities and the Eastern Mediterranean: Sound, Media and Power, 1789–1922, approved by the European Research Council and funded by UK Research and Innovation (EP/X032833/1). I am grateful to a number of people who helped to improve the chapter by commenting on earlier drafts and presentations, including Peter McMurray, David Trippett, Onur Engin, Nazan Maksudyan, Hande Betül Ünal, Elif Yumru, Carmel Raz, Francesca Brittan, David Cohen, and an anonymous reviewer. Thanks are also due to Vera Wolkowicz for inviting me to present the research as part of the Royal Musical Association Research Colloquium series at the University of Glasgow in February 2025.

2. The intellectual and historical background of this theory has been explored in previous literature, though without a detailed discussion of Salih Zeki's contributions or its relation to experimental acoustics. See Jacob Olley, "Measuring Progress: The Ottoman Revival of Systematist Music Theory, c. 1900," *Oriens* 51, nos. 1–2 (2023): 127–54; and Okan Murat Öztürk, "Türk Müziğinde Yekta, Ezgi ve Arel Teorilerinin Pozitivist İnşası: Kısa Fakat Eleştirel Bir Tarihçe," *Eurasian Journal of Music and Dance* 16 (2020): 171–215.

3. For general discussions of *samāʿ*, see Leonard Lewisohn, "The Sacred Music of Islam: *Samāʿ* in the Persian Sufi Tradition," *British Journal of Ethnomusicology* 6 (1997): 1–33; Arthur Gribetz, "The *Samāʿ* Controversy: Sufi vs. Legalist," *Studia Islamica* 74 (1991): 43–62. On the connections between Sufism and Ottoman court music, see Walter Feldman, *From Rumi to the Whirling Dervishes: Music, Poetry, and Mysticism in the Ottoman Empire* (Edinburgh: Edinburgh University Press, 2022).

4. See, e.g., *Seydī's Book on Music: A 15th Century Turkish Discourse*, ed. and trans. Eugenia Popescu Judetz in collaboration with Eckhard Neubauer (Frankfurt: Institute for the History of Arabic-Islamic Science at the Wolfgang Goethe University, 2004), 20–25.

5. Nina Ergin, "The Soundscapes of Sixteenth-Century Istanbul Mosques: Architecture and Qur'an Recital," *Journal of the Society of Architectural Historians* 67, no. 2 (2008): 204–21; Nina Ergin, "Multi-Sensorial Messages of the Divine and the Personal: Qur'an Inscriptions and Recitation in Sixteenth-Century Ottoman Mosques in Istanbul," in *Calligraphy and Architecture in the Muslim World*, ed. Mohammad Gharipour and İrvin Cemil Schick (Edinburgh: Edinburgh University Press, 2013), 105–118. For wider reflections on the aural reception of the Quran, see Navid Kermani, *God Is Beautiful: The Aesthetic Experience of the Quran*, trans. Tony Crawford (Malden: Polity, 2015).

6. Avner Wishnitzer, *Reading Clocks,* Alla Turca: *Time and Society in the Late Ottoman Empire* (Chicago: University of Chicago Press, 2015), chapter 1.

7. For a discussion of the ways in which the sonic dimensions of modern technologies transformed religious sensibilities, see Adam Mestyan, "Upgrade? Power and Sound During Ramadan and *'Id al-Fitr* in the Nineteenth-Century Ottoman Arab Provinces," *Comparative Studies of South Asia, Africa and the Middle East* 37, no. 2 (2017): 262–79.

8. On the creation and dissemination of international standards of pitch, see Fanny Gribenski, *Tuning the World: The Rise of 440 Hertz in Music, Science, and Politics, 1859–1955* (Chicago: University of Chicago Press, 2023). For more general discussions of the rise of modern acoustics, see Viktoria Tkaczyk, *Thinking with Sound: A New Program in the Sciences and Humanities Around 1900* (Chicago: University of Chicago Press, 2023); Viktoria Tkaczyk, Mara Mills, and Alexandra Hui, eds., *Testing Hearing: The Making of Modern Aurality* (New York: Oxford University Press, 2020); Alexandra Hui, *The Psychophysical Ear: Musical Experiments, Experimental Sounds, 1840–1910* (Cambridge: MIT Press, 2013); Benjamin Steege, *Helmholtz and the Modern Listener* (Cambridge: Cambridge University Press, 2012). On global practices and experiences of temporality in the modern era, see Vanessa Ogle, *The Global Transformation of Time: 1870–1950* (Cambridge: Harvard University Press, 2015); On Barak, *On Time: Technology and Temporality in Modern Egypt* (Berkeley: University of California Press, 2013); and Wishnitzer, *Reading Clocks.*

9. Talal Asad, *Secular Translations: Nation-State, Modern Self, and Calculative Reason* (New York: Columbia University Press, 2018).

10. For a discussion of synchronization in the semi-autonomous Ottoman province of Egypt, see On Barak, "Synchronization and Its Discords: Calendric Reform and Imperial Politics at the End of the Nineteenth Century," in *Egypt and Empire: The Formation of Religious Identity After Rome*, ed. Elisabeth R. O'Connell (Leuven: Peeters, 2022), 133–48.

11. On science and political ideology in the late Ottoman period, see M. Alper Yalçınkaya, *Learned Patriots: Debating Science, State, and Society in the Nineteenth-Century Ottoman Empire* (Chicago: University of Chicago Press, 2015); Şeyma Afacan, "Idle Souls, Regulated Emotions of a Mind Industry: A New Look at Ottoman Materialism," *Journal of Islamic Studies* 32, no. 3 (2021): 317–53; Şükrü Hanioğlu, "Blueprints for a Future Society: Late Ottoman Materialists on Science, Religion, and Art," in *Late Ottoman Society: The Intellectual Legacy*, ed. Elisabeth Özdalga (London: Routledge, 2005), 28–116.

12. For Salih's biography, see Şeref Etker, "Salih Zeki Bey: Üç Boyutlu Bir Biyografi İçin," *Osmanlı Bilimi Araştırmaları* 7, no. 1 (2005): 137–54; Alper Atasoy, "Türk Bilim Tarihçiliğinde Salih Zeki: Hayatı, Çalışmalar ve Eserleri," in *Şimdiyi Bilmek ve Geçmişi Anlamak: Bilim ile Bilim Tarihi Arasında Salih Zeki*, ed. Elif Baga (Istanbul: Ketebe Yayınları, 2021), 17–46; Remzi Demir and İnan Kalaycıoğulları, "Büyük Bir Matematik Tarihçisi ve Felsefecisi: Salih Zeki Bey (1864–1921)," *Kutadgubilig Felsefe-Bilim Araştırmaları* 6 (2004): 195–211.

13. Feza Günergün, "Salih Zeki ve Astronomi: Rasathane-i Amire Müdürlüğü'nden 1914 Tam Güneş Tutulmasına," *Osmanlı Bilimi Araştırmaları* 7, no. 1 (2005): 97–122. Coumbary is often mistakenly identified as French, but he was actually an Ottoman subject (of Greek heritage) born in Istanbul. See Aykut Kar, "Arşiv Belgeleri Işığında Rasathane-i Amire'nin Müdürü Coumbary Efendi'nin Çalışmaları," in *Yeditepe Üniversitesi Tarih Bölümü I. Uluslararası Türk Kültürü ve Tarihi Sempozyumu 19-20-21 Nisan 2018 Bildiri Kitabı*, ed. Ahmet Taşağıl, Aykut Kar, Ahmet Büyükaksoy, and Ayşen Müderrisoğlu (Istanbul: Yeditepe Üniversitesi Tarih Bölümü, 2018), 610–25.

14. Though, as a number of scholars have shown, such international agreements were subject to dispute and took several decades to come into widespread effect. See Ogle, *The Global Transformation of Time* and Charles W. J. Withers, *Zero Degrees: Geographies of the Prime Meridian* (Cambridge: Harvard University Press, 2017). For musical parallels, see Gribenski, *Tuning the World.*

15. Günergün, "Salih Zeki ve Astronomi," 104–5. For details of Salih's involvement in time reform, see Tuğçe Kancı Karagedikli, "Horological Connections: Salih Zeki and the Standardization of Clocks in the Late Ottoman Empire" (Master's thesis, Middle East Technical University, 2024).

16. The exact causes of Salih's death in 1921 when he was only in his late fifties are unclear, though most sources report a neurological condition of some sort. See Atasoy, "Türk Bilim Tarihçiliğinde Salih Zeki," 27; Etker, "Salih Zeki Bey," 139.

17. For a preliminary bibliography, see Alper Atasoy, "Salih Zeki'nin Makaleleri: Bir Bibliyografya Denemesi," *Osmanlı Bilimi Araştırmaları* 23, no. 2 (2022): 335–94.

18. On Lacoine, see Günergün, "Salih Zeki ve Astronomi," 117–18.

19. For further references and analysis, see Kancı Karagedikli, "Horological Connections."

20. One of Salih's articles on Comte was written and published together with his second wife, the prominent novelist, feminist, and Turkish nationalist Halide Edib (1884–1964). See Atasoy, "Salih Zeki'nin Makaleleri," 383, and Demir and Kalaycıoğulları, "Büyük Bir Matematik Tarihçisi," 203.

21. For Salih's own account of this research, see the transcription provided in Demir and Kalaycıoğulları, "Büyük Bir Matematik Tarihçisi," 197–99. For an early assessment of his historical scholarship, see the review by A. Adnan, "Ṣāliḥ Zekī [*sic*], *Athār-i-Bāqiya*, a history of Arabic mathematics," *Isis* 19, no. 3 (1933): 506–15.

22. Ṣāliḥ Ẕekī *Ās̱ār-i baḳīye*, 2 vols. (Istanbul: Maṭbaʿa-i ʿāmire, 1329/1911).

23. Ṣāliḥ Ẕekī, *Ḥikmet-i ṭabīʿīye* (Istanbul: Ḳarabet maṭbaʿası, 1311/1894), 477–607.

24. For further discussion, see Feza Günergun, "Derviş Mehmed Emin pacha (1817–1879), serviteur de la science et de l'État ottoman," in *Médecins et ingénieurs ottomans à l'âge des nationalismes*, ed. Méropi Anastassiadou-Dumont (Paris: Maisonneuve & Larose, 2003), 171–83; Meltem Akbaş, "The Military March of Physics II: Teachers and Textbooks of Physics and Mechanical Sciences of the 19th Century Ottoman Military Schools," *Osmanlı Bilimi Araştırmaları* 14, no. 1 (2012): 89–108.

25. Ṣāliḥ Ẕekī, *Mebḥās̱-i ṣavt* (Istanbul: Maṭbaʿa-i ʿāmire, 1326/1910). For a partial transcription and commentary, see Ayşegül Başar, "Salih Zeki Bey'in 'Hikmet-i Tabîiyye-i Umûmiyye'den Mebhâs-ı Savt' İsimli Eseri ve 'Mûsikî Sesleri' Bölümünün İncelenmesi" (Master's thesis, Sakarya University, 2021). See also İrem Aslan Seyhan, "Salih Zeki'nin Hikmet-i Tabîʿiyye-i 'Umûmiyyeden Mebhas-i Savt Eseri," in *Ölümünün 100. Yılında Bilim Tarihçisi Salih Zeki'yi Anmak ve Anlamak Paneli Bildiri Kitabı*, ed. Vural Başaran (Ankara: Atatürk Kültür Merkezi Başkanlığı, 2022), 45–59.

26. David Pantalony, *Altered Sensations: Rudolph Koenig's Acoustical Workshop in Nineteenth-Century Paris* (New York: Springer, 2009), 9–11; Paolo Brenni, "19th Century French Scientific Instrument Makers. VI: The Triumph of Experimental Acoustics: Albert Marloye (1795–1874) and Rudolph Koenig (1832–1901)," *Bulletin of the Scientific Instrument Society* 44 (1995): 13–17.

27. Pantalony, *Altered Sensations*, 262–65.

28. For descriptions of experiments using the sonometer, see C. L. Barnes, *Lessons in Elementary Practical Physics*, vol. 3, part 1: *Practical Acoustics* (London: Macmillan and Co., 1897),

18–22. Salih explains the construction of the sonometer in Ṣāliḥ Ẕekī, *Ḥikmet-i ṭabīʿīye*, 564–5, and briefly describes his experimental methods in Ṣāliḥ Ẕekī, "ʿAṭūfetlü Aḥmed Midḥat Efendi ḥażretlerine," *İḳdām*, no. 1638 (January 29, 1899), 2–3.

29. For Yekta's account of this research, see Raouf Yekta, "La musique turque," in *Encyclopédie de la musique et dictionnaire du conservatoire. Première partie: Histoire de la musique*, vol. 5, ed. Albert Lavignac and Lionel de la Laurencie (Paris: Librairie Delagrave, 1922), 2945–3064, at 2982–84. For discussion, see Olley, "Measuring Progress," and Öztürk, "Türk Müziğinde Yekta, Ezgi ve Arel."

30. Yekta, "La musique turque," 2984; Raʾūf Yektā, "Fisaġoras ġāmı 2," *İḳdām*, no. 1668 (February 28, 1899), 3–4, at 3, translit. in Rauf Yekta, *Türk Müzikolojisinin Kurucusu Rauf Yekta Bey*, vol. 3: *Polemik Yazıları*, ed. Mehmet Öncel and Muhammet Ali Çergel (Istanbul: Ketebe Yayınları, 2024), 119–27, at 121. According to Yekta, before Ataullah began using a sonometer it was also used by Prince Abdulhalim Pasha (1830–94), a member of the Egyptian khedivial family who was banished to Istanbul in 1868: cited in İbnülemin Mahmut Kemal İnal, *Hoş Sadâ: Son Asır Türk Musikişinasları* (Istanbul: Maarif Basımevi, 1958), 85. Both Ataullah and Abdulhalim were high-ranking Freemasons, in common with Idrīs Rāghib Bey (1862–1931), another Egyptian intellectual who conducted experiments with the sonometer, as documented in Muḥammad Kāmil al-Khulaʿī, *Kitāb al-musīqā al-sharqī* ([Cairo]: Maṭbaʿat al-taqaddum, 1322/1904), opposite 32 (my thanks to Salah Eddin Maraqa for this reference). On the links of these figures to Freemasonry, see Thierry Zarcone, "Şeyh Mehmed Ataullah Dede (1842–1910) ve Galata Mevlevîhânesi: Doğu-Batı Arasında Entelektüel ve Tinsel Bir Köprü," in *Saltanatın Dervişleri Dervişlerin Saltanatı: İstanbul'da Mevlevîlik*, ed. Ekrem Işın (Istanbul: Suna ve İnan Kıraç Vakfı İstanbul Araştırmaları Enstitüsü, 2007), 58–75; Karim Wissa, "Freemasonry in Egypt 1798–1921: A Study in Cultural and Political Encounters," *Bulletin (British Society for Middle Eastern Studies)* 16, no. 2 (1989): 143–61.

31. Yekta, "La musique turque," 2984; Raʾūf Yektā, "Rūm kilīselerinde mūsīḳī," *İḳdām*, no. 1930 (November 17, 1899), 4, translit. in *Türk Müzikolojisinin Kurucusu*, 151–60, at 158. Although Yekta states that his first article appeared in *İkdam* on April 6, 1898 ("La musique turque," 2382), he began publishing in *Resimli Gazete* on January 11, 1897 (Olley, "Measuring Progress," 137–38).

32. For simplified transliterations of Ahmed Midhat's articles on music, see Fazlı Arslan, *Ahmet Midhat Efendi ve Mûsikî: Batılılaşma Döneminin İlk Metinleri* (Istanbul: VafıkBank Kültür Yayınları, 2020). Yekta's responses are transliterated in *Türk Müzikolojisinin Kurucusu*; he summarizes the exchange in "La musique turque," 2982–84. On the broader context of the debate, see Jacob Olley, "An Orient of One's Own: Music and Islamic Modernism in the Late Ottoman Empire," *Royal Musical Association Research Chronicle* 54 (2023): 21–36.

33. İsmāʿīl, "Raʾūf Yektā Beğ'e iḫṭār," *Ṣabāḥ*, no. 3276 (January 5, 1899), 3.

34. Raʾūf Yektā, "'İsmāʿīl' imżālı varaḳa ṣāḥibine," *İḳdām*, no. 1615 (January 6, 1899), 3, translit. in *Türk Müzikolojisinin Kurucusu*, 107.

35. Ṣāliḥ Ẕekī, "İsmāʿīl Beğ Efendi'ye," *İḳdām*, no. 1616 (January 7, 1899), 3, translit. in Samet Tengiz, "İkdam Gazetesi'nde Mûsikî Makaleleri (Transkribe ve İnceleme)" (Master's thesis, Ankara Yıldırım Beyazıt University, 2020), 67–73.

36. Ṣāliḥ Ẕekī, "İsmāʿīl Beğ," 70.

37. Ibid.

38. Ibid.

39. Ṣāliḥ Ẕekī, "İsmāʿīl Beğ," 71–73.

40. Raʾūf Yektā, "Üstād-i muḥterem saʿādetlü Ṣāliḥ Ẓekī Beğ Efendi ḥażretlerine," *İḳdām*, no. 1621 (January 12, 1899), 3, translit. in *Türk Müzikolojisinin Kurucusu*, 108–10.

41. Aḥmed Midḥat, "Saʿādetlü Ṣāliḥ Ẓekī Beğ Efendi ḥażretlerine," *Tercüman-i ḥaḳīḳāt*, no. 6363 (January 11, 1899), 2–3, translit. in Arslan, *Ahmet Midhat Efendi*, 195–202, at 196.

42. Ṣāliḥ Ẓekī, "ʿAṭūfetlü Aḥmed Midḥat Efendi ḥażretlerine," *İḳdām*, no. 1627 (January 18, 1899), 2–3, at 2, translit. in Arslan, *Ahmet Midhat Efendi*, 283–87, at 283. For the following installments, see: Ṣāliḥ Ẓekī, "ʿAṭūfetlü Aḥmed Midḥat Efendi ḥażretlerine," *İḳdām*, no. 1632 (January 23, 1899), 2–3; no. 1638 (January 29, 1899), 2–3; no. 1639 (January 30, 1899), 2–3; no. 1647 (February 7, 1899), 3; no. 1649 (February 9, 1899), 3. The contents of the articles are summarized in Arslan, *Ahmet Midhat Efendi*, 185–95.

43. Ṣāliḥ Ẓekī, "ʿAṭūfetlü Aḥmed Midḥat Efendi" (January 29, 1899), 2.

44. Aḥmed Midḥat, "Saʿādetlü Ṣāliḥ Ẓekī Beğ," 3, translit. in Arslan, *Ahmet Midhat Efendi*, 201.

45. Ṣāliḥ Ẓekī, "İsmāʿīl Beğ," translit. in Tengiz, "İkdam Gazetesi'nde Mûsikî," 70.

46. Yekta, "La musique turque," 2990. The tetrachords are 9:8 + 9:8 + 256:243 and 10:9 + 10:9 + 27:25 respectively. For details of al-Urmawī's theoretical system, see Owen Wright, *The Modal System of Arab and Persian Music, A.D. 1250–1300* (Oxford: Oxford University Press, 1978).

47. Ṣāliḥ Ẓekī, "ʿAṭūfetlü Aḥmed Midḥat Efendi" (January 23, 1899), 2.

48. For further discussion, see Olley, "Measuring Progress."

49. On the role of attention in acoustic experiments, see Steege, *Helmholtz and the Modern Listener*, chapter 3.

50. See David Trippett's contribution in this volume.

51. An argument along these lines had been made in 1885 by Alexander Ellis, "On the Musical Scales of Various Nations," *Journal of the Society of Arts* 33, no. 1688 (March 27, 1885): 485–527. For further discussion of the relationship between concepts of tuning and colonial difference, see Parkorn Wangpaiboonkit, "On Offering Oneself to Music History: Perspectives and Positionalities from Colonial Siam," *Journal of Musicology* 40, no. 3 (2023): 275–83.

52. Ṣāliḥ Ẓekī, "ʿAṭūfetlü Aḥmed Midḥat Efendi" (January 23, 1899), 2.

53. Ṣāliḥ Ẓekī, "ʿAṭūfetlü Aḥmed Midḥat Efendi" (January 29,1899), 2.

54. William Thomson [Lord Kelvin], "The Six Gateways of Knowledge," in *Popular Lectures and Addresses*, vol. 1 (London: Macmillan and Co., 1889), 253–99, at 270–71. Cf. Ṣāliḥ Ẓeḳī, *Mebḥāṣ-i ṣavt*, front matter, translit. in Başar, "Salih Zeki Bey'in 'Hikmet-i Tabîiyye-i Umûmiyye'den Mebhâs-ı Savt'," 56.

55. Asad, *Secular Translations*.

56. For further discussion, see Wishnitzer, *Reading Clocks*, chapter 1; Ogle, *The Global Transformation of Time*, chapter 6; Mestyan, "Upgrade?"

57. For accounts of listening to the chimes of clock towers, see Mehmet Bengü Uluengin, "Secularizing Anatolia Tick by Tick: Clock Towers in the Ottoman Empire and the Turkish Republic," *International Journal of Middle East Studies* 42, no. 1 (2010): 17–36. On other sonic technologies, see Onur Engin, "Church Bells, Cannons, and Talking Machines: Sounding Devices and the 'Modernization' of Late Ottoman Istanbul" (PhD diss., Koç University, 2021).

58. Ṣāliḥ Ẓekī, "Baḥrīye Nāẓırı Paşa ḥażretlerine," *Ṭanīn*, no. 167 (January 18, 1909), 2–3. See Kancı Karagedikli, "Horological Connections," 50–55.

59. Ṣāliḥ Ẓekī, "Baḥrīye Nāẓırı Paşa," 2; Kance Karagedikli, "Horological Connections," 52.

60. Muhammad U. Faruque, "Attention, Consciousness, and Self-Cultivation in Sufi-Philosophical Thought," *Journal of the Institute for Sufi Studies* 2, no. 2 (2023): 300–316.

61. Ines Weinrich, "Sensing Sound: Aesthetic and Religious Experience According to al-Ghazālī," *Entangled Religions* 10 (2019), https://doi.org/10.13154/er.10.2019.8437, para 13. On Mevlevism and music, see Feldman, *From Rumi to the Whirling Dervishes* and Anders Hammarlund, Tord Olsson, and Elisabeth Özdalga, eds., *Sufism, Music and Society in Turkey and the Middle East: Papers Read at a Conference Held at the Swedish Research Institute in Istanbul, November 27–29, 1997* (Istanbul: Svenska Forskningsinstitutet Istanbul, 2001).

62. The book on music and listening from al-Ghazālī's *Iḥyā ʾ ʿulūm al-dīn* (Revival of the religious sciences) is translated in Duncan B. MacDonald, "Emotional Religion in Islām as Affected by Music and Singing," *Journal of the Royal Asiatic Society of Great Britain and Ireland* (April 1901): 195–252; (October 1901): 705–48; (January 1902): 1–28. The quote is from April 1901, p. 199. For further discussion, see Weinrich, "Sensing Sound" and Lewisohn, "The Sacred Music of Islam."

63. MacDonald, "Emotional Religion" (January 1902), 3.

64. Jean-Jacques Thibon, "Sufi Views on Time and History," in *Sufi Cosmology*, ed. Christian Lange and Alexander Knysh (Leiden & Boston: Brill, 2023), 89–108, at 91–97. See also Gerhard Böwering, "Ideas of Time in Persian Sufism," *Iran* 30 (1992): 77–89.

65. Thibon, "Sufi Views," 96.

66. Böwering, "Ideas of Time," 86; Feldman, *From Rumi to the Whirling Dervishes*, 81–84.

67. Jonathan Crary, *Suspensions of Perception: Attention, Spectacle, and Modern Culture* (Cambridge: MIT Press, 1999); Michael Hagner, "Toward a History of Attention in Culture and Science," *MLN* 118, no. 3 (2003): 670–87.

68. Barak, *On Time*, 252–65; Wishnitzer, *Reading Clocks*. See also Vanessa Ogle, "Whose Time Is It? The Pluralization of Time and the Global Condition, 1870s–1940s," *American Historical Review* 118, no. 5 (2013): 1376–1402.

69. Crary, *Suspensions of Perception*, 65–71.

70. Afacan, "Idle Souls"; Şeyma Afacan, "Searching for the Soul in Shades of Grey: Modern Psychology's Spiritual Past in the Late Ottoman Empire," *European Journal of Turkish Studies* 32 (2021), https://doi.org/10.4000/ejts.7248; Özgür Türesay, "Between Science and Religion: Spiritism in the Ottoman Empire (1850s–1910s)," *Studia Islamica* 113, no. 2 (2018): 166–200.

71. Henri Bergson, *Time and Free Will: An Essay on the Immediate Data of Consciousness*, trans. F. L. Pogson (Abingdon: Routledge, 2013 [1910]), 11–18. See also Alexandra Kieffer, "Bergson, Debussy, and Early Twentieth-Century *Psychologie*," *Journal of the American Musicological Society* 72, no. 2 (2019): 563–69; Suzanne Guerlac, *Thinking in Time: An Introduction to Henri Bergson* (Ithaca: Cornell University Press, 2006), 47–52.

72. On Bergson's understanding of attention, see Crary, *Suspensions of Perception*, 316–28; Guerlac, *Thinking in Time*, 130–38.

73. Martin Jay, *Downcast Eyes: The Denigration of Vision in Twentieth-Century French Thought* (Berkeley: University of California Press, 1994), 113–14. Cf. Bergson, *Time and Free Will*, 161–62.

74. On the perception of the sound of a pendulum and its spatial representation, see Bergson, *Time and Free Will*, 104–10.

75. Bergson, *Time and Free Will*, 44–46.

76. Nazım İrem, "Bergson and Politics: Ottoman-Turkish Encounters with Innovation," *The European Legacy* 16, no. 7 (2011): 873–82; Afacan, "Searching for the Soul," 12–16. See also Damian A. Howard, *Being Human in Islam: The Impact of the Evolutionary Worldview* (London & New York: Routledge, 2011), chapter 2. On Bergsonism in global anticolonial thought, see

Souleyman Bachir Diagne, *Postcolonial Bergson*, trans. Lindsay Turner (New York: Fordham University Press, 2020) and Andrea J. Pitts and Mark William Westmoreland, eds., *Beyond Bergson: Examining Race and Colonialism Through the Writings of Henri Bergson* (Albany: State University of New York Press, 2019).

77. Donna V. Jones, *The Racial Discourses of Life Philosophy: Negritude, Vitalism, and Modernity* (New York: Columbia University Press, 2010); Mark Antliff, *Inventing Bergson: Cultural Politics and the Parisian Avant-Garde* (Princeton: Princeton University Press, 1993).

78. Larry S. McGrath, "Bergson, Servant of Colonialism: Education and Empire in North Africa Under the French Third Republic," *Bergsoniana* 2 (2022), https://doi.org/10.4000/bergsoniana.920.

79. M. Sait Özervarlı, "Ottoman Perceptions of al-Ghazālī's Works and Discussions on His Historical Role in Its Late Period," in *Islam and Rationality: The Impact of al-Ghazālī*, vol. 2, ed. Frank Griffel (Leiden: Brill, 2016), 253–82, at 260–62 and 277.

80. See e.g. M. Brett Wilson, "The Twilight of Ottoman Sufism: Antiquity, Immorality, and Nation in Yakup Kadrı Karaosmanoğlu's *Nur Baba*," *International Journal of Middle East Studies* 49, no. 2 (2017): 233–53; and Afacan, "Idle Souls."

81. Feza Günergun, "Timekeepers and Sufi Mystics: Technical Knowledge Bearers of the Ottoman Empire," *Technology and Culture* 62, no. 2 (2021): 348–72. On Sufi engagements with modern technologies and institutions, see Brian Silverstein, "Sufism and Governmentality in the Late Ottoman Empire," *Comparative Studies of South Asia, Africa and the Middle East* 29, no. 2 (2009): 171–85.

82. M. Sait Özervarlı, "Alternative Approaches to Modernization in the Late Ottoman Period: İzmirli İsmail Hakkı's Religious Thought Against Materialist Scientism," *International Journal of Middle East Studies* 39, no. 1 (2007): 77–102, at 81.

CHAPTER 11

The Attention Wave: Sound and Sense in Cycles ca. 1905

D. Graham Burnett

The English-born, German-trained scientist Edward B. Titchener presided rather grandly over the most prolific American center of experimental psychology in the early twentieth century—at Cornell University. His textbook and lab manual, *Experimental Psychology: A Manual of Laboratory Practice* of 1901, was used all over the United States and beyond.[1] One of the early experiments in that volume taught students how to establish the existence of an "attention wave," a psychological phenomenon to which this dean of American science was quite deeply committed. Let us begin, then, with a close look at just exactly how one caught a glimpse of this important feature of the human mind—at least as it was understood circa 1900.

The experiment proceeded as follows: Students were assigned to work in pairs, and one sat in front of a "kymograph"—an electrical recording device that used a recording needle to inscribe a line on a sheet of smoked paper on a slow-revolving drum; then this student pressed a little key-switch, the needle jumped, and the line marked the change. With that student at the ready, the *other* student carried a ticking watch away from the kymograph table until the sound of the timepiece could no longer be heard by the student at the switch; then, slowly, inching a bit closer together, the pair found the absolute "supraliminal" threshold of audibility for the student on the kymograph. At that distance, this latter student proceeded to press the key *when the sound could not be heard*, and release it when it was again perceived. When the experiment "worked," even if both students stayed perfectly still, the sound seemed to come and go for the student on the kymograph, whose recordings of the phenomenon were supposed to take on an evenly oscillating curve form.

This activity, in fact, reproduced a classic psychophysical experiment: In 1875, the Austrian ear surgeon Viktor Urbantschitsch had noticed that his patients experienced subjective fluctuations in their ability to hear a faint sound without moving their heads, despite there being no measurable change in the intensity of the sound itself. It seemed clear that this was happening in the biology of the listener, because while various subjects would report the same effect, they did not experience it "in phase" together, even when they were all listening to the same tuning fork hum or trickle of water. Urbantschitsch ascribed the phenomenon to a (hypothetical) cyclical change in the receptivity of the auditory nerves themselves.[2] Perhaps, he thought, they went through periods of exhaustion and recuperation along the lines of other nervous responses studied by Gustav Fechner and others. Urbantschitsch explicitly analogized the appearance and disappearance of the sound to a visual phenomenon discussed by Hermann von Helmholtz in the second volume of his *Handbuch der Physiologischen Optik* (originally published in 1860), where the faintest visible rings on a rotating Masson disk seemed to appear and disappear to the eye.[3]

About a decade later, both the acoustic and the visual phenomenon of fluctuations of liminal perception (together with parallel effects in a third sensory modality—haptic discrimination) came under the scrutiny of Nicolai Lange, who worked closely with Wilhelm Wundt at Leipzig. Across a barrage of experiments, Lange satisfied himself that these cycles were a robust manifestation of something much deeper than the fatiguing of peripheral receptors: He made a claim that these discernable patterns of perceptual sensitivity were actually "central," and hence were better understood as fluctuations not of mere perception, but of *active apperception* itself, which he would call "Aufmerksamkeit"—*attention*.[4]

In the last decade of the nineteenth century, the emergent sense that oscillations in the experience of perception at liminal thresholds reflected some sort of general cyclicity in the apperceptive capacities of the human organism gave rise to the formal idea of an "Attention Wave." This idea was controversial. Titchener, for one, came down firmly on the side of those who believed that these various rhythmical experiences of faint perception (in vision, touch, and hearing) indeed reflected an *underlying organic cycle*: He believed that the minds of human beings could be shown to have an attentional rhythm not unlike the rhythm of the breath.[5] In fact, by 1899, Titchener was persuaded

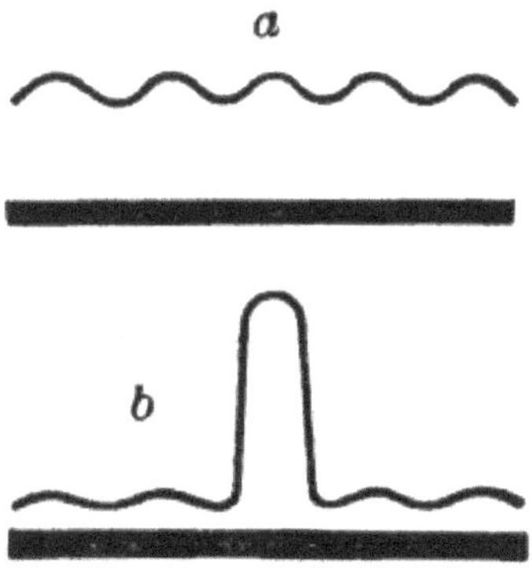

FIG. 10

Figure 11.1. Figure 10 from Edward Bradford Titchener, *A Primer of Psychology*, revised edition. New York: Macmillan, 1899, illustrating the rhythmic dynamic of attention.

that experimental evidence pointed to discrete "attention-waves" of variable but distinct periodicities:

> The attention-wave rises and falls at short intervals: attention fluctuates. The longest stretch of attention recorded is a stretch of 24 sec., and the average length of attention is no more than 5 or 6 secs.[6]

He even offered an illustration of the rhythmic dynamic of attention, an image meant to depict not only the frequency, but also the amplitude of an attention wave under changing conditions (Figure 11.1).[7]

In sum, from Titchener's perspective, "continuous attention" was, as he put it, "a contradiction in terms"—because attention *came and went*, perpetually, in a series of sinusoidal oscillations.[8]

* * *

This core claim—that attention itself was subject to an organic rhythm operating at the scale of two to thirty seconds—was hotly debated by a number of leading experimental psychologists in Europe and the United States. Already, in the immediate wake of Urbantschitsch and Lange, a half-dozen published experiments on both sides of the Atlantic contested *both* the factual findings *and* the interpretations at stake. Some denied that there was any real subjective fluctuation at all, provided you set the experiments up just so.

Others acknowledged perceptual fluctuations, but insisted that they were best explained by peripheral (or modally particular) accommodations of the eyes, ears, or sensory faculties of the skin.

Bracket all this debate (which I discuss in detail in a longer study of this problem). What is relevant for our purposes in this essay is that Titchener, in 1900, was a partisan of the full-on "Attention Wave" theory.[9] Which is to say, he believed that what was fluctuating was indeed a central apperceptive capacity, and one that corresponded to what he wanted to call "Attention." In his view, the specific threshold effects in each of the senses were all properly attributed to the cyclical nature of human attention itself. Thus, his laboratory manual for *Experimental Psychology* gave clear instructions about how a student of psychology was to be taught to discern this attentional dynamic.

For instance, in Chapter 8, the manual explained that it was important, when setting up the basic fluctuation experiments, to put the student doing the threshold listening *right in front of the revolving drum of the kymograph itself*—so he or she could see the rise and fall of the indicator needle *that he or she was causing*. This was key. It allowed "the attention wave" to take shape right before the eyes of the experimental subject. The object of the experiment? *To learn to see/feel the attention wave.* The assignment? *To produce a strip of kymograph indicator showing sinusoidal attentional fluctuations.*

What better way to help the student learn to "introspect" these fluctuations than to position him or her in front of the recording device, so that there could be a visible "check" and reinforcement of the auditory dynamics being experienced? A telling note to the lab instructor went on to explain that if the drum was *set to rotate too slowly* the subject, "(at any rate, in this stage of practice), grows inattentive, and the results are therefore untrustworthy."[10]

This is an extraordinary moment in the text, one that speaks a small volume on the intricate nexus of attention (as a problem) and introspection (as a method) in psychology at the start of the twentieth century. Here, the leading figure in the field in the United States provides instructions to young psychologists as to how to *be attentive to their own attention*. It was a learning process, to be sure. It took a while to get the hang of it. And this was what the laboratory practices of early psychology explicitly taught. Titchener, who had very specific ideas about the nature of attention (he defined it, ultimately, as a kind of mental "clearness") and how it worked (in *waves*), laid out, in his laboratory manual, *exceedingly precise instructions* for how everyone working in psychology would learn to experience, in themselves, and document, in the lab, the wave-nature of human attention. Accordingly, the "clarity" of things—one's core capacity to see, hear, feel, and think—was in perpetual, rhythmic, and very subtle flux. Successful introspectors acquired, through disciplined psycho-technical pedagogy, a heightened somatosensory *attunement*: They learned to *see past* the illusion of a stable world of thoughts and perceptions. These expert scrutinizers of mental life could discern what most of us miss: Our minds, too, "beat," just like our hearts; consciousness *pulses* in a systole–diastole of attention–distraction; there is a respiratory rhythm to what Titchener would come to call "attensity."[11]

Becoming an expert psychological introspector thus entailed doing hundreds of pages of exercises across weeks and months of laboratory study, and (for those who went on to pursue master's and doctoral degrees) following such practice with *years* of careful reinforcement. The result was a cadre of certified experimental subjects for new psychological experiments. It is perhaps little surprising, then, that these individuals were able to generate relatively large amounts of handsomely consonant experimental results. While not a "shell game," exactly, it was one of the most remarkably refined, elaborate, intensive, and ultimately abandoned regimes of observation ever devised in the modern sciences.[12] Moreover (at least for quite a while), it *worked*—producing vast numbers of professional publications and securing teaching berths for hundreds of newly minted PhDs in an emergent field that would hold a successful place in the modern research university.

By 1900, then, Titchener had created a whole system of "formation," a process of disciplinary acculturation, an elaborate socio-technical regime of sequential exercises and regular "practices"—in short, a *dispositif*, in the sense invoked in late Foucault—by which to instantiate his theory of attention in the eyes, ears, hands, and minds of those who would pursue the problem for a full generation in North America.[13] In this way, Attention Waves were "baked in" to the dominant process of disciplinary professionalization in the United States at the turn of the century. Titchener not only believed that human attention operated in identifiable waveform cycles, he and his colleagues had anchored that finding in a distinctive complex of laboratory experiments and trained observation. Competent subjects in the psychology labs of Cornell, Ann Arbor, and elsewhere had learned how to discern the rhythmical nature of their own attention through carefully drilled introspective exercises. Empirical technologies of graphic inscription turned this expert knowledge into objective, time-sequenced data, suitable for display and further analysis.

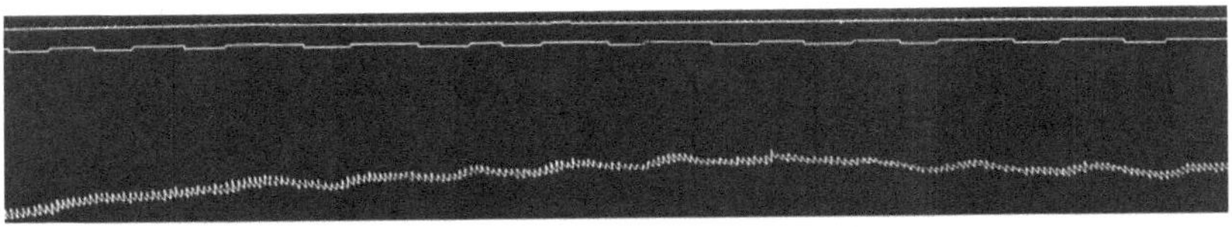

Figure 11.2. Figure 3 from John W. Slaughter, "The Fluctuations of the Attention in Some of Their Psychological Relations," *American Journal of Psychology* 12, no. 3 (April 1901), leaf between pages 328 and 329.

Figure 11.2 shows an image from an article entitled "The Fluctuations of the Attention in Some of Their Psychological Relations," published in 1901 in the *American Journal of Psychology* as part of a series of research reports contributed by Walter Pillsbury on work being done on attention waves in his Michigan laboratories.[14] The recording was made by John W. Slaughter as part of his PhD work under Pillsbury, who had just recently arrived at the university from his days in Ithaca with Titchener (and who was already on his way to becoming the leading attention scientist in the United States).

We can think of the upper part of this recording as a paradigmatic inscription of the attention wave in the period. Unlike the Titchener graphic shown above (which was merely an "illustration" for pedagogical purposes), this is a photostatic reproduction of an actual sheet of smoked paper peeled off the drum of a kymograph after an active experiment. In that experiment, the author of the paper, Slaughter, serving as his own expert subject, attempted to discern the faintest ring on a rotating Masson disk for a little over half a minute. The top line depicts the tick-marks made every fifth of a second by a Jacquet-Verdin indicator clock on the kymograph sheet. The crenellating curve below records Slaughter's pressing and releasing of a telegraph key, as (through the introspective scrutiny of his own attention) he saw, and then failed to see, that lightest gray bit of the Masson disk. In other words, that stepwise, oscillating curve depicts the "attention wave" itself, as it was then understood. A full cycle (of visibility and invisibility, which translated to "attention" and "inattention") varies, but runs about 3 to 5 seconds—the shorter end of the "normal" spectrum that had come to be established in the course of traditional laboratory exercises like those sketched in Titchener's *Manual*.

If the upper lines in the diagram, then, establish this document as an inscription of "normal science" for the study of human attention at the turn of the century (the kind of thing that everyone who had done a laboratory course in psychology knew how to make), it is the *lower* line that accounts for Slaughter's publication of the research in a professional journal. Because that line shows something different. It depicts the delicate tracing of a "finger plethysmograph," a highly precise mechanical piston-sensor that could be fitted over the finger and that was capable of recording a number of physiological rhythms—most notably the pulse and the cyclical perturbations in blood pressure known as "Traube-Hering" (or sometimes "Mayer") waves.[15] And what Slaughter claimed that he could show, from this inscription and others made over the course of his doctoral research, was that *attention waves could be correlated with Traube-Hering waves*. There seemed to be a relationship between the cycles of human attention and the cycling of blood pressure in the body.[16] Not only that, respiration, too, appeared to play a role. The rhythms of attention thus seemed to be embedded in a series of fundamental somatic cycles.

I explore this moment in attention wave research circa 1901 in order to give a feel for the kind of work being done on the topic during its heyday, especially the richness of the research program and the dynamism of the larger project of investigation. Literally *dozens* of intricate, data-intensive papers exploring the workings of attention waves were published in the United States in the first decade of the twentieth century, many of which clashed with one another in sharp arguments—over measurements, equipment, and interpretation. I will not belabor the point by attempting to offer a résumé of all this work, or the details of the minor point-scoring by junior scholars that went on in those exchanges. But it will perhaps help the reader get a sense of the field if I sketch a few more instances of the work being done by the circle around Pillsbury, who established himself as the leading exponent of attention-wave work—more so even than his mentor, Titchener.

So, for instance, while John Slaughter was working to correlate attention waves with the physiological cycles of circulatory phenomena, another University of Michigan scientist, R. W. Taylor, applied a very similar experimental setup to run nearly one thousand attention wave observations on three different experimental subjects (all with "previous psychological training," all "careful, interested observers").[17] His experiments had two basic aims: (1) to examine changes in the attention waves under the influence of various pleasant and unpleasant "stimuli" (the warm odor of balsam, the soothing effects of smoking a cigarette, a sudden galvanic shock); and (2) to run these tests while closely monitoring the subject's breathing by means of a "pneumograph"—a device that recorded the rise and fall of respiratory activity.

The results were mixed: It was hard to distinguish the effects of pleasant and unpleasant stimulation; but *any* stimulation did seem to have a direct effect on the attention wave (lengthening its cycle). Moreover, confirming Slaughter's findings, there seemed to be excellent evidence that respiratory rhythms, too, were affected, thus suggesting that respiration was also involved in the dynamic of attention waves.[18]

Related work in other labs followed. Frederick G. Bonser, working at the University of Illinois, expanded attention-wave research by looking systematically at the relationship between Traube-Hering waves and a wide range of sensory modalities in which attention waves were thought to be identifiable. These included the visual modality, using a Masson disk; the auditory modality, using faint audition (as in the original psychophysical work of Urbantschitsch); and then a pair of related phenomena—retinal rivalry (the oscillation of eye dominance in a split presentation of different hues on the left and right side of a stereoscope), and the "flipping" of ambiguous figures like a Necker cube.[19] Bonser developed a set of composite images that showed the tight correlation across these manifestations, shown in Figure 11.3.

So robust did these correlations come to seem (to some, anyway), that a good few physiologists decided to dispense with the finicky business of trying to monitor Traube-Hering waves (always very hard to detect) and instead simply to use the much easier to generate attention waves as a workable proxy.[20] And for his part, Pillsbury began to investigate, in conjunction with Wundt's protégé Oswald Külpe in Würzburg, the idea that diurnal patterns in the attention wave could become a useful index for mental fatigue, a topic of considerable interest in the period.[21] By 1904, attention waves felt so solid that they could serve as a means for investigating other psychological and physical phenomena.[22]

This sense of the attention wave as a fundamental feature of the embodied mind extended beyond laboratory research in those years and enjoyed wider cultural resonance. The Hungarian-born social activist Gustav Spiller, one of the early leaders of the Ethical Culture movement (and a convener of the 1911 anticolonial "First Universal Races Congress" in London), picked up on the theme of attention waves in his popular writings on the human mind in the early twentieth century and went on to elaborate what might be thought of as an "attention wave theory" of aesthetics, education, and human possibility more generally. In *The Mind of Man* (1902), he mined the experimental literature on the dynamic nature of attention to sketch "attention"

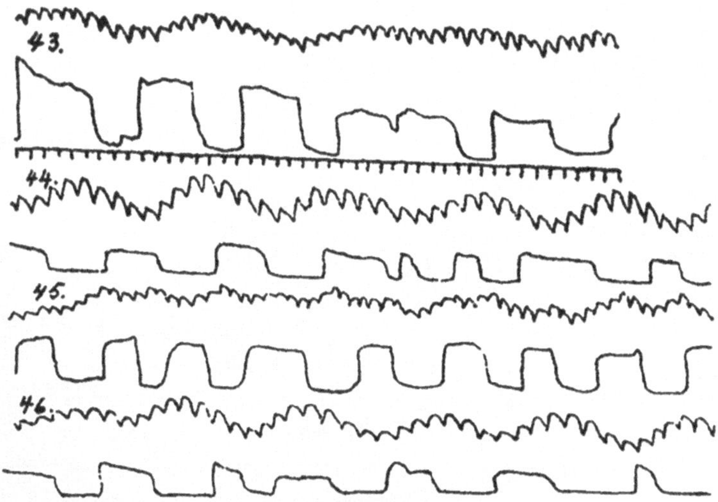

Fig. 14. Parallelism of Traube-Hering and Attention Waves. Curve 43, Masson Disc ; 44, 45, 46, geometrical figure.

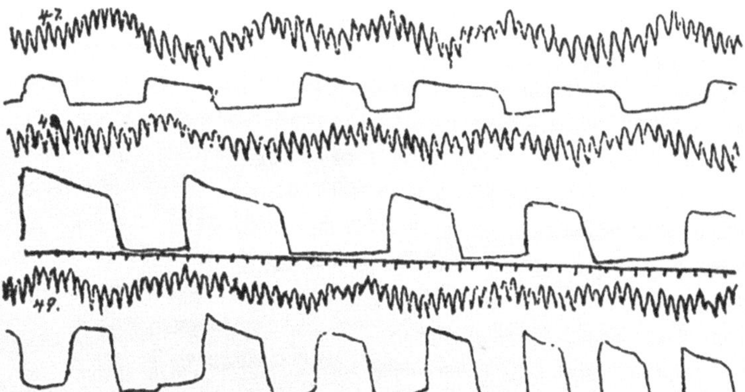

Fig. 15. Parallelism of Traube-Hering and Attention Waves. 47, Retinal-rivalry; 48, Audiometer ; 49, Masson disc.

Figure 11.3. Figures 14 and 15 from Frederick G. Bonser, "A Study of the Relations Between Mental Activity and the Circulation of the Blood," *Psychological Review* 10, no. 2 (1903), 135.

as something like the essential "delta" or "dynamic of change" in the human organism, understood as a sensory and cognitive being.[23] Waves operated at every scale. There were the grand swells:

> Instead of imagining attention in time as a smooth sea, we have to look upon it as a stormy one where the surface consists of huge waves. These waves, which form the ocean of thought, represent the several subjects which constitute the field of attention in time, and the largest waves are constantly broken up so as allow others to form.

And then there were the continuous *pulsings*, which he invoked in a sentence the actual cadences of which reproduced the very dynamic at issue:

> Or, we may say, that as the blood is propagated, not in a steady stream, but in waves, so in attention, or neural functioning, advance proceeds by pulsations.

For Spiller, more than for Titchener, Pillsbury, and others, static attention was authentically a contradiction in terms—because static attention meant inattention, even *vacancy*:

> Attention is like a river; it cannot rest; it must report progress. If the larger waves represent the subjects of thought, then the minute ripples which cover the whole surface of the stormy sea represent the essentially dynamic property of attention. Attention, like all work, is movement.[24]

When attention "sat still" it was no longer attention at all: "In normal life we ceaselessly pass from detail to detail, for persistent attention to one detail, as in hypnotism, produces vacancy or non-attention."[25]

Again, the rhythmic nature of the prose—the pacing of these sentences, in little comma-marked units—may well have been intentional, in that Spiller was ultimately committed to the idea that the wave-nature of attention was constitutive of human understanding and aesthetic experience alike. In a remarkable synthetic chapter at the end of his study, Spiller summarized his whole volume under the ambitious heading "Systems as Attention-Determined" and proceeded to lay out the way that the "waves" and "ripples" in human attention determine nearly every aspect of our experience: beauty, poetry, prose, music, humor, play—ultimately, the nature of learning itself. Aesthetic experiences, in his account, amount to a *coordination of the rhythms of phenomena in the world with the rhythms of our native attentional being*, hence we experience both pleasure and "progress" when we are, effectively, "in time," attentionally, with the stuff of existence. Not for nothing did Spiller turn his own attention, in his work with Felix Adler and the Ethical Society, to the collation of (secular) songs and hymns by which a new generation of well-tempered souls might be gathered in the enacted harmony of chorale.[26]

Nor was Spiller alone in linking the attention wave to the wider dynamics of language and aesthetic experience. Carl Seashore, the Iowa-based psychologist who would go on to develop the "Meier-Seashore Art Judgement Test" and the "Seashore Tests of Musical Ability," devoted six pages of his own popular textbook, *Elementary Experiments in Psychology* (1908), to attention waves, depicting them graphically in a particularly striking way (Figure 11.4).

He also insisted, like Spiller, that "the psychological explanation for our enjoyment of rhythm in all sorts of mental activity is to be found largely in this basal fact of a natural, biological rhythm of attention."[27]

We should pause for a moment on his uncanny image. We are told that this figure gives a sense of what a "detailed record of attention to a constant stimulus" might look like across about ten seconds. The "rugged" nature of the crests and troughs alike indicates that the "rhythmic fluctuation" operates continuously on multiple frequencies. Here is Seashore's explanation: "There are infinitely short ripplets of attention, hardly perceptible; these

Figure 11.4. Attention waves depicted in Carl Seashore, *Elementary Experiments in Psychology* (New York: Henry Holt, 1908), 160.

form the surface of ripples, which in turn form the surface of wavelets; and these wavelets in turn form the surface of waves, and so on." Summing up, he sketched a kind of fractal topology of attention, which undulated across all possible scales: "Thus, all attention is rhythmic, and there is rhythm within rhythm from the infinitesimally short to the very long, even daily and annual periodicities."[28]

But how strange, then, to read on and discover, in a footnote, that, in fact, the figure in question *isn't an actual image of attention waves at all.* While Seashore offers that it "might represent the distribution of attention . . . schematically," he owns, in the note, that the picture itself is meant as "merely a hypothetical diagram." Why? Because what the image shows is something quite different: As he concedes, his picture of "attention waves" is actually an engraved copy of "a photograph of the manometric flame for a vowel."

A *what?* A quick turn into turn-of-the-century "phonometry" may be in order. The "manometric flame" was the nineteenth-century technical term for a kind of "dancing candle" demonstration pioneered by Paris-based Prussian instrument-maker Karl Rudolph Koenig. It converted sound-waves into a series of vibrations in a gas-fed flame, the oscillations of which could then be seen stretched into a kind of time-lapse strip by means of a rotating mirror. Located somewhere between a parlor-trick and an early instance of "sound-writing" (such as would later give rise to phonographs and sound spectrographs), elaborate manometric flame apparatuses were developed by Helmholtz and other distinguished physicists and engineers interested in music and speech in the second half of the nineteenth century. The instrument got a new lease on life when high-speed photographic technique permitted the near-instantaneous documentation of the flame effects (creating a newly concrete inscription of sonic phenomena). The whole tradition was just passing from currency by the end of the first decade of the twentieth century, when Seashore used this imagery as an (unexpected) reference for the complex nature of attention waves.

But his "hypothetical" substitution is telling. By using the graphic representation of the sound-patterns of a human voice articulating a vowel to represent (if "schematically") the layered dynamics of the attention waves, Seashore was gesturing at something he thought was real: Like Spiller, Seashore too believed that when the object of attention was itself periodic ("as in music, poetry, or rhythmic movements"), then "the attention-waves adapt themselves to the rhythm so that the essentials of the group [i.e., the "units" of periodicity in the object] are grasped during the crest of the attention-wave and each momentary effort is well timed and is followed by a period of rest."[29] Both understanding and aesthetic experience, by these lights, were a kind of *phase conjunction*, a coordination between attention waves and those of sense. Indeed, Seashore likely felt that the inherent sonic rhythms of the voice (like those of music, to which he would give much of his professional attention in subsequent decades) were closely related to the attention waves that apprehended them.

This very idea—that the rhythms of human language were inextricable from the rhythms of attention waves—was brought to its most elaborate and systematic exposition in the same years by the American applied psychologist, adman, and general dignitary of commercially viable learned enterprise, Walter Dill Scott. Another Midwestern farm boy who eventually took a PhD (circa 1900) with Wundt in Leipzig, Scott had played a central role in establishing experimental psychology at Northwestern University in the early twentieth century—and he would go on not only to be president of the American Psychological Society, but also of Northwestern itself. He was an early and energetic exponent of the utility of psychological testing and psychological theories to the worlds of commerce, administration, and education, and his first book, *The Theory of Advertising; A Simple Exposition of the Principles of Psychology in Their Relation to Successful Advertising* of 1903, can be thought of as inaugurating the particular unholy alliance of science and capital that comes down to us as the "attention economy"—a dubious distinction.[30]

Scott's successful 1906 volume, *The Psychology of Public Speaking*, gives an entire chapter to attention wave phenomena and elaborates a remarkably detailed vision of the ways that the successful use of language depends on a mastery of the organic rhythms of the human attentional capacity. Reproducing several kymographic strips that depict the fundamental oscillation of attention (such as had been derived from Masson disk work and other threshold perceptual experiments), Scott went on to abstract a full-spectrum attention wave—one showing the layered attentional temporalities revealed in the laboratory (Figure 11.5).

From this diagram, Scott proceeded to elaborate a host of practical observations concerning the nature of human communication. The most fundamental point was simply that "all our thinking is done in 'spurts' which are uniformly followed by periods of inactivity." And for this reason, the work of "public address" must be tailored to the reality of the attention waves—at *both ends* (meaning both speaker and listener). And it must take into account the

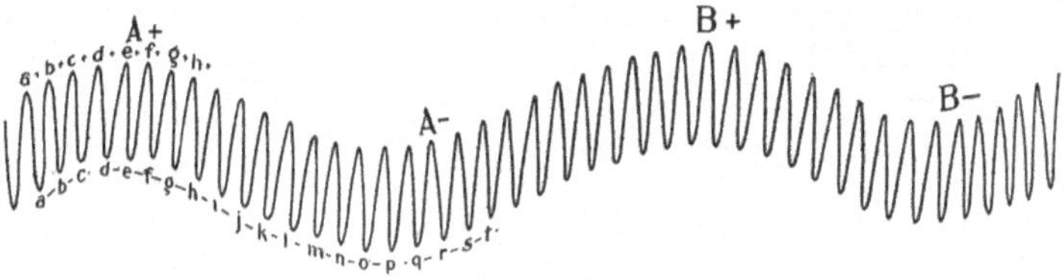

Figure 11.5. Figure 7 from Walter Dill Scott, *The Psychology of Public Speaking* (New York: Hinds, Noble, and Eldredge, 1906), leaf between pages 112 and 113.

multiple vibratory cycles at play in every act of attention. Indeed, it would not be wrong to think in terms of a kind of "harmonics" at play in every effort to communicate. Dill actually used the musical language of "partials" to characterize the subunits of the larger attentional waves, drawing on the theory of musical resonance laid out in Alexander John Ellis's important translation of (and commentary on) Helmholtz's *On the Sensations of Tone as the Physiological Basis for a Theory of Music.*[31]

Scott appealed to the experiences of his readers to support his general claim concerning the waveform experience of attentive consciousness: "Every public speaker has doubtless had the experience of finding himself unable to attend to the matter at hand for a moment, and then he seems to lose all control of himself and what he is trying to accomplish." This unsettling episode of "struggle" corresponds, Scott explained, to a dip in the major attention wave, such as can be seen between "A+" and "A-." Scott went further, noting that when the trough of the major wave coincides with a local minimum in the "minute waves" (which he identified as "A-q"), the result is an attentional nadir:

> The speaker may forget his part or he may speak on from the familiarity of the part, but with a minimum of efficiency. The hearer will not understand that which is being spoken.

All of this is often happening at a speed that militates against clear awareness (which is anyway what is somatically compromised). If the amplitude and frequency of the waves are sufficiently high, "when the crest of the second wave and of the minute wave coincide at a later time, the view of the whole may be so complete that the loss sustained by the previous lack of attention may be wiped out and not become known to the individual at all."[32] This meant that high peaks of attentiveness had the capacity, in effect, *retrospectively to supersede* upon the inevitable and continuous "gaps" created by inattention. One senses a theory of consciousness that is fundamentally analogous to the phenomenology of cinema: The experience of smooth, conjoint perception is an artifact of the physiology and psychology of the perceiving subject and does not correspond with the underlying discontinuities revealed by techno-scientific analysis.[33]

This had direct implications for the work of communication—implications of material importance to anyone (like an adman) who needed to get a message through to an audience. And Scott drove the point home: "No one can attend to what we are saying or doing unless we are presenting a new aspect of the subject every few seconds." Not only that, it stood to reason that the way to manage the inevitable series of local minima of attention in any time sequence was to make sure that the peaks of attentional engagement were sufficiently full and engaging that they could cast a little time-defying ring of ripples capable of widening across the stone-skipping lacunae in any listener's capacity to attend.

In this effort to "tune" a message for the human receiver, it was the actual structure of language that provided the key tool. In a remarkable section of his analysis of attention waves entitled "The Sentence, or Clause, as the Unit of Thought," Scott turned to a delicate discussion of the concordance between syntactical structure and the dynamics of human attention itself. While the verbal expression of a given "conception or idea" may require a number of seconds, Scott again contended that laboratory study had shown definitively that "we do not sustain a maximum of attention to it during the entire time." And this was, of course, true for both speaker and listener. Under certain circumstances, "as we pronounce a sentence our minds may hardly be on it at all, but we may be thinking of the sentences which are to follow, and the sentence which is being pronounced almost 'says itself.'" Similarly, when we are listening to a discourse, "we do not hold ourselves to a continuous strain, but we comprehend the sentences by very short pulses of attention, which are followed by periods of comparative rest."[34]

This meant that understanding itself proceeded by a kind of saccadic rhythm: "As a sentence is pronounced we may give but little attention to it and may not comprehend any of it till the very last word is spoken, and then by a single momentary activity of the mind we get the import of the entire sentence." And this could go farther. Now and again, the lacuna may endure "till after the following sentence has been commenced, and this later sentence may give the cue to the former, and it may then be grasped by a single pulse of thought."[35]

What the laboratory study of the attention wave had established was that the "normal" periodicity of attentional peaks seldom exceeded a few seconds, and thus that "to be compelled to attend without pause longer than this time is exhausting and very unpleasant." At the same time, "vibrations of attention that are shorter than a full second" are equally problematic, as they, too, are out of sync with the natural rhythm of our attentional capacities. Scott insisted that this "psychological law" had direct "application" in the work of effective and artful expression: The rhythm of the mind needed to be brought into harmony with the rhythm of the body.[36] Scott pointed out that normal human respiration could be shown to involve exhalations of about three seconds' duration: "Sound may be produced during an entire exhalation; or several 'breath groups' or 'exhalation groups' of sounds may be produced by a single exhalation." Citing physiological and psychological research done on the poet James Whitcomb Riley as he declaimed his thirty-two-line poem "Old-Fashioned Roses," Scott reported a total of twenty-nine exhalations, of which twenty-two fell at the end of a line: "Each expiration period expressed a single idea and could be grasped as a single pulse of attention," with an average period of 3.13 seconds. The point? "Here we have the expiration period usually identical with the line of poetry and the time of the enunciation of the line (or expiration period) . . . well within the time stated as the fluctuation of attention."[37]

Here was the modal optimum for the syncopation of human breath, human attention, and English syntax. Yes, longer sentences were possible. And shorter ones too. But the former were "exhausting" and the latter felt "choppy." The reason for this lay in the essential periodicity of the faculty of attention itself. There was some elasticity in all of this, naturally, in that "the nature of the discourse and the mental development of the auditors seem to make a difference," but the divergence fell in a narrow zone, given by Scott, on the basis of laboratory research as "from a minimum of approximately a single second to a maximum of a very few seconds." He summarized his discussion with concrete advice to those who wished to deliver an effective message to an audience:

> In general it may be said that each single sentence requires a 'spurt' of the attention, and the presentation must conform to this psychological fact. Every lecturer or entertainer must also expect a rise and

> fall of the intensity and duration of those recurrent 'spurts' of attention, and so [must] adjust his lecture to the mental limitations of his hearers. By skillful adjustments to the attention waves of the hearers the maximum effect is produced with the minimum of strained attention.[38]

* * *

The "Attention Wave" rose across the first decade of the twentieth century. But it fell before the First World War. The full story of that collapse is beyond the scope of this chapter, but suffice it to say that new experiments made it hard to hold on to the dream. Yes, there were fluctuations in sensory experience at the threshold of perception; but it became harder and harder to insist that what was fluctuating could properly be described as "attentional." On the contrary, a spate of younger researchers, working on the cusp of the second decade of the century, gradually established that what was "fluctuating" were the neural signals being generated at the sense organs themselves. This was hardly "attention" in any meaningful sense; it was simply "perception." Along the way, the primary means by which the scientists of the attention wave did their work—relying on trained "introspection"—also slumped into the dustbin of history. A new cohort of psychology researchers, eventually termed "behaviorists," would move away from the kind of self-scrutiny on which the science of Edward Titchener and his acolytes depended. Treating the mind as a "black box," these new researchers focused on inputs and outputs, and mostly deemed actual "thinking" as simply beyond scientific scrutiny. Sure, stuff happened in the mind, but there was no way to get at it from the first-person perspective. This had been the error of the introspectors, who spent so many hours feeling the rise and fall of the attention wave inside their persons.

But the sense of synchrony of voice and mind—the sense of a self understanding itself as a kind of rhythmic coordination—was a dream that died hard. As a coda to this story, we can do no better than return to one of the champions of the attention wave at its peak: Walter Pillsbury, one of Titchener's earliest and most faithful students, and himself the author of a distinguished treatise on attention, published in 1908. But the decade that followed was tough—globally, personally, and in the learned precincts of psychology. The First World War had sundered the Germanophilic world of American mind science and devastated the laboratory meccas of Old Europe to which that project had long oriented itself. Pillsbury himself, though always prolific and still the éminence grise of Ann Arbor, had teetered through an episode of psychological and physical collapse just before the outbreak of the war, slipping from the podium mid-lecture in a sudden rictus of epilepsy or stroke—to be caught by an attentive student in the front row. Convalescence ensued. And doggedly, the old warhorse of the attention wave returned, even in his depleted condition, to the pursuit of that essential periodicity he still posited, despite the ever-mounting evidence to the contrary. In a touching exercise of what might be called "absent-minded introspection" (or is it the "psychophysics of scholarship"?), Pillsbury had both himself *and his typewriter* wired up to a kymographic recorder in order that, as he wrote up a new psychology textbook in the postwar period, the rhythms of his scholarly productivity could be correlated to his respiration rate, his blood pressure, and, yes, his "attention waves," which were also monitored. The results were ambiguous. He presented them at the eighth international conference of psychology, held in Groningen in 1926. By the time the abstract saw print, the following year, Titchener himself was dead, and so was introspective/structuralist psychology.

The attention wave was a flatline.

Notes

1. For a recent not entirely unfavorable assessment of all of this by a philosopher interested in the ways philosophy and psychology orient to introspection, consider: Eric Schwitzgebel, "Introspective Training Apprehensively Defended: Reflections on Titchener's Lab Manual," *Journal of Consciousness Studies* 11, no. 7–8 (2004): 58–76.

2. Viktor Urbantschitsch, "Ueber eine Eigenthümlichkeit der Schallempfindungen geringster Intensität," *Centralblatt für die medicinischen Wissenschaften* 37 (21 August 1875): 625–28.

3. The Masson disk is a circular card marked with a single radius of black hachures. When rotated sufficiently quickly each of these marks appears as a gray circle. Although the marks are all identical, the concentric gray circles they generate in rotation are evenly graded from dark (in the middle) to very pale (at the margin)—on account of the faster speed of the turning disk as one moves out toward the circumference from the center. This was a standard instrument of the psychophysical and optical laboratory in the period. See Hermann von Helmholtz, *Treatise on Physiological Optics*, edited by James P. C. Southall (New York: Dover, 1962). This publication presents the three original volumes in two, with Helmholtz's volumes one and two bound together with separate numeration in a single book. The relevant pages for the Masson disk work are thus volume one (of the Dover), but volume two (of the work), and they are numerated pp. 178–180.

4. Nicolai Lange, "Beitrige zur Theorie der Aufmerksamkeit," *Philosophische Studien*, 4 (1888): 390–422. Interestingly, Lange also tried to demonstrate that these perceptual fluctuations

correlated to what would later be thought of as "gestalt shifts" in bivalent imagery (like a Necker cube). For a discussion of Lange and the fate of the idea of fluctuating attention in Wundt's work, see: Sven Luders, "The 'Fluctuations of Attention' Between Physiology, Experimental Psychology and Psycho-Technical Application," in *Psychology's Territories*, edited by Mitchell Ash and Thomas Sturm (Mahwah, NJ: Lawrence Erlbaum Associates, 2007): 31–50.

5. In the United States, Titchener's primary adversary in this matter was Hugo Münsterberg, who already by 1889 had defined the core elements of an alternative view by focusing on the role of specific ocular adaptations (blinking, averted vision) on perceptual fluctuations as they were experienced at sensory thresholds.

6. Edward Bradford Titchener, *A Primer of Psychology*, revised edition (New York: Macmillan, 1899), 88–89.

7. Ibid., p. 75. The work that this graphic does in the *Primer* should not be underestimated, as the simple image is actually deployed in a rhetorically intricate way. Yes, Titchener references this figure as an "attention-wave" on page 89, in the course of an analogy that likens the dynamics of attention to the carnival game in which an egg, bouncing atop a little jet of water, offered a difficult target for sharpshooters. But on page 75, where the image originally appears, Titchener glosses it as depicting the flow of consciousness coming out of the page at the reader and suggests that each peak be understood as an idea or perception in the field of awareness (i.e., apprehended at the same time); as one perception or idea seizes the attention, it can be understood to be "amplified," and the others correspondingly diminished. This is, obviously, not at all the same thing as the "natural" and continuous oscillation of attention in time—though Titchener uses the same image to illustrate both phenomena.

8. He asserts that continuous attention is a contradiction in terms in his *Experimental Psychology: A Manual of Laboratory Practice*, vol. 1, 196.

9. The specific formulation "attention wave" arises in English, in the relevant sense, for the first time (that I have yet discovered) in *Science* in 1889 in the course of a discussion of the publication, in German, of a series of experiments conducted in Strasbourg by the pioneering Swiss Sanskrit scholar Ernst Leumann. Leumann, inspired by experimental practices in physiology and psychology, monitored the pulse and respiration in *readers*, and believed he had noted correspondences between the rates of bodily functioning and textual expression/comprehension. Coincident patterns suggested rhythms of attention. See: Anonymous, "Mental Activity in Relation to Pulse and Respiration," *Science* 14, no. 355 (22 November 1889): 347–48. Leumann's work became part of the general argument for "attention waves" in the early twentieth century, though it was sometimes noted in the psychological literature as a "lay" contribution (e.g., in the *American Journal of Psychology*).

10. Titchener, *Experimental Psychology*, 195.

11. On "attensity," see Edward Bradford Titchener, "The Term 'Attensity,'" *The American Journal of Psychology* 35 (1924): 156.

12. One thinks of how richly Titchener's regime of introspection could be assimilated to the analytics laid out in Lorraine Daston and Elizabeth Lunbeck's *Histories of Scientific Observation* (Chicago: University of Chicago Press, 2011).

13. Kenneth Rogers, in *The Attention Complex: Media, Archaeology, Method* (New York: Palgrave Macmillan, 2014), plays out a fully "Foucauldian" treatment of attention from William James to John Conway, although his study really gets started with an analysis of behaviorism. The first chapter, for instance, examines how a behaviorist account of attention was inextricable from the "subject" of regimes of labor, training, management, and education in the period

1900–1945. The work on attentional cyclicity and mental fatigue in the early part of the century makes for an interesting counterpoint to his study (he does not discuss "attention waves").

14. John W. Slaughter, "The Fluctuations of the Attention in Some of Their Psychological Relations," *American Journal of Psychology* 12, no. 3 (April 1901): 313–34.

15. Traube-Hering-Mayer phenomena are genuinely complex and remain imperfectly understood to this day. But the consensus is that they relate to changes in the "tone" of the blood vessels, changes effected biochemically. They came to the attention of physiologists in the 1860s and 1870s, and their relationship to respiration and other physiological patterns was a subject of active research in the late nineteenth century. Continuous blood pressure recording sensitive enough to see them and stable enough to permit them to be distinguished from other factors was not easy to achieve. The finger-plethysmograph used by Slaughter is described in detail in: Warren P. Lombard and W. B. Pillsbury, "A New Form of Piston Recorder and Some of the Changes of the Volume of the Finger Which It Records," *American Journal of Physiology* 3 (1900): 186–200.

16. It is important to note that the English physician-physiologists Charles S. Roy and Charles S. Sherrington had suggested, in 1890, that threshold perceptual experiences (like the fluctuations of the visibility of faint rings on a Masson disk) might be attributable to "Traube-Hering undulations of the arterial pressure—with the wavelength of which they agree very well." But these medical doctors were not engaged in the study of attention and did not conceive liminal perceptual fluctuations as "attentional." See C. S. Roy and C. S. Sherrington, "On the Regulation of the Blood Supply of the Brain," *Journal of Physiology* 11 (1890): 85–108, at p. 108. Because modern neurophysiology continues to ascribe threshold perceptual fluctuations to circulatory rhythms, it is interesting to consider a more recent reappraisal: Robert P. Friedland and Constantino Iadecola, "Roy and Sherrington (1890): A Centennial Reexamination of 'On the Regulation of the Blood-Supply of the Brain,'" *Neurology* 41 (1991): 10–14.

17. R. W. Taylor, "The Effect of Certain Stimuli upon the Attention Wave," *American Journal of Psychology* 12, no. 3 (1901): 335–45, at p. 333.

18. Though not in any simple way. Taylor deferred to Slaughter's hypothesis that the Traube-Hering waves were tightly aligned to the attention waves. But Taylor thought that an inspiration could provide a "respiratory reinforcement" that could "keep the [Masson] rings in consciousness." Taylor saw his work as complementing "Slaughter's conclusions that the attention waves are due to overflows from the medulary centers to the central nervous system in general"—a process that heightened activity in the nervous system and that was itself, in all likelihood, "reinforced" by physiological processes of circulation and respiration. See Slaughter, "The Fluctuations of Attention in Some of their Psychological Relations," 333.

19. Frederick G. Bonser, "A Study of the Relations Between Mental Activity and the Circulation of the Blood," *Psychological Review* 10, no. 2 (1903): 120–38.

20. See Taylor, "The Effect of Certain Stimuli upon the Attention Wave," 345.

21. Pillsbury was resident in Würzburg in the summer of 1902, where he became acquainted with the closely related work of E. D. Wiersma (Wiersma, E. "Untersuchungen über die sogenannten Aufmerksamkeitsschwankungen," *Zeitschrift für Psychologie und Physiologie der Sinnesorgane* 26 [1901]: 168–200, and two subsequent supplements), then working with Külpe. More generally on the study of fatigue, the classic study remains Anson Rabinbach, *The Human Motor: Energy, Fatigue, and the Origins of Modernity* (Berkeley: University of California Press, 1992).

22. W. B. Pillsbury, "Attention Waves as a Means of Measuring Fatigue," *American Journal of Psychology* 14, no. 3/4 (1903): 277–88.

23. "Delta" and "dynamic of change" are my phrases. But here is his theory *in nuce*: "We have identified attention with cerebral change. The meaning of our key-word is readily distinguishable from the only other term, besides activity, with which it might be confounded. Willing is divided from attention, and attention from willing, in that attention *always is* change and nothing else, while volition *never is* change, but only points to it. The relation is that between being and becoming, between *sein* and *werden*." Gustav Spiller, "The Dynamics of Attention," *Mind* 10, no. 40 (1901): 498–524, at 523. Emphases in the original. It is an extraordinary passage, suggesting as it does that "attention" is properly thought of as "being"—*understood dynamically*. But it is precisely a dynamic conceptualization of "being" that has proved elusive within the history of Greek-derived philosophical inquiry. About a decade later, Heidegger himself would be preoccupied with this very problem, as it emerged from a neo-Kantian "psychologism" not unlike that of Spiller.

24. This and the block quote above, from Gustav Spiller, *The Mind of Man* (London: Sonnenschein, 1902), 69–70.

25. Ibid.

26. The wider project of linking physiological rhythms to artistic form in the period is the subject of Michael Golston, *Rhythm and Race in Modernist Poetry and Science: Pound, Yeats, Williams, and Modern Sciences of Rhythm* (New York: Columbia University Press, 2007). Golston mentions attention waves briefly in connection with the work of Thaddeus Bolton, but the book could have made more of the centrality of specifically *attentional* rhythms.

27. Carl E. Seashore, *Elementary Experiments in Psychology* (New York: Henry Holt, 1908), 163. The image above is found on p. 160.

28. Ibid., p. 161.

29. Ibid., p. 163.

30. Walter Dill Scott, *The Theory of Advertising; A Simple Exposition of the Principles of Psychology in Their Relation to Successful Advertising* (Boston: Small, Maynard & Company, 1903). The longest chapter in this book is the one that immediately follows the introduction, and it is on "Attention." Interestingly, Scott is overlooked by Tim Wu in his lively and wide-ranging book on the history of the buying and selling of attention, *The Attention Merchants* (New York: Knopf, 2016). Wu takes up "scientific advertising" a quarter-century later. But Scott was writing a regular column for *Mahin's Magazine*, the Chicago-based ad-industry newsletter from the first issue (in 1902).

31. Walter Dill Scott, *The Psychology of Public Speaking* (New York: Hinds, Noble, and Eldredge, 1906), 112–113. Helmholtz's book went through multiple editions, the first in 1875. I have consulted that of 1895, published in London by Longmans.

32. Scott, *The Psychology of Public Speaking*, 113–114.

33. See Jimena Canales, *A Tenth of a Second: A History* (Chicago: University of Chicago Press, 2009).

34. Scott, *The Psychology of Public Speaking*, 115.

35. Scott, *The Psychology of Public Speaking*, 115. The sense that understanding itself was effectively "saccadic" (in the way I am invoking here) wants to be examined in the context of the actual identification of the saccadic movements of the eye in reading. I strongly suspect a relationship. There is a notable convergence in the timing of the inquiries, as well as in some of the research personnel. It is the French optical scientist Émile Javal who is usually credited with first observing the jittery skipping of the eye along a line of text. But much of the mechanical/experimental early eye-tracking work was done by the idiosyncratic American psychologist

Edmund Burke Huey. See his *The Psychology and Pedagogy of Reading* (New York: Macmillan, 1908). It is important to note that the term "saccade" would not come to be used in English in the relevant sense until later, when Raymond Dodge applied the Francophone neologism to the jumpy, horizontal eye-movements associated with reading.

36. Scott, *The Psychology of Public Speaking*, 117. Interestingly, the dramatist, radical, and Victorian man of letters George Henry Lewes, writing in his *Problems of Life and Mind* (London: Trübner, 1879), offered a surprising early invocation of the "syncopation" of attention and breath-rhythms. Scott does not cite him, but Lewes's decidedly "nonprofessional" essay on attention was taken up by a number of scientific psychologists in the later nineteenth century, including Ribot. Here is Lewes's striking footnote (which appears in a note that runs from pages 188 to 189 of the work cited above): "The acquisition of the power of attention is the learning how to alternate mental adjustments with the rhythmic movements of respiration."

37. Scott, *The Psychology of Public Speaking*, 117.

38. Here and above, ibid., pp. 116–18.

CHAPTER 12

Attention in Musical Action: Theories of Attention in Early Music Psychology and Piano Pedagogy

Youn Kim

The nineteenth century famously saw the ascendence of various new forms of auditory attention. These new modes of listening responded to emerging musical aesthetics—such as the dramatic increase in the prestige of instrumental music—as well as to innovations in the domain of medicine, which saw the emergence of mediate auscultation and a bevy of associated techniques of attending to bodily sounds.[1] This chapter examines conceptions of attention in the context of psychology and music pedagogy at a late-century moment when the boundaries between these fields were still largely undefined.

Investigations of auditory attention took place in a variety of research laboratories toward the end of the nineteenth century. But the topic was also studied within the context of an active speculative tradition that had long attempted to understand the human mind in context. These approaches collectively formed what later historians have termed "second psychology."[2] Attention was a prominent topic across these various psychology programs, with rich and intricate conceptual layers. Many contemporary music pedagogy writings, and especially piano treatises that related attention to concepts of muscular sensation, automaticity, bodily effort, proprioception of the performing body, and the unconscious, can thus be regarded as also representing this "second" psychological tradition. Situating attention as a form of agency rather than discussing it as a quality of experience, they advanced broad, multimodal concepts of musical attention with significance well outside pedagogy. Here, I examine these multiple dimensions of attention in musical action, focusing on the reciprocal relationship between psychological research and piano pedagogy.

Late Nineteenth-Century Psychologies of Attention

With the emergence of scientific psychology in the late nineteenth century, many attempts were made to systematically discuss the topic of attention.[3] Edward Bradford Titchener, who studied with the founder of scientific psychology, Wilhelm Wundt, even wrote that "the discovery of attention" was one of the principal achievements of the new experimental psychology:

> What I mean by the discovery of attention is the explicit formulation of the problem; the recognition of its separate status and fundamental importance; the realization that the doctrine of attention is the nerve of the whole psychological system and that as men judge of it, so shall they be judged from the tribunal of psychology.[4]

However, the study of attention within the context of psychology remained complicated due to a lack of clarity as to the nature of the phenomenon. The reality was far from William James's famous assertion that "everyone knows what attention is."[5] On the contrary, as historian Kurt Danziger has aptly pointed out, the boundaries between intra-scientific and lay usage of some terms have long been blurry in the human sciences, partly because specialized language "continues to seep into the languages of everyday life" (and vice versa).[6] This was particularly true for attention. Though today the term is generally associated with the selective process of perception, historically, it had much broader and more comprehensive implications.

The most widely discussed attentional models in early psychology were based on analogies to vision. Wundt deemed the focus of attention the *Blickpunkt* ("focal point") and the wider field of awareness the *Blickfeld* ("field of view").[7] This approach influenced several structural psychologists who investigated the basic elements of consciousness in order to understand the structure of the human mind. Titchener, for instance, delineated different "levels or degrees of clearness [that] may coexist in the same consciousness." In his influential *Lectures on the Psychology of Feeling and Attention* (1908), he writes: "Increased clearness of any one part-contents of consciousness implies the decreased clearness of all the rest."[8] This model of attention prevailed even in writings by those who disagreed with the foundational tenets of structural psychology. For example, psychologist Oswald Külpe, who emphasized higher mental processes rather than the structure of consciousness, provided a similar explanation of attention as a "state of consciousness," writing, "We think of the attention as a state of mind which may be realized in greater or less degree, and may embrace at any given time a larger or smaller number of conscious processes."[9] A similar view is found in American psychologist and philosopher James Mark Baldwin's graphic representation of consciousness as a series of concentric circles that illustrate gradations of awareness, ranging from the unconscious to active acts of apperception. This further emphasizes the analogy with vision (see Figure 12.1).[10]

British ethologist and psychologist Conwy Lloyd Morgan also proposed that "in any moment of consciousness, there are in addition to and alongside the dominant elements constituting the summit of full clear, consciousness,

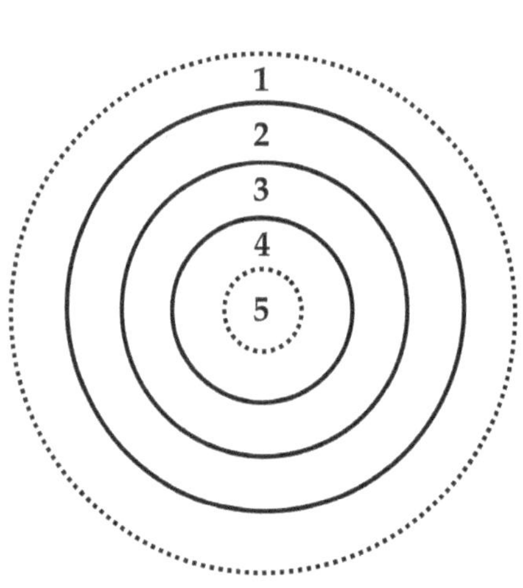

1. The Unconscious (physiological)
2. The Subconscious
3. Passive (or diffused) Consciousness
4. Active Consciousness or Attention
5. Apperception

Figure 12.1. Baldwin's "graphic representation of area of consciousness, after analogy with vision."

dimly felt elements that may have little or no direct connection with those dominant elements."[11] And American psychologist and educator James Rowland Angell made a similar contention about "a central focal point" and "fading gradually away to vaguer and vaguer consciousness."[12]

The aforementioned visual models typically consider attention as a psychological state that reflects the limits of our perceptual capacity. Here, attention is "sensory clearness."[13] Such conceptions have been frequently discussed in the history of psychology and, as historian Christopher Mole observes, became "the central explananda for a theory of attention."[14]

In nineteenth-century discussions of audition, however, a subtly yet significantly different conceptualization of attention emerges. In his monumental work, *Die Lehre von den Tonempfindungen als physiologische Grundlage für die Theorie der Musik* (1863), Hermann von Helmholtz proposed two kinds of ears: the "bodily ear" (*das körperliche Ohr*), which analyzes the compound musical tone into its constituent simple sinusoidal components, and the "mental ear" (*das geistige Ohr*), which synthesizes them back into a single representation. As a result, many constituent aspects of a given sound are overlooked or ignored in our everyday listening. We are used to hearing a musical tone as a whole. In order to perceive the details of a sound, such as its constituent overtones, an "exercise of attention" (*Uebung der Aufmerksamkeit*) is necessary.[15] As Benjamin Steege puts it, "Helmholtz sought to produce a listener who could choose and analyze its objects at will and thereby makes its perceptions its own." As a result, Helmholtz's listener is "endowed with considerable perceptual agency."[16]

Titchener was profoundly influenced by Helmholtz's theory of auditory attention. In *Experimental Psychology: A Manual of Laboratory Practices* (1901), he provided detailed guidelines for instructors on how to conduct experiments on auditory attention. Proposing that "under certain conditions, the process attended to becomes more intensive as well as more clear and distinct" as the "second law of attention," Titchener suggested an experiment using a piano (Figure 12.2):

> (a) Strike on the piano the chord here indicated, and hold the keys down while the tones ring off. When the sound is quite weak, *single out by attention* some one of the constituent tones. Is the selected tone intensified? . . . (c) Strike the chord *c-e-g* strongly upon the piano keyboard, *directing the attention to* the *c*. Is it intensified? Strike the chord again, directing the attention to the *e* or *g*. Is the tone attended-to intensified?[17]

Figure 12.2. Experiment suggested by Titchener to verify his second law of attention.

Here, a listener directs their attention to one of the constituent tones of the C-major chord, allowing it to occupy the focal point of consciousness. As in the preceding model, this explanation of attention proposes distinct levels of consciousness with varying degrees of clarity; as such, it remains grounded in the vision-based models discussed above. However, it also suggests a different connotation of attention as an action and invokes the agency of an attentive listener.

Attention and the Sense of Agency

Attention, conceptualized as an action of the psychological subject rather than a psychological state of consciousness, thus involves a sense of agency. The etymology of the word itself also reinforces this point. "To attend" in Latin is *ad tendere*, derived from "ad" (toward) and "tendere" (stretch), meaning *the act of stretching toward*. Attention has long played a significant role in the traditional philosophical discourse on volition.[18] Given its close link to the concept of an active subject, how could attention be investigated within the new framework of physiological and experimental psychology? In his 1889 lecture for the Neurological Society of London titled "The Psycho-Physical Process in Attention," English psychologist James Sully addressed this challenge directly: "In its higher form of voluntary selection and fixation of particular presentations of sense and ideas," he writes, attention is often thought of as "a mysterious emanation of spiritual activity that interferes with and controls the mechanism of mind."[19] As a result, attention calls the "possibility of a true science of psychology" into question. In contrast to this notion of an "occult Ego that asserts itself in the activity of attention," Sully advocated for attention as a psycho-physical process "of which we may, by properly scientific methods, determine the co-operant factors and conditions."[20] In his lecture, he not only discussed attention as a psychological phenomenon but also explored sensory and muscular changes involved in acts of focus, thereby presenting a biological conception of attention.

A similar emphasis on action and agency, as well as on the link between attention and motor processes, is found in William James's *The Principles of Psychology* (1890). Here, James emphasizes the significance of attention in the production of behaviors, stating that our "experience is what [we] agreed to attend to."[21] Theorizing the nature of the attentive process, he observes that the sensory organs and the bodily muscles accommodate and adjust themselves in our attention to not only sensory stimuli but also to the "idea" of a sensible object. As he puts it, "The two processes of sensorial adjustment and ideational preparation probably coexist in all our concrete attentive acts."[22] In his chapter on attention, James does not discuss listening in detail but provides a long footnote in which he cites Ernst Mach's discussion of auditory attention, highlighting the view of attention as a "bodily activity":

> We 'search' among the sounds. This hearkening search is *very observably a bodily activity*, just like attentive looking in the case of the eye. . . . If . . . we understand by attention nothing mystical, but a bodily disposition, it is most natural to seek it in the variable tension of the muscles of the ear. . . . it seems to me a very plausible view that quite generally *Attention has its seat in the mechanism of the body.*[23]

Motor Theories of Attention

A more substantial and methodical examination of attention in relation to action and motor processes emerged throughout the late nineteenth and early twentieth centuries, giving rise to the "motor theory of attention."[24] The Scottish philosopher Alexander Bain was a leading member of this school of thought. The major contribution of his two treatises on physiological psychology—*The Senses and the Intellect* (1855) and *Emotions and the Will* (1859)—is an emphasis on action and movement in all psychological matters. Bain regarded the muscular sense as primary and emphasized awareness of the body's movement and force as exerted by the muscles:

> In treating of the Sense, besides recognizing the so-called muscular sense as distinct from the five senses, I have thought proper to assign to Movement and the feeling of Movement a position preceding the Sensations of the senses; and have endeavored to prove that the exercise of active energy originating in purely internal impulses, independently of the stimulus produced by outward impressions, is a primary fact of our constitution.[25]

Bain treats music extensively in both treatises. In his discussion of touch, he links action and muscular sense: "In playing on finger instruments, the piano, guitar, organ, etc., the touch must measure the stroke or pressure that will yield a given effect on the ear."[26] The expression of feeling is hence conditioned by the nature of the action itself. According to Bain, attentive listening involves the sensations of both sounds and muscular activities. The gradual increase and decrease of loudness, for example, can evoke intense emotion as it involves muscular effects as well as the stimulation of the auditory nerves:

> I think it not unlikely, that *a muscular effect enters into this sensation*: the gradually increased or relaxed tension of the muscles of the ear being a probable accompaniment of the increase or diminution of loudness. On this supposition, the influence on the mind would be one of the very delicate effects of movement.[27]

Noting that attention involves many deliberate bodily actions, Bain deemed it a kind of motor phenomenon.

Contemporaneous developments in neurology provided support for Bain's notion of attention as action. In his work *The Functions of the Brain* (1876), Scottish neurologist David Ferrier claimed that "when engaged in the attentive consideration of some idea or ideas, we are in reality throwing into action, but in an inhibited or suppressed manner."[28] Several other psychologists continued to adopt the notion of action as central to attention. American psychologist Walter B. Pillsbury, for example, authored a 346-page monograph on attention in which he proclaimed, "One of the [strongest] facts in connection with the attention is its close connection with bodily movement and posture. . . . There is no act of attention that is unaccompanied by some motor process."[29] For Pillsbury, the "motor concomitants of attention" included the movements of the sense organs involved, as well as the effects on other voluntary muscles, on posture, and on the involuntary mechanisms of pulse, respiration, and vaso-motor activities. Arguing along similar lines, German-American psychologist Hugo Münsterberg classified "attention" as an "activity" in his *Psychology: General and Applied* (1914), claiming that attention consists of both "mental and physical activities."[30] He emphasized the fact that "the body adjusts itself to the center of attention, and this adjustment sends kinesthetic sensations to consciousness." In his view, the body and mind are an inextricable unit: Bodily processes occur in tandem with brain processes, and they are not the "result of attention, but are themselves the attention."[31]

Physiologically grounded motor theories of attention thus described how the attentional process was connected not only to the activation but also to the inhibition of physical movements. French psychologist Théodule Ribot's conceptualization of attention is particularly noteworthy in this regard. In his monograph *La psychologie de l'attention* ("The Psychology of Attention"), Ribot famously asserted, "Every intellectual state is accompanied by definite physical manifestation."[32] His claim emphasized the close connection between our mental experiences and the biological processes that underlie them. When it comes to attention, the significance of the bodily movements and modifications becomes more crucial. Accordingly, "The fundamental role of the movements in attention is to *maintain* the appropriate state of consciousness and to *reinforce* it."[33] They are not "simply effects, outward marks"; they are "*the necessary conditions, the constituent elements, the indispensable factors of attention*." To Ribot, physical movements are not mere side effects but rather the essential component of attention.

Claims such as this placed attention within the framework of experimental psychology, linking attentiveness to muscle contraction/relaxation and thus rendering it observable and measurable. This idea was soon applied to music. In an experimental study of musical enjoyment, psychologist Harry P. Weld drew on plethysmography, the technique of measuring *changes in the volume of body parts* and observed that "the act of listening to music was almost invariably attended by a decrease in the volume of the fore-arm." He concluded that "volumetric changes are essentially phenomena of attention;

it proves to be possible, in every instance, to correlate these characteristic changes with changes in the nature or condition of attention."[34] Weld's research highlighted the role of bodily movements and muscular senses in music listening. The *New York Times* showcased his research under the title "Develop Your Muscle So You Can Enjoy Music," highlighting its potential challenge to the prevailing view that "a strictly auditory enjoyment of music is the only way in which melody pleases us."[35]

Muscular Attention and Music-Making

These discussions of attention as a motor process proved highly pertinent to practical discussions of music-making; indeed, a noticeable shift in piano pedagogy at the turn of the twentieth century coincided with emerging motor theories of attention. New treatises on piano playing brought attention–action correlates to the realm of music psychology by considering attention—both in auditory perception and in music-making—as manifested through bodily action. While a directly causal relationship between motor theories of attention and piano pedagogy cannot be established here, it is nevertheless instructive to explore resonances between the two fields.

The end of the long nineteenth century saw the rise of a new pedagogical approach that often incorporated contemporary physiological findings. The "physiological turn" in the study of piano playing is particularly evident in the work of the French performer-pedagogue Marie Jaëll. Central to Jaëll's teaching was the principle of touch (*toucher*).[36] She was especially well known for her fingerprint experiments, undertaken in collaboration with the physician Charles Féré, which involved applying ink to pianists' fingertips and then recording their prints on the piano keys.[37] Her keyboard print-maps served as evidence of proper or improper pianistic technique. More significant here is the particular emphasis she placed on attention in her piano pedagogy, which, she maintained, was "not particularly aimed at those who are gifted or those who are not; it is aimed at those capable of *an effort of will*."[38] Conscious attention to physical processes, Jaëll argued, was necessary: The pianist must continuously attend to the relationship between haptic and aural sensations, and such attention is closely linked to muscular tension. Chapter 2 of her *La musique et la psychophysiologie* ("Music and psychophysiology," 1896) is entirely devoted to the topic of "attention and muscular sense." According to her, the "mechanism of attention resides in the muscles, which, like rubber bands, warm up as they contract."[39]

Jaëll opposed the mechanistic "finger school" of piano pedagogy, which advocated for the independence of individual fingers and stillness in other parts of the upper body.[40] However, she recognized the importance of muscular tension or "fixity" (*la fixité*) in piano playing. In her piano pedagogy, the concept was equated with attention itself: "Without perfection of the muscular tension, there is neither a perfection of attention nor a perfection of timbre." Such fixity allows students to "think about the movements, the sounds they hear, and the tactile sensations they are feeling."[41]

Whereas the concept of fixity was important for attention in Jaëll's teaching, other piano pedagogues approached the issue in different ways. For example, attention featured frequently in writings by the "arm weight" school. The treatises in this tradition were mostly theoretical and explanatory, and often drew on contemporary physiology and psychology, as opposed to the finger school, which focused on providing a wealth of practical exercises.[42] German piano pedagogue Rudolf Maria Breithaupt, for example, considered attention as a characteristic of finger school piano pedagogy, with its focus on ideals of "immobility" and "fixation." "To play 'fixed,'" he wrote, "is perhaps the same as to play 'with concentrated attention.'" Here, attention was identified as the cause of the rigidity, not desirable in piano playing. By contrast to this artificial fixation, Breithaupt foregrounded muscular relaxation by emphasizing the importance of relaxed forearms, upper arms, and even shoulders.[43] He advocated for the development of the body's "natural playing-faculties," arguing that "the arm must be weighted and yet loose in all its movements."[44] While sharing a similar emphasis on arm weight, British piano pedagogue Tobias Matthay provided a different, fuller account of attention. Instead of likening attention and inhibition to action alone, he identified different kinds of attention, thereby proposing a kind of attentional psychology. In *The Act of Touch in All Its Diversity: An Analysis and Synthesis of Pianoforte Tone-Production* (1903), which underwent multiple editions by the mid-twentieth century, Matthay conceptualized attention multimodally, deeming it *necessary* for natural pianism. As he wrote, "The act of *Attention* needed in playing, although it seems but a single flash of consciousness, is really a compound act, and that it must embrace two distinct purposes."[45] He diagrammed "the intelligence needed in piano-playing" as follows (Figure 12.3):

The Act of INTELLIGENCE *In Playing,*
implies:

Attention Inwards—
to enable us to obey Music:
This implies use of our Musical-Imagination
so that by perceiving
Music-Emotion and Musical-Shape,
we shall be able to judge the
TIME-SPOT and TONE-KIND
due for each note.

Attention Outwards—
to enable us to obey the Key:
This implies Attention
Muscularly and Aurally,
to enable us to
Feel key-resistance before and during descent,
and to
Hear the completion of key-descent.

Or finally, to put the case as a positive direction:
I: *We must try to SEE the Emotion, and the Musical-material;* —
— the latter being the material-shapes of Time (rhythm) and Tone, through which the former is exhibited;
and II: *We must try to OBTAIN this mentally-pictured Music from the Keys;*
— by employing our knowledge of the requirements of Key-treatment, and our muscular-adaptations to these.
The latter moreover involves, that we must give:
a) *Muscular-Attention* to the key-Resistances, and
b) *Aural-Attention* to the beginning of each sound-cote consummation of key-descent.

Figure 12.3. Matthay's diagram of "the intelligence needed in piano-playing."

Matthay identified two different levels of attention: (1) the "inward" or "aural" attention to music as a sonic phenomenon, and (2) the "outward" or "muscular" attention to "key-resistance." These claims require some unpacking. The notion of "inward" or "aural" attention extends beyond hearing to include a vivid visualization of sonic imagery.[46] Hence, attention is directed toward the intended object of our musical action. The idea of "outward" and "muscular" attention, on the other hand, includes attention to our bodily parts in action as well as the interaction between our body and the instrument, which, in Matthay's words, refers to "the resistance the key offers *before* it will move, and *during* its descent." Muscular-sense is the "medium through which we thus become aware of key-resistance."[47] In this view, instead of merely acting upon the sounding mechanism of the instrument, the performer's body is also perceived by the performer in connection with their own position. Consequently, muscular attention is not equated with mere inhibition of movement during an action but embraces a more nuanced physicality. Although Matthay's treatise did not include the terms "kinesthesia" or "proprioception," he was aware of their sense and importance.[48] His treatises describe both awareness of the position and movement of body parts (*proprioception*) and a sense of the strength of effort required to move the body (*kinesthesia*). Figure 12.4 shows the procedure of Matthay's "'Aiming'

Fig. 3.

The "Aiming" Exercise.

a)
"Before the drop:"

b)
"Down with the keys:"

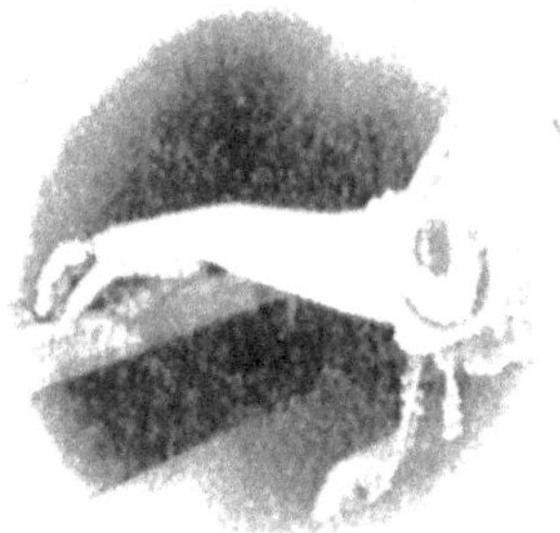

c)
"Let go:"

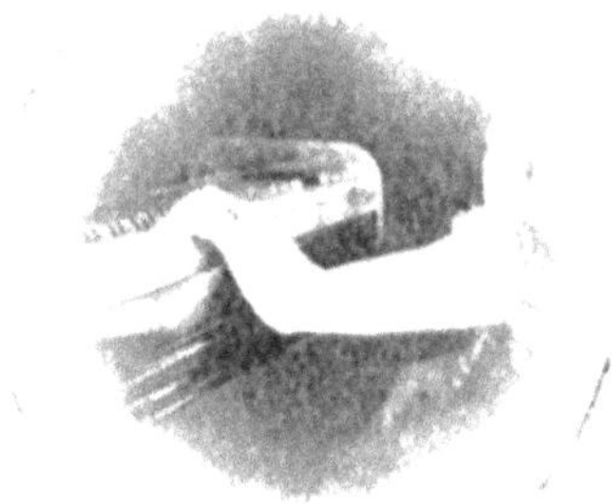

Figure 12.4. Matthay's illustration of the "Aiming" exercise.

Exercises," by which the student could bring their mind "properly on *Key-attention*," with both their "muscular-sense and aural-sense now keenly applied to each other."[49]

Matthey's treatise can be evaluated from the viewpoint of psychological theory as well as in the context of practical piano pedagogy. As his approach demonstrates, early twentieth-century piano pedagogues identified different kinds of attention—including attention to the object and to the attending agent—and contributed to expanding attentional discourse across different modalities.

Unconscious Attention and Virtuosic Pianism

In addition to describing pianistic attention as a psychological and physiological phenomenon, late nineteenth-century treatises also proposed integrated attention models related to action. This concept emerged particularly clearly in discussions surrounding virtuosity.[50] Historically, the term virtuosity denoted not just performers but also composers and music theorists. Eighteenth-century French music theorist Sébastien de Brossard, for example, argued that "virtuoso" should be applied to excellent musicians who "apply themselves to the theory or the composition of music."[51] It was only in the nineteenth century that the term became associated primarily with extremely skilled performers. One of its key features was physical velocity—the speed at which musicians move their fingers. Intrigued by pianists' sheer motor abilities, several late nineteenth-century physiologists empirically investigated the speed of pianists' muscle movements, often in collaboration with players themselves. These studies call into question the role of attention during rapid, complicated passages.

To take one example: In 1902, James Paget, a leading medical figure of the time, collaborated with pianist Natalia Janotha. He was particularly interested in the rapidity with which she performed Mendelssohn's *Perpetuum Mobile*, Op. 119, clocking her at 4 minutes and 3 seconds.[52] Janotha did not leave a recording of this particular piece, thus no verification of her speed is possible, yet it would seem to fall well within the range of plausibility (for context, some twentieth-century recordings have a duration of less than three minutes).[53] Certainly, her rendition was fast enough to impress Paget. According to his calculation, Mendelssohn's piece contained 5,595 notes, which meant that Janotha had executed at least 72 bimanual finger movements per second, even without considering the movements of the hand, forearm, arm, and foot.[54] Later, physiologist Homer William Smith continued this line of investigation, working with American pianist David Saperton. On studying Saperton's playing, Smith concluded that, even focusing on finger actions alone (i.e., leaving aside the movements of other body parts involved in performance), "a speed of 20 to 30 notes per second may involve 400 to 600 separate motor actions."[55] Overall, Smith concluded that effortful attention had little influence on virtuosic motor execution, maintaining that "in the most complicated patterns of voluntary [motor] activity, consciousness may play a very limited role."[56]

This point can be made concrete by examining Saperton's performance of one of Leopold Godowsky's piano pieces. Figure 12.5 shows the beginning of Godowky's Study No. 47, "Badinage," from *53 Studies on Chopin's Études*, which combines Chopin's études Op. 10, No. 5 ("Black Key") and Op. 25, No. 9 ("Butterfly"), both in G flat major. Each étude requires complex finger actions on its own, and performing the combination of the two études requires even more complex motor coordination.

Godowsky was, of course, a significant pianist and pedagogue, as well as Saperton's father-in-law. Like Breithaupt, whom we met earlier, Godowsky was a proponent of the relaxed "arm weight" school. He was also known to be a "wizard of technique," and yet appeared to exert very little effort while playing.[57] His student Heinrich Neuhaus observed that while executing his "super-acrobatic task," Godowsky's face muscles demonstrated "enormous concentration" while his body appeared "terribly natural" and "completely effortless," with no concentrative traces at all.[58] Perhaps, then, extreme virtuosity stood in opposition to directed muscular attention?

This conclusion seemed attractive, yet it was irreconcilable with theories of attention *as* effort, particularly those of American psychologist Thaddeus L. Bolton. In his extensive study on rhythm (1894), Bolton described "efforts of attention" and more specifically, mental "efforts to attend to sound."[59] Similarly, in "Attention: A Historical Summary of the Discussion Concerning the Subject," Bolton's compatriot Lemon Leander Uhl wrote, "Physiologically, attention is a motor effort or a sensory impulse combined with motor innervation."[60] These notions of attention in action—that is, attention as equated with muscular effort—seemed at odds with observations of virtuosic pianism and with the "arm weight" school of piano pedagogy. How could the two be reconciled?

Among those who recognized and attempted to solve this conundrum was German physiologist Friedrich Adolf Steinhausen. In his treatise *Die*

Figure 12.5. The beginning of Godowsky's Étude No. 47 from *53 Studies on Chopin's Études*, "Badinage" (New York: Schirmer, 1899), combining Chopin's Étude Op. 10, No. 5 ("Black Key") and Étude Op. 25, No. 9 ("Butterfly").

physiologischen Fehler und die Umgestaltung der Klaviertechnik (1913), Steinhausen argued for unconscious attending in action, writing that "only the goal of the movement, the intention or purpose, is continually conscious to us."[61] He claimed that "the muscular position and motion senses" that play a great role in piano playing are "totally unconscious (*alles unbewußt*)."[62] The natural or passive motions (promoted by the arm weight school) are generated by weight and elasticity, not active focus. In fact, in order to learn to play the piano, an unlearning (*das Verlernen*) of disturbing and unnecessarily controlling forms of attention was required, this being "in the real sense[,] a mental process."[63]

Attention, in Steinhausen's formulation, thus plays a significant role in virtuosic performance, but differs from the conscious attention accompanied by motor innervation, constituting instead a kind of automaticity. It is also distinguished from the more widely known Helmholtzian notion of inattention, which dictates, according to James's characterization, that "we leave all impressions unnoticed which are valueless to us as signs," as in the perception of upper partial tones.[64] In Steinhausen's framework, attention is an unconscious proprioceptive sense, a form of muscular *intention* manifested on and through the body.

Conclusions

This chapter has examined various aspects of attention in the context of late nineteenth- and early twentieth-century psychology and music pedagogy. It reveals that theories of auditory, rather than visual, attention contributed to the emergence of an account of attention as action. The physiological consequences of this approach toward attention were explored and expanded in piano pedagogy, particularly by "arm weight" pianists active in the early 1900s. Histories of music pedagogy often credit advocates of this pedagogical school with introducing a "naturalistic" style of playing that engaged the entire body rather than focusing solely on the fingers. I have highlighted a different contribution made by these treatises by reading them through the lens of psychological theory. As demonstrated, the piano pedagogues in the arm weight tradition distinguished between various kinds of attention and thus recognized musical activity as a fertile arena for observing the relationship between attention and muscular action. They also considered the role of effort in attention and the possibility of attention not only *to* but also *in* action, as a muscular (rather than purely mental) phenomenon.

Historians of psychology often point out that attention has long been studied as a facet of perception, focusing on internal mental processes.[65] Especially post-1950, when definitions of attention as "selection-for-perception" were dominant, auditory focus was primarily studied in response to external sounding stimuli and "*in the absence of actions.*"[66] Not until the late 1980s did certain scholars advance an action-related approach,[67] arguing that "viewing attention and action in isolation would not offer a realistic representation of either of the two types of processes."[68] Their approach was not entirely novel, however. As demonstrated in this chapter, an attempt to integrate attention and action was already present in various forms in the late nineteenth and early twentieth centuries, especially in the discussions of musical activities such as piano playing. Because keyboard performance involves a complex interplay between the intended object of attention (the music) and the performing/attending subject (the player), it necessitates a theory of attention that goes beyond "the auditory self" or the "attentive ear," casting the pianist as the agent of embodied attending. The piano treatises examined in this chapter, when viewed in relation to physiological and psychological theories of their day, thus represent an important part of the history of attention.

Notes

1. See, e.g., Daniel Chua, *Absolute Music and the Construction of Meaning* (Cambridge: Cambridge University Press, 1999); Lydia Goehr, *The Imaginary Museum of Musical Works: An Essay in the Philosophy of Music* (Oxford: Clarendon Press, 1992); Mark Evan Bonds, *Absolute Music: The History of an Idea* (Oxford: Oxford University Press, 2014); Jonathan Sterne, *The Audible Past: Cultural Origins of Sound Reproduction* (Durham: Duke University Press, 2003).

2. On various nineteenth-century "second psychology" programs, see Emily D. Cahan and Sheldon H. White, "Proposals for a Second Psychology," *The American Psychologist* 47, no. 2 (1992): 224–235.

3. As philosopher Christopher Mole puts it, when psychology and philosophy parted, "it was psychology that got custody of attention." Christopher Mole, "The Metaphysics of Attention," in *Attention: Philosophical and Psychological Essays*, ed. Christopher Mole, Declan Smithies, and Wayne Wu (New York: Oxford University Press), 60–77, xi.

4. Edward Bradford Titchener, *Lectures on the Elementary Psychology of Feeling and Attention* (New York: The Macmillan Company, 1908), 173.

5. William James, *The Principles of Psychology*, vol. 1, rev. ed. (New York: Dover Publications, 1950; originally published 1890), 404.

6. Kurt Danziger, "Psychology and Its History," *Theory & Psychology* 23, no. 6 (2013): 829–39, 836.

7. Wilhelm Max Wundt, *Grundzüge de physiologischen Psychologie* (Leipzig: W. Engelman, 1874), 85–97.

8. Titchener, *Lectures on the Elementary Psychology of Feeling and Attention*, 220.

9. Oswald Külpe, "The Problem of Attention," *The Monist* 13, no. 1 (1902), 38–68, 41.

10. James Mark Baldwin, *Handbook of Psychology: Senses and Intellect* (London: Macmillan, 1890), 68.

11. C. Lloyd Morgan, *An Introduction to Comparative Psychology* (London: Walter Scott Publishing Co., 1894), 14.

12. James Rowland Angell, *Psychology: An Introductory Study of the Structure and Function of Human Consciousness* (New York: Holt, 1908), 65.

13. For such a discussion of attention in relation to sensation, see Edward Bradford Titchener, "Attention as Sensory Clearness," *Journal of Philosophy, Psychology and Scientific Method* 7, no. 7 (1910): 180–82 and Ernest Glen Wever, "Attention and Clearness in the Perception of Figure and Ground," *The American Journal of Psychology* 40, no. 1 (1928): 51–74.

14. Christopher Mole, "Attention," in *The Stanford Encyclopedia of Philosophy*, edited by Edward N. Zalta (Stanford, CA: Metaphysics Research Lab, Stanford University, 2017), 61.

15. Hermann von Helmholtz, "Über Combinationstöne," *Annalen der Physik und Chemie* 99 (1856): 497–540, 526.

16. Benjamin Steege, *Helmholtz and the Modern Listener* (Cambridge: Cambridge University Press, 2012), 81.

17. Edward Bradford Titchener, *Experimental Psychology: A Manual of Laboratory Practices*, vol. I (New York: The Macmillan Company, 1906), 111.

18. See, for example, Lorraine J. Daston, "British Responses to Psycho-Physiology, 1860–1900," *Isis* 69, no. 2 (1978): 192–208.

19. James Sully, "The Psycho-Physical Process in Attention," *Brain* 13, no. 2 (1890): 145–64, 146, emphasis added.

20. Sully, "The Psycho-Physical Process in Attention," 146–147. Here, we observe the status of attention study during the formative years of psychology when the focus shifted from the soul to the mind. See Edward S. Reed, *From Soul to Mind: The Emergence of Psychology, from Erasmus Darwin to William James* (New Haven, CT: Yale University Press, 1997).

21. William James, *The Principles of Psychology*, vol. 1, rev. ed. (Dover Publications, 1950; originally published 1890), 402.

22. James, *The Principles of Psychology*, 434.

23. Ernst Mach, "Zur Theorie des Gehörorgans," *Sitzungsberichte der kaiserlichen Akademie der Wissenschaften in Wien. Mathematisch-naturwissenschaftliche Classe* 48, no. 2 (1863): 283–300. Translation cited in James, *The Principles of Psychology*, 436; emphasis added.

24. Myra O. Smith, "History of the Motor Theories of Attention," *The Journal of General Psychology* 80, no. 2 (1969): 243–57.

25. Alexander Bain, *The Senses and the Intellect* (London: J. W. Parker, 1855), v–vi.

26. Ibid., 194.

27. Ibid., 207, emphasis added.

28. David Ferrier, *The Functions of the Brain* (New York: G. P. Putnam's Sons, 1876), 285.

29. Walter Bowers Pillsbury, *Attention* (London & New York: Swan Sonnenschein; The Macmillan Co., 1908), 12.

30. Hugo Münsterberg, *Psychology: General and Applied* (New York: Appleton, 1914), 190, emphasis in the original.

31. Ibid., 190. Münsterberg was one of the pioneering psychologists who postulated a "second psychology," previously discussed at the beginning of this chapter. He distinguished two psychologies, "causal" and "purposive." Each has different approaches, one focusing on explaining "mental objects" and the other on understanding "acts." Both are needed for "complete psychology." Although he still maintained attention as a topic of causal psychology, his emphasis on acts, activities, and agency is apparent in his discussion of attention. His approach to attention illustrates his endeavors to "modernize the soul" by "reinterpreting it for a world in which a natural scientific worldview, one without a place for the soul, dominated psychology." Robert Kugelmann, *The Soul in Soulless Psychology* (Cambridge: Cambridge University Press, 2023), 57.

32. Théodule Ribot, *The Psychology of Attention* (Chicago: Open Court Publishing Company, 1903), 12; originally published as *La psychologie de l'attention* (Paris: F. Alcan, 1889).

33. Ibid., 19, emphasis in the original. Ribot's claim has predecessors. Psychiatrist Henry Maudsley wrote, "Attention is the arrest of the transformation of energy for a moment—*the maintenance of a particular tension.*" Maudsley, *The Physiology of Mind*, 3rd ed. (New York: D. Appleton, 1883), 308; emphasis added. Philosopher George Stout also stated that the attention process "often involves a noticeable tension of the muscles, similar to that which arises when we make an effort to move and at the same time by a counter effort arrest the intended movement in its first inception." Stout, "Apperception and the Movement of Attention," *Mind* 16, no. 61 (1891): 23–53, 25.

34. Harry Porter Weld, "An Experimental Study of Musical Enjoyment," *The American Journal of Psychology* 23, no. 2 (1912): 245–308, 251.

35. Harry Porter Weld, "Develop Your Muscle So You Can Enjoy Music: Dr. H.P. Weld Makes Some Curious Experiments That Show the Connection Between the Strenuous Life and Aesthetic Gratification," *New York Times*, May 5, 1912, sec. Magazine.

36. Marie Jaëll, *Le Toucher, Nouveaux Principes Élémentaires Pour l'Enseignement Du Piano* (Paris: Heugel & Cie Editeurs, 1894); *La Musique et La Psychophysiologie* (Paris: Alcan, 1896); *Le Toucher: Enseignement Du Piano Basé Sur La Physiologie* (Paris: Costallat, 1899).

37. See Julia Kursell, "Visualizing Piano Playing, 1890–1930," *Grey Room*, no. 43, 2011: 66–87 and James Q. Davies, *Romantic Anatomies of Performance* (Berkeley: University of California Press, 2014).

38. "L'étude du Piano ne s'adress spécialment ni ceux qui son doués ni à ceux qui le son pas; elle s'adress à ceux capables d'un effort de volonté." Jaëll, *Le Toucher: Nouveaux Principes*, i.

39. "Le mécanisme de l'attention réside dans les muscles, qui, de même que des fils de caoutchouc, s'échauffent en se contractant." Jaëll, *La Musique et La Psychophysiologie*, 17.

40. The "finger school" of piano pedagogy, as the name implies, focused on finger actions only. It advocated for independence of individual fingers, and for stillness in other parts of the upper body. Piano pedagogues, including Muzio Clementi, Johann Nepomuk Hummel, and Carl Czerny promoted the importance of "training all the fingers of both hands to an equal degree of power and independence of action" (Johann Nepomuk Hummel, *A Complete Theoretical and Practical Course of Instructions on the Art of Playing the Piano Forte* (London: T. Boosey, 1829), 34; originally published Vienna: Tobias Haslinger, 1827.

41. "sans perfectionnement de la tension musculaire n'y a ni perfectionnement de l'attention, ni perfectionement du timbre de la sonorité." Jaëll, *Le Toucher: Enseignement Du Piano,* 10; translation adopted from Laure Struber, "The Education of Musical Thinking Through the Hand

According to Marie Jaëll (1846–1925)" (DMA diss., University of Washington, 2017), 171; "cette fixité d'attitude correspond à l'intensité avec laquelle l'éleve arrive à penser: 1° aux mouvements réalisés; 2° aux sons-entendus; 3° aux sensations tactiles éprouvées." Jaëll, *Le Toucher: Enseignement Du Piano*, 27; translation from Struber, "The Education of Musical Thinking," 173.

42. Rudolf Maria Breithaupt, *Natural Piano-Technic*, trans. John Bernhoff, vol. 2 (Leipzig: C. F. Kahnt Nachfolger, 1909), 99.

43. For a discussion of this paradigm shift as an expansion of the perception-action interface in piano playing, see Youn Kim, "Music Psychology of the Piano-Playing Hands in Historical Discourse," *Journal of Musicology* 28, no. 2 (2021): 32–66.

44. Breithaupt, *Natural Piano-Technic*, 99.

45. Tobias Matthay, *The Act of Touch in All Its Diversity: An Analysis and Synthesis of Pianoforte Tone-Production* (London: Longmans, Green and Co., 1903), 32–33. Diagram in Figure 12.3 appears on p. 34.

46. In the field of motor learning, the external focus of attention is *"where the performer's attention is directed to the effect of the action"* while an internal focus of attention means *"where attention is directed to the action itself."* Gabriele Wulf, Barbara Lauterbach, and Tonya Toole, "The Learning Advantages of an External Focus of Attention in Golf," *Research Quarterly for Exercise and Sport* 70, no. 2 (1999): 120–26, 120.

47. Matthay, *The Act of Touch in All Its Diversity*, 120, note 1; emphasis in original. Matthay's conceptualization of "inward and outward attention" is remarkably well in line with recent theories of attention and action in the cognitive sciences, for instance, Wayne Wu's two ways of understanding the relationship between attention and action. On one level, we can attend to "target of action, the object we intend to manipulate with our bodies," which corresponds to Matthay's inward attention. On another level, we can attend to our body deployed in action, which may not be "salient to us when we act," consistent with Matthay's outward attention: This kind of attention "must both select the target to manipulate and the part of the body manipulate with . . . attention to action, so understood, is a form of attention in action." See Wayne Wu, "Attention as Selection for Action," in *Attention: Philosophical and Psychological Essays*, Christopher Mole, Declan Smithies, and Wayne Wu, eds. (Oxford: Oxford University Press, 2011), 104.

48. The two terms first appeared in 1887 and 1906, respectively. H. Charlton Bastian, "The 'Muscular Sense'; Its Nature and Cortical Localisation," *Brain* 10, no. 1 (1887): 1–89; Charles S. Sherrington, *The Integrative Action of the Nervous System* (New Haven, CT: Yale University Press, 1906), 129–130.

49. Tobias Matthay, *Relaxation Studies in the Muscular Discriminations Required for Touch, Agility and Expression in Pianoforte Playing* (Leipzig: Bosworth & Co., 1908), 23; original italics. Illustration in Figure 12.4 is taken from p. 12.

50. Several intricate aspects related to virtuosity are addressed in Thomas Carson Mark, "On Works of Virtuosity," *The Journal of Philosophy* 77, no. 1 (1980): 28–45; Susan Bernstein, *Virtuosity of the Nineteenth Century: Performing Music and Language in Heine, Liszt, and Baudelaire* (Stanford, CA: Stanford University Press, 1998); Antoine Hennion, "'As Fast as One Possibly Can . . .': Virtuosity, a Truth of Musical Performance," in *Critical Musicological Reflections. Essays in Honour of Derek B. Scott*, ed. Stan Hawkins (Farnham, Surrey; Burlington, VT: Ashgate, 2012), 125–38; Jane Ginsborg, "'The Brilliance of Perfection' or 'Pointless Finish'? What Virtuosity Means to Musicians," *Musicae Scientiae* 22, no. 4 (2018): 454–73.

51. "Plûtôt à ceux qui s'appliquent à la Théorie, ou à la Composition de la Musique." Sébastien de Brossard, *Dictionaire de musique* (Paris: Christophe Ballard, 1703), 248–249.

52. James Paget, *Memoirs and Letters of Sir James Paget* (London: Longmans, Green, 1902), 205.

53. Benjamin Frith's 1995 Naxos recording performance, for example, lasts 2 minutes and 47 seconds.

54. Physiologists of our own time have contended that the motor actions involved in executing these complex piano passages are considerably more complicated than Paget's calculation. See, for example, Shinichi Furuya et al., "Secrets of Virtuoso: Neuromuscular Attributes of Motor Virtuosity in Expert Musicians," *Scientific Reports* 5, no. 1 (2015): 1–8.

55. Homer William Smith, *From Fish to Philosopher* (Boston: Little, Brown and Co., 1953), 209.

56. Ibid., 211.

57. Heinrich Neuhaus, *The Art of Piano Playing*, ed. K. A. Leibovitch (London: Kahn & Averill, 1993), 12.

58. Neuhaus, *The Art of Piano Playing*, 105.

59. Thaddeus L. Bolton, "Rhythm," *The American Journal of Psychology* 6, no. 2 (1894): 145–238, 211.

60. Lemon Leander Uhl, *Attention, a Historical Summary of the Discussion Concerning the Subject* (Baltimore: Johns Hopkins Press, 1890), 124.

61. Friedrich Adolf Steinhausen, *Die physiologischen Fehler und die Umgestaltung der Klaviertechnik* (Breitkopf & Härtel, 1913). Translations are from Donald Rupert, "The Physiological Errors and the Reshaping of Piano Technique" (PhD diss., University of Rochester, 1963), 49.

62. "Dem Muskel-, Lage- und Bewegungssinn (33) fällt eine weit wichtigere Rolle zu, was jedoch, weil alles unbewußt, genauer kaum abzuschatzen ist." Steinhausen, *Die physiologischen Fehler*, 68–69, Rupert, "The Physiological Errors," 52.

63. "Das Verlernen des störenden und dann ja unnötig kontrollierenden Aufpassens ist im eigentlichen Sinne ein seelischer Vorgang." Steinhausen, *Die physiologischen Fehler*, 77, trans. Rupert, "The Physiological Errors," 61.

64. James, *The Principles of Psychology*, 456.

65. As historian Gary Hatfield notes in a historical overview of attention in early scientific psychology, attention was presented as a kind of internal sense in most standard psychological textbooks, along with memory and imagination. Bodily motion was discussed separately. Gary Hatfield, "Attention in Early Scientific Psychology," in *Visual Attention*, vol. 8 (New York: Oxford University Press, 1998), 3–25.

66. Jay Pratt, Eric Taylor, and Davood Gozli, "Action and Attention," in *Handbook of Attention*, edited by E. F. Risko, J. Fawcett, and A. Kingstone, 325–48 (Cambridge, MA: MIT Press, 2015), 326; emphasis added.

67. Donald A. Norman and Tim Shallice, "Attention to Action," in *Consciousness and Self-Regulation: Advances in Research and Theory*, vol. 4, edited by Richard J. Davidson, Gary E. Schwartz, and David Shapiro (Boston, MA: Springer, 1986), 1–18. Alan Allport, "Selection for Action: Some Behavioral and Neurophysiological Considerations of Attention and Action," in *Perspectives on Perception and Action*, edited by H. Heuer and H. F. Sanders (Hillsdale, NJ: Lawerence Erlbaum, 1987), 395–419.

68. Pratt, Taylor, and Gozli, "Action and Attention," 326.

CONTRIBUTORS

Francesca Brittan is Associate Professor of Music and Affiliated Faculty in the Department of Cognitive Science at Case Western Reserve University. She is the author of *Music and Fantasy in the Age of Berlioz* (Cambridge, 2017) and co-editor of *Berlioz and His World* (Chicago, 2024). She has published widely on topics at the intersection of sound, science, and medicine, most recently, "The Neural Orchestra: Cognitive Instrumentalities" in the *Journal of the American Musicological Society* (2024). Her work has been recognized by a Junior Research Fellowship (University of Cambridge), a Senior Research Fellowship (University of Amsterdam), the W.P. Jones Presidential Faculty Development Award (CWRU), the Dimitrije Pivnicki Award in Neuro and Psychiatric History (McGill University), and the Alfred Einstein Award (American Musicological Society). She co-directs the *Brain Media Lab* at CWRU.

D. Graham Burnett is the Henry Charles Lea Professor of History and History of Science at Princeton University, where he is affiliated with the Interdisciplinary Doctoral Program in the Humanities (IHUM). He is the author of a number of books on technology, nature, and society from the seventeenth through the twentieth centuries. Recent collaborative works include *Attensity! A Manifesto of the Attention Liberation Movement* (Crown, 2026); and *Scenes of Attention: Essays on Mind, Time, and the Senses* (Columbia, 2023), edited with Justin E. H. Smith. Burnett is associated with the ESTAR(SER) collective and was a 2023 visiting artist at the Academy of Fine Arts in Helsinki.

Tony Day holds a PhD in Southeast Asian History from Cornell University (1981); has taught at the University of Sydney, Wesleyan University, and Yale-NUS College; and now lives in Graz, Austria, where he is working on a collection of translations and interpretations of the early nineteenth-century Javanese Sufi narrative poem *Sĕrat Cĕnthini* with Nancy Florida. His publications include *Fluid Iron: State Formation in Southeast Asia* (2002); "Stepping on a Wulu: Minor Characters and Narrative Possibilities in the *Sĕrat Cĕnthini*," in Ronit Ricci, ed., *Storied Island: Explorations in Javanese Literature* (2023); and with Nancy Florida, "A Sufi Traveler on the Road in Nineteenth-Century Java," *Philological Encounters* 9, 1–2 (2024).

Céline Frigau Manning is a Professor of Italian and Performance Studies at Université Lyon 3, France, and an honorary member of the Institut universitaire de France. A graduate of the École Normale Supérieure, she was a researcher at the Bibliothèque Nationale de France from 2006 to 2010 and resident scholar at the Villa Medici in 2011. Her articles have appeared in English, French, and Italian, in *19th-Century Music*, *Opera Quarterly*, and *L'Avant-Scène Opéra*. Her latest monograph, *Ecstatic Sounds: Music, Hypnosis, and Medicine in the Nineteenth Century*, is forthcoming with The University of Chicago Press in the series New Material Histories of Music.

Ewan Jones is Associate Professor in English at the University of Cambridge. He is the author of *Coleridge and the Philosophy of Poetic Form* (Cambridge, 2014), and *The Turn of Rhythm: How Victorian Poetry Shaped a Concept* (Virginia, 2023), along with numerous articles on poetics, intellectual history, and digital humanities. *Close Reading as Attentional Practice*, a series of essays co-edited with Marion Thain, was published by Edinburgh University Press in 2025.

Brian Kane is Professor of Music and Affiliated Faculty in Film and Media Studies at Yale University. He is the author of *Sound Unseen: Acousmatic Sound in Theory and Practice* (Oxford, 2014) and *Hearing Double: Jazz, Ontology, Auditory Culture* (Oxford, 2024). His interdisciplinary work focuses on issues in sound studies, voice studies, jazz, and the philosophy of music.

Alexandra Kieffer is Associate Professor of Musicology at Rice University. Her work on nineteenth- and twentieth-century music and intellectual culture in France has appeared in *19th-Century Music*, the *Journal of Musicology*, the *Journal of the American Musicological Society*, *Music Theory Spectrum*, and *Ethnomusicology*; and her book *Debussy's Critics: Sound, Affect, and the Experience of Modernism* appeared from Oxford University Press in 2019. Her current book project explores ideas about music, religion, and ritual in the social sciences in the twentieth century.

Youn Kim obtained her PhD from Columbia University and is currently Associate Professor at the University of Hong Kong. Her research interests include history of music theory, music psychology, and the interrelationship between music theory and the science of the mind. She has published in the *Journal of Musicology*, *Psychology of Music*, *Journal of Musicological Research*, and *Musicae Scientiae*, and co-edited *The Oxford Handbook of Music and the Body* (Oxford, 2019). Kim's monograph *Body and Force in Music* (Routledge, 2022) was published as a part of SEMPRE Studies in the Psychology of Music, and she co-edited *Music and the Cognitive Humanities* (Oxford, in press).

Nicholas Mathew is Professor of Music and Richard and Rhoda Goldman Distinguished Professor in the Humanities at the University of California, Berkeley. He is the author of *Political Beethoven* (Cambridge, 2012) and *The Haydn Economy* (Chicago, 2022) and, as a pianist, has recorded the music of the Haitian diaspora in collaboration with Jean-Bernard Cerin and the Lisette Project.

Jacob Olley is Career Development Fellow in Music at Durham University, having previously held research posts at the University of Cambridge and at the University of Münster. His work has appeared in *Journal of the American Musicological Society*, *Journal of the Royal Musical Association*, and *World of Music (new series)*. His article "Evliya's Song: Listening to the Early Modern Ottoman Court" (2023) won the American Musicological Society Alfred Einstein Award and the British Forum for Ethnomusicology Early Career Prize.

Carmel Raz is an assistant professor of music at Cornell University. She is the author of *Hearing with the Mind: Proto-Cognitive Music Theory in the Scottish Enlightenment* (Oxford, 2025). Other projects include *Sound and Sense in British Romanticism*, co-edited with James Grande (Cambridge, 2023), which won the 2025 Diana McVeagh Prize for Best Book on British Music, and *Thinking Music: Global Sources for the History of Music Theory*, co-edited with Thomas Christensen and Lester Zhuqing Hu, a digital collection of annotated primary sources which will be published by the University of Chicago's online research and press service (CORPUS) in 2027. Her current book project explores theorizations of habit, automaticity, and attention in musical performance from antiquity to the present.

Benjamin Steege is Associate Professor of Music at Columbia University. His research centers on problems in the history of listening, intersections with the history of science, music phenomenology, as well as film music. He is the author of *Helmholtz and the Modern Listener* (Cambridge, 2012), *An Unnatural Attitude: Phenomenology in Weimar Musical Thought* (Chicago, 2021), and numerous articles. His research has been supported by the Max Planck Institute for the History of Science, the Alexander von Humboldt Foundation, the National Endowment for the Humanities, and the Radcliffe Institute for Advanced Study at Harvard University.

David Trippett is Professor of Music at the University of Cambridge and a Fellow of Christ's College. His research focuses on nineteenth-century intellectual history, Richard Wagner, and the role of sound within the history of science. He is author of *Wagner's Melodies* (Cambridge, 2013), and co-edited *The Cambridge Companion to Music in Digital Culture* (Cambridge, 2019) and *Nineteenth-Century Opera and the Scientific Imagination* (Cambridge, 2019), and *Wagner in Context* (Cambridge, 2024). Between 2015 and 2020 he was PI for a European Research Council Grant on "Sound and Materialism in the Nineteenth Century." A monograph arising from this, *Sound in Mind and Body*, will center on composers and scientists embroiled in the nineteenth-century clash of idealism and materialism. In more hands-on work, he produced the first critical and performing editions of Liszt's unfinished Italian opera *Sardanapalo* (1851/2018), which has now been performed in six countries. As pianist and conductor, he recently recorded his first three CDs of contemporary music with the Philharmonia Orchestra.

Sarah Weiss is University Professor for the Institute for Ethnomusicology at the University of Music and Performing Arts, Graz (KUG). She is an ethnomusicologist who specializes in Southeast Asian performing cultures, gender studies, postcoloniality and hybridity, analysis of musical mode, and ritual studies. Among her publications are two monographs—*Ritual Soundings: Women Performers and World Religions* (Illinois, 2019) and *Listening to an Earlier Java: Aesthetics, Gender, and the Music of Wayang in Central Java* (KITLV, 2007)—several edited volumes including *Is There Such a Thing as Singaporean Performance?* (Shaker Verlag, 2023), and many articles. An active musician, Sarah Weiss is the artistic director for the university's Central Javanese ensemble, Gamelan Nyai Rara Saraswati.

Richard David Williams is Reader (Associate Professor) in Music and South Asian Studies at SOAS University of London. A cultural historian of music, he is especially interested in literary engagements with music and sound in early modern and colonial north India. His first book, *The Scattered Court: Hindustani Music in Colonial Bengal*, was published with the University of Chicago Press in 2023. He is currently writing a book on the intellectual history of *rāgamālā*, the early modern north Indian art of imagining music through poetry and painting. His research primarily engages with musical archives in Brajbhasha (old Hindi), Bengali, and Urdu.

INDEX

ACKNOWLEDGMENTS

This volume emerged over the course of several years as the fruit of sustained conversation across three continents, several symposia, and innumerable pandemic Zooms. In short, it has been an act of shared attention nourished by precisely the forms of concentration, generous collectivity, wired alertness, and creative distraction that it investigates. We extend our thanks to the Max Planck Institute for Empirical Aesthetics for providing a Frankfurt home for the Music and Attention Working Group, and to Diana Gleiss for her tireless logistical support. We are grateful to Walter Biggins and Emma Dillon at the University of Pennsylvania Press, whose enthusiasm for the project carried us through to the end. We would like to thank the Salina Public Library in Kansas for generously allowing us to reproduce an image from their collection on our cover, and Nir Cohen-Shalit, who virtuosically proofread the volume and compiled our index. Our appreciation goes also to scholars who contributed to discussions at our workshops and study sessions dating back to 2017, including Deirdre Loughridge, Nina Sun Eidsheim, Danny K. S. Walden, Michael Auer, Erin Johnson-Williams, Brian Hyer, Brian O'Connor, Zeynep Bulut, Annette Richards, Davinia Caddy, and Jonathan de Souza. Finally, and most sincerely of all, we thank our contributors. Your imagination, conviviality, and willingness to think with us about the complexity and promise of sound attentions made this work possible.

FSC
www.fsc.org
MIX
Papier | Fördert gute Waldnutzung
FSC® C083411

Kolibri 360 GmbH
Ferdinand-Jühlke-Straße 7
99095 Erfurt, Deutschland
produktsicherheit@kolibri360.de